Michael

INTRODUCTION TO QUALITY CONTROL

KAORU ISHIKAWA

INTRODUCTION TO QUALITY CONTROL

3A CORPORATION

Originally printed in Japan as "*Dai-3-pan Hinshitsu Kanri Nyūmon*" (Introduction to Quality Control 3rd Edition) by Kaoru Ishikawa. ©Kaoru Ishikawa 1989, published by JUSE Press Ltd.

INTRODUCTION TO QUALITY CONTROL

Shoei Bldg., 6-3, Sarugaku-cho 2-chome, Chiyoda-ku, Tokyo 101, Japan

ISBN 4-906224-61-X C0034

First Printing.....October 1990

Translated by J.H. Loftus

Printed in Japan

Contents

CHAPTER 3
THE PREPARATION AND USE OF CONTROL CHARTS 147

CHAPTER 4
PROCESS ANALYSIS AND IMPROVEMENT 201

CHAPTER 5
PROCESS CONTROL 287

CHAPTER 6
QUALITY ASSURANCE AND INSPECTION 335

Figures and Tables

Preface

Forty years have passed since I and my colleagues formally introduced the quality control (QC) movement to Japan in 1949. Since that time, Japanese QC methods have changed considerably. QC began with statistical quality control (SQC) and statistical process control (SPC) and progressed through supplier QC (1960), QC in new product development (1961) and QC in the sales department and distribution networks to QC in the construction and service industries. During this process, SQC has evolved into TQC (Total Quality Control) and CWQC (Companywide Quality Control), and recently into GWQC (Groupwide Quality Control, i.e. quality control covering a whole corporate group including its suppliers and distribution organizations).

The first edition of this book was published in 1954 and the second in 1964, and a total of about a hundred printings of these editions have been issued. Minor corrections were of course made to these editions, but the plates are beginning to wear out, and QC and its methods have changed. I have therefore decided to issue this revised edition. The basic principles described in the original editions remain the same, but new developments and advice have been added. The result is an introduction to QC with over 430 pages.

To understand QC, we need to know not only what QC itself is but also what SQC, SPC, TQC, CWQC and GWQC are. Every person in a company should know this, from the president down to the lowest worker. To achieve this understanding, we should first grasp what QC is and then move on to learn about methods such as quality assurance methods, the statistical approach, control methods and improvement methods. In this book, I have tried to explain these matters in down-to-earth, easily-understood language. Although computer-based statistical methods are becoming common, I have not dealt with them here. I have concentrated on pencil-and-paper methods, because analysis and control using such methods are still essential in the workplace, particularly in service industries. I have also found that everyone studying QC methods needs the experience of plotting and analyzing data manually before moving on to using computers.

Top managers, department managers, administrative staff and workplace supervisors should study Chapters 1 and 2 by discussing the contents of these chapters thoroughly over a two-day period, while younger department managers, section managers, under-managers and technical staff should take six to eight days to study whole. When the book is used in undergraduate courses, case studies should also be included, since university students have no work experience. If one lecture is given per week, the book should be covered in one year.

I strongly urge readers who learn about QC through this book to put it into action. Theory and scholarship are also needed in QC, but it only yields tangible results if actually practiced. Some readers may think that what the book says is obvious. They are right. Before QC, what should obviously have been done inside and outside companies was either not being done at all or only being done piecemeal. An alternative definition of QC could in fact be, 'Everyone doing what should be done, in an organized, systematic way.' The machine industry was slow in adopting QC, and some in the construction and service industries at first claimed that it could not be applied to them because they were different from other industries. However, when it was actually tried, almost all the basic principles were found to be the same. Saying that 'QC won't work in our industry because we're different from the rest' is just an excuse for lack of motivation.

I hope readers will study this book carefully line-by-line, digest its contents well, and use them to enable their companies to survive in today's competitive free economy.

QC starts and ends with education.

In conclusion, I should like to express my warmest thanks to all at JUSE (The Union of Japanese Scientists and Engineers) who have assisted in preparing this book for publication since the first edition.

October 1988

Kaoru Ishikawa

Acknowledgment

This book is an English translation of the late Dr. Kaoru Ishikawa's classic work "*Dai-3-pan Hinshitsu Kanri Nyūmon*" ("Introduction to Quality Control 3rd Edition"), originally published in Japanese by JUSE Press Ltd., the publishing arm of the Union of Japanese Scientists and Engineers. The publication of this English translation was planned with Dr. Ishikawa's ready consent before his death. Regrettably, he fell ill just at that time, and finally passed away in April, 1989. I cannot help but feel a deep sense of sorrow at his loss.

For many years, Dr. Ishikawa lectured and counseled managers and engineers from overseas in his capacity as a Principal Lecturer and Director of the Association for Overseas Technical Scholarship, 3A Corporation's parent organization. Many of the trainees who received his guidance are now playing key roles in their respective countries' business and industrial circles.

"*Hinshitsu Kanri Nyūmon*" was widely acclaimed on its publication in 1954 and has since been revised and reprinted many times. After particularly extensive rewriting and expansion to cover the use of computers and other modern developments, this third edition was published under the title "*Dai-3-pan Hinshitsu Kanri Nyūmon*" in 1988, making it Dr. Ishikawa's posthumous legacy to all students and practitioners of quality control.

In translating and publishing this book, we received exceptionally generous guidance and editorial assistance from Dr. Hitoshi Kume, Professor of the Faculty of Engineering of the University of Tokyo, Dr. Noriaki Kano, Professor of the Faculty of Engineering of the Science University of Tokyo, and Dr. Yoshinori Iizuka, Associate Professor of the Faculty of Engineering of the University of Tokyo. The publication of this book would have been extremely difficult without their help; and I am deeply indebted to them.

I am also grateful to them for approaching the American quality control pioneer Dr. J. M. Juran with a request to write some words of recommendation for this book. He was kind enough to comply with this request, and I would like to thank him sincerely for adding the finishing touch.

Finally, I would like to thank all those at JUSE Press Ltd. for their valuable cooperation over the long period from the planning stage to final publication.

October 1990 Publisher

Modern quality control constitutes a revolution in management thinking, and implementing it companywide can dramatically improve a compamy's corporate culture.

As industry advances and society modernizes, quality control becomes more and more important.

It is my sincere hope that quality control will achieve the following aims:

Strengthen a country's economic base by making it possible to export high-quality, reasonably-priced products in large volumes.

Establish reliable industrial technology and enable technology transfer to other countries to flourish.

Secure a solid economic foundation for the future.

Finally, to enable companies to share their profits fairly among consumers, employees and company investment and raise their nation's standard of living.

If every nation plays its part in promoting quality control, the world will find peace, and its people will be able to live together harmoniously and happily.

We should all strive to create a lively, cheerful atmosphere within our companies and to build happy lives for our countries and the world.

Chapter 1
WHAT IS QUALITY CONTROL?

1.1 WHAT IS QUALITY CONTROL?

The top management of a company of course bears overall responsibility for that company's products and services, but factory managers, department managers, section managers, supervisors, and foremen are all responsible for the quality of the products and services produced by their respective factories, departments, sections, groups, and teams. Meanwhile, the duty of engineers and technical specialists is systematically and methodically to prepare, revise, and improve standards that will enable their companies to supply society with products as economically as possible.

Controlling quality does not simply mean studying statistics or preparing control charts. I believe that the aims of quality control should be first to strengthen a country's economic base by enabling it to export large volumes of high-quality, reasonably priced products, and finally to secure a firm economic foundation for the future by establishing and actively exporting industrial technology. The ultimate aims of quality control should be to enable companies to share their profits sensibly and fairly among consumers, employees, and shareholders, to raise the country's standard of living, and to make life better for the world as a whole.

1.1.1 The Definition of Quality Control

In the JIS (Japanese Industrial Standards) terminology standard Z8101–1981, quality control is defined as follows:

"A system of methods for the cost-effective provision of goods or services whose quality is fit for the purchaser's requirements.

Quality control is often abbreviated to QC.

Since modern quality control uses statistical methods, it is sometimes referred to as statistical quality control (abbreviated to SQC). The effective implementation of quality control requires the participation and cooperation of all the employees of a company from top management through middle management and supervisors down to ordinary workers at every stage of the company's activities, from market research through research and development, product planning, design, production preparation, purchasing and subcontracting, production, inspection, sales, and after-sales service, as well as in the financial, personnel, and education functions. Quality control carried out in this way is known as companywide quality control (abbreviated to CWQC) or total quality control (abbreviated to TQC)."

Quality control is a new way of thinking about and viewing management. The following is my personal definition:

"Quality control consists of developing, designing, producing, marketing, and servicing products and services with optimum cost-effectiveness and usefulness, which customers will purchase with satisfaction. To achieve these aims, all the separate parts of a company (top management, head office, factories, and individual departments such as production, design, technical, research, planning, market research, administration, accounting, materials, warehousing, sales, servicing, personnel, labor relations, and general affairs) must work together. All the company's departments must strive to create cooperation-facilitating systems, and to prepare and implement standards faithfully. This can only be achieved through full use of a variety of techniques such as statistical and technical methods, standards and regulations, computer methods, automatic control, facility control, measurement control, operations research, industrial engineering, and market research."

Since real quality control can only be achieved by marshalling all of a company's strengths, this kind of quality control is called companywide quality control (CWQC), or total quality control (TQC). Implementing CWQC/TQC requires the following:

1. All departments must participate, with the head of each department taking the lead. Each department must take the initiative in liaising with related departments.
2. Every employee must become involved; in other words, all members of the company, from the Chairman of the Board through the Chief Executive, senior officers, directors, department and section managers, and

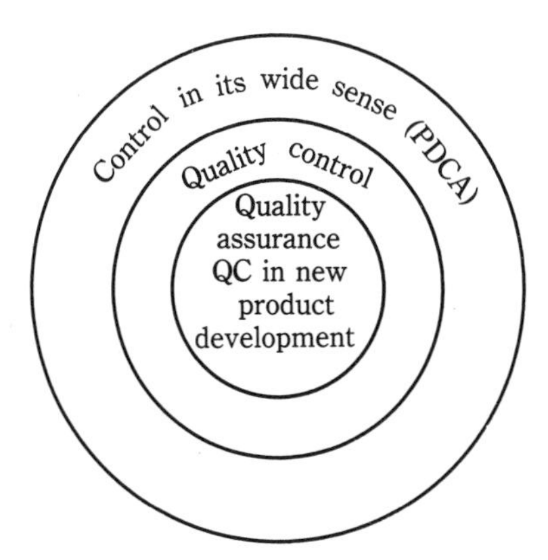

Figure 1.1 What is TQC?

technical and administrative staff down to QC circle members (i.e., shop foremen, full-time workers, sales staff, and part-timers) must participate in implementing quality control.

3. QC must be implemented comprehensively. To produce products that consumers and society will buy happily, quality (Q) must come first, but at the same time, costs (C) (i.e., sales price and profit), delivery (D) (i.e., production volumes, sales volumes, and inventories), and safety (S) (including social and environmental factors) must be comprehensively controlled. This is why the term "total quality control" (TQC) is used.

Groupwide quality control (GWQC) means quality control and total quality control implemented throughout an entire company group, i.e., not just within a company, but throughout a complete organization, including the company's suppliers and subcontractors, distribution organizations, and affiliates.

Quality control is related to quality assurance in that QC consists of activities intended to assure quality for customers and end-users; in other words, quality assurance is the aim and essence of quality control.

At present, the terms "companywide quality control" and "total quality control" appear to be used interchangeably and are defined identically in the Japanese Industrial Standards. In this book, we will mainly use the term "total quality control."

Note: What is TQC in its broad sense? (Refer to Figure 1.1):
As explained above, total quality control consists essentially of developing, controlling, and assuring the quality of products and services. This is shown by the innermost ring of the figure. However, when we understand what good quality means in terms of products and services, we can expand the definition of

TQC to mean improving the quality of everything, i.e., creating a high-quality company, high-quality executives and department managers, high-quality sales departments, personnel departments, factories and laboratories, high-quality sales staff and supervisors, high-quality suppliers, high-quality distributors, etc. This is represented by the second ring of the figure. Some companies use TQC in an even wider sense to mean applying quality-control methods rigorously to all their work (the outermost ring of the diagram) and rotating the PDCA cycle (Section 1.5).

In implementing TQC, companies are free to choose the standpoint from which they define it, according to the nature of the company and top-management policy. This means that when a company introduces TQC, its top management must clearly announce its aims in introducing it, and their particular definition of it. However, the essence of TQC must not be forgotten: the "quality-first" principle, quality assurance, and quality control in new-product development. QC circle activities were also begun in 1962 as one aspect of TQC; a more detailed explanation of this is presented in Section 1.10 (Figure 1.19).

1.1.2 Some Misunderstandings about Quality Control and Total Quality Control

Some common misconceptions about QC and TQC are:

- QC consists of tightening up inspection.
- QC means making standards.
- QC consists of preparing control charts.
- QC is statistics.
- QC means studying something difficult.
- QC can be left up to the inspection section.
- QC is something the QC section does.
- QC can be left up to the factory.
- QC can be left up to the workplace.
- QC is nothing to do with the administration department.
- QC costs money.
- We are making a profit at the moment, so we don't need anything like QC.
- We are doing QC circle activities, so we must be doing TQC.
- A QC campaign consists of QC circle activities.
- As long as we are doing QC circle activities, we are doing all right.
- Our company doesn't need QC circle activities.
- QC is nothing to do with me.

1.1.3 The Benefits of Companywide Quality Control

What benefits are obtained when a company implements quality control in earnest throughout its organization? The following are some that have already been demonstrated in Japanese companies:

- Quality (in its narrow sense) is raised, and the number of defective products decreases.
- Quality becomes more uniform and the number of complaints decreases.
- Reliability increases, confidence in the products improves, and customers' trust is obtained.
- Costs decrease.
- Products can be sold at higher prices.
- A quality assurance system is established, and the trust of consumers and customers is obtained.
- Complaints are dealt with more quickly, and effective action is taken to prevent their recurrence.
- Unit costs improve, and value-added productivity increases.
- Production volumes increase, and it becomes possible to prepare rational production plans.
- Wasteful work disappears, rework decreases and efficiency improves.
- Technology is established, engineers can be employed in their true capacity, and technology improves. Ways of employing people, particularly engineers, become more rational.
- Inspection and testing costs decrease.
- Contracts with suppliers, subcontractors, and consumers can be rationalized.
- Sales routes expand.
- Relationships and the flow of information within the company organization become smoother.
- Research and development is speeded up and made more effective.
- Research investment becomes more rational.
- Employees' humanity is respected, personnel development becomes possible, and workplaces become more cheerful.
- Talent-spotting becomes possible, and people are able to exercise their full capacities.
- Human relations improve, and barriers between departments are broken down.
- People begin to speak a common language and to understand each other better.

- The whole of the company organization can be rationalized, and department managers, section managers, supervisors, and foremen become able to work more effectively.
- Good market information is received more quickly.
- New-product development speeds up and improves. Products of world-beating quality can be made.
- People become able to talk frankly and openly.
- Meetings go more smoothly.
- Plant and equipment repair and expansion can be done rationally according to priority.
- The entire company works together, and a system of cooperation is established.
- Decision making is speeded up, and policy deployment and management by objectives improve.
- The corporate culture is improved.
- The company becomes trusted.
- All departments understand the idea of dispersion and become able to utilize QC techniques.
- The company and its factories cease to issue false data.

In addition, introducing total quality control helps to rationalize all aspects of company management; and consumers, employees (including top management), and shareholders all profit.

As with many other things, there is a surprising amount of prejudice against quality control, but the proof of the pudding is still in the eating. Quality control only succeeds when top management feels responsible for the quality of its company's products and takes up quality control as a matter of policy, and everyone—not just middle management and technical staff, but also administrative staff and front-line workers, and even further, subcontractors, distribution organizations, subsidiaries, and affiliates—bands together to implement it. It will not usually succeed if it consists merely of a handful of engineers studying statistics in a corner of a factory. This is why the understanding, enthusiasm, and leadership of top management, and the accompanying action, are all so important.

The prerequisite for a company to act as a single unit in promoting quality control is to improve human relations, i.e., to construct a system of cooperation covering the whole company.

1.2 The History and Current Status of Quality Control

After the end of World War II, modern quality control blew like a breath of fresh air through Japan's ravaged industries. It was a major force in helping rationalize the country's manufacturing, and it revolutionized the management policies and organizational structures of Japanese companies.

The benefits are obvious. Some companies in Japan and abroad have already earned huge profits through total quality control and QC circle activities, while countless others have succeeded in cutting costs and saving energy. Even medium-sized businesses and small businesses with as few as fifteen employees are using these methods to produce high-quality products at unbeatable prices. It has already been shown that the methods can produce good results in any kind of industry.

Company executives and engineers often claim that they have always taken good care of the quality of their products, and that their products are fine without going to the bother of practicing QC or TQC. However, companies who make such claims will almost certainly be destroyed by the competition if their competitors start implementing the methods described in the following pages. Japanese industry is particularly vulnerable to changes in the economic climate, and its management base is weak. Companies that do not adopt these methods and do not rationalize their management will probably disappear from the industrial scene as a result of rising costs and lack of confidence in their products. I hope everyone will heed the following warning: *Any enterprise that does not practice quality control will not be around for long.*

(1) Statistical quality control

Statistical methods are extremely helpful and are often used in quality control. For this reason, quality control is often called "statistical quality control" (SQC).

Although statistics are very useful in quality control, many people meeting them for the first time—particularly managers and administrative staff—feel apprehensive. If, however, one understands the ideas behind statistical methods, actually using them is very simple; all one needs is a primary-school knowledge of arithmetic—addition, subtraction, multiplication, and division. While the science of statistics is making rapid progress, and some of its methods are extremely complex, the "seven QC tools" are now being extensively used in all industries by supervisors and ordinary workers, male and female, regular or part-time.

As well as being used to make process control charts, to design experiments, and for sampling inspection, modern statistics have a wide variety of social uses, such as in opinion polls, cost-of-living surveys, agricultural production surveys, tax surveys, market research and operations research, production planning, transport planning, stock control, equipment control, and management research.

However, while statistics are important, it is more important to understand the quality control approach and to follow it faithfully.

Modern quality control was started in America in the 1930s as a result of advances in measurement technology and the application to industry of the control chart (invented in 1924 by Dr. W. A. Shewhart of Bell Telephone Laboratories) and other statistical methods. Shewhart's classic *Economic Control of Quality of Manufactured Product* was published in 1931, and quality control was subsequently adopted in the United Kingdom. It continued to develop in both the U.S. and the U.K., but was only applied in earnest in all industries when World War II appeared imminent. When planning its increased industrial production in preparation for war, the U.S. set its sights on turning out good-quality products cheaply and in large quantities. This was in marked contrast to the unscientific approach of Japan's wartime military and government authorities, who told industry to step up production even if they had to do so at higher costs.

By this time, some extremely simple forms of control chart had already been investigated and had given good results when used in some U.S. factories. To have the munitions industry adopt them, they were promulgated in the form of standards in 1941 and 1942. These were the well-known American War Standards Z1.1–Z1.3, as follows:

Z1.1: Guide for Quality Control (1941)
Z1.2: Control Chart Method of Analyzing Data (1941)
Z1.3: Control Chart Method of Controlling Quality During Production (1942), American Standards Association.

(2) The development of QC in America and Europe

The American Society for Quality Control (ASQC) was founded in 1946. Control charts came into widespread use, and, in 1958, when I visited the U.S. for the first time, I observed that the 3,000-strong workforce at Western Electric's Allentown, Pennsylvania plant was using ap [illegible] e 5,000 employees working in Eastman-Kodak' [illegible] icilities had prepared 35,000 charts (including [illegible]). Both companies achieved remarkable result [illegible] department stores starting to introduce qualit [illegible] npanies commencing actively to introduce it to [illegible] ufacturing company, it then began to develop i [illegible] n example of the development of quality contr [illegible] control in its wide sence.

Meanwhile, QC research in Great Britain proceeded rapidly, as might have been expected from the country that gave birth to modern statistics. In 1935, papers on quality control by E. S. Pearson and others were used as the basis for British Standard BS 600. America's Z1 Standards were later adopted without modification as BS 1008, and many other QC standards were established and followed.

Other European countries, including France, Switzerland, Czechoslovakia, Sweden, Italy, and West Germany, also started using statistical quality control methods, and began to implement quality control in earnest in 1953, when instructors were invited from America. The European Organization for Quality Control (EOQC) was established in 1965.

(3) The introduction of QC to Japan

The British Standards mentioned above had already been brought to Japan before World War II; a Japanese translation was published during the war. The war ended as some mathematicians were studying these standards and attempts were being made to implement them. Meanwhile, academic specialists were studying modern statistics, and their research had reached world-class levels. The methods, however, were explained in mathematical terms, which conveyed to some the idea that the methods were extremely difficult, and they were not generally adopted. After the war, it gradually became apparent that the methods were achieving great successes in the U.S. The occupation army found that Japan's telephone communication system at that time was virtually useless, and it was proposed that Japan's communications equipment manufacturers should implement quality control in order to eliminate defects and nonuniformity in the quality of the equipment. This happened in May 1946.

Since before the war, some companies had been using the old Taylor method (regarded as modern at that time) as a method of management.

In 1946, the Union of Japanese Scientists and Engineers (JUSE) was set up, as a private, non-profit organization. In 1949, interested parties from academic establishments, industry, and government met at JUSE and formed a group called the Quality Control Research Group (QCRG) with the aim of carrying out quality-control research and education and promoting it in Japan. The Quality Control Research Group had no connection with the government; it was simply a collection of volunteers who thought they would try to promote quality control and thereby help to rationalize Japanese enterprises, improve the quality of Japanese products (at that time, "Made in Japan" meant "cheap and nasty"), and improve exports. In 1949, JUSE started its first Basic QC Course (BC), a 36-day course given at the rate of three days per month over twelve months.

Meanwhile, in 1950, the Japanese Industrial Standards (JIS) marking system was instituted based on the industrial standardization law. Under this system, companies had to implement statistical quality control and quality assurance in order to be able to place the JIS mark on their products. Japan can be considered fortunate in starting to promote industrial standardization and quality control at the same time.

In 1950, JUSE invited Dr. W. E. Deming from the U.S. to conduct a QC seminar for top management, department and section managers, and engineers. This seminar was extremely enlightening; with the proceeds donated by Dr. Deming from the sale of the seminar transcripts, the Deming Prize for Quality Control was inaugurated in 1951. This scheme has contributed immeasurably to the progress of quality control in Japan.

Initially, however, Japanese quality control also suffered from various problems. The first was that statistical methods were overemphasized, and this fostered the mistaken impression that quality control and statistical quality control were difficult. Second, the emphasis on standardization led to a tendency for quality control to be carried out merely formally. The third problem was that top management and department and section managers did not develop much enthusiasm for quality control.

To help solve these problems, Dr. J. M. Juran was invited to Japan in 1954 to give a seminar for executives and department and section managers. Quality control at last started to be used as a management tool. This marked the beginning of a gradual transition from statistical quality control to total quality control, and in turn led to the promotion of quality control in which all departments and all employees participated—in other words, total or companywide quality control.

This required the involvement of the workplace, and, in 1956, a quality-control course for foremen was started on Japanese shortwave radio. This course was later repeated on the Japan Broadcasting Corporation's radio and television channels. In 1960, JUSE issued a publication entitled *Shokukumichō no Tame no Hinshitsu Kanri Tekisuto* ("A Quality Control Text for Foremen"). Then, in April 1962, JUSE launched a journal called *Genba to QC* ("QC and the Workplace"). The name of this journal was later changed to *FQC* ("QC for the Foreman") and then to its present *QC Sākuru* ("QC Circles"). At the same time, JUSE started group activities in the workplace under the title "QC Circles." These QC circle activities, started in Japan as an integral part of TQC, are now being copied in countries all over the world.

(4) Japanese quality control

Throughout this time, Japanese QC researchers and users began to realize that while disciplines such as physics, chemistry, mechanical engineering, and electrical engineering are common to all countries throughout the world, disciplines such as quality control that talk in terms of control and management involve cultural differences and human factors. It became apparent that American and European QC methods could not be applied to Japan without modification, and that a Japanese form of quality control suitable for use in that country would have to be developed.

Table 1.1 lists some of the relevant cultural and social differences between Japan and the West. The total quality control practiced in Japan today was developed by promoting a Japanese style of quality control that took these differences into account.

At the QC Symposium held in Hakone, Japan, in 1968, Japanese-style quality control was distinguished from the quality control practiced in the West. The following six main characteristics were identified, some desirable and others not so desirable:

1. TQC: QC with the participation of all departments and the involvement of all employees; total quality control.
2. Enthusiasm for QC education and training.
3. QC circle activities.
4. QC audits: the Deming Application Prize and company presidents' audits.
5. Use of statistical methods: dissemination of the Seven QC Tools and use of advanced methods.
6. National QC promotion campaigns: Quality Month, various QC symposia, the QC Circle Headquarters.

Table 1.1 Some Cultural Differences between Japan and the West

	The West	Japan
1. Professionalism	Strong	Weak
2. Organization	Strong staff role	"Vertical society"
3. Labor unions	Mainly trade-based	Mainly company-based
4. The Taylor system	Prevalent	Rare
5. Graduate elitism	Strong	Not so strong
6. Salary system	Merit-based	Seniority-based
7. Job transfer rate	High (layoffs common)	Low (lifetime employment)
8. Writing system	Phonetic	Pictorial (ideographic)
9. Educational level	Depends on country	Particularly high
10. Ethnic characteristics	Multiracial	Single race
11. Religion	Judeo-Christianity	Buddhism, Confucianism
12. Relationship with subcontractors /buying-out rate	Adversarial/50-60%	Friendly/70%
13. Capitalism	Old-fashioned	Democratic
14. Government control	Depends on country	Not very strong

During this time, the postwar "seller's market," with all commodities in short supply, turned into a buyer's market. As Japan's industry continued to grow, the government's trade liberalization policies, introduced in 1960, made quality control even more important. We devised the slogan "Trade Liberalization through QC," and started activities designed to help companies cope with trade liberalization by producing products of high enough quality and low enough cost to make them suitable for export. This campaign succeeded to the point where Japan can now manufacture goods of world-beating quality capable of being exported all over the globe.

Japanese-style TQC continued to evolve; the following ten characteristics were identified in 1987:

1. QC activities with the participation of all departments and the involvement of all employees led by top management.
2. Widespread acceptance of the quality-first principle in management.
3. Policy deployment and management by policy.
4. The QC audit and its application.
5. Quality assurance programs extending from planning and development right through to sales and service.
6. QC circle activities.
7. QC education and training.
8. The development and application of QC methods.
9. The extension of QC from the manufacturing industry to other industries.
10. National QC promotion campaigns.

The Japanese Society for Quality Control (JSQC), the formation of which had been discussed since about 1950, was finally established in 1970. While the American Society for Quality Control (ASQC) is a professional society, the JSQC is an academic one.

Quality control is not a passing fad. As long as a company is selling products or services, it must continue to control quality. I keep repeating the truism that *total quality control consists of doing what should be done as a matter of course.* Moreover, *TQC is not a fast-acting drug like penicillin, but a slow-acting herbal remedy that will gradually improve a company's constitution if taken over a long period.* Quality control means doing what ought to be done in all industries, and Japan has already proved that implementing it gives remarkable results.

Some people say that there is no way that quality control can be applied to their particular companies or factories, but this is because they do not understand the true meaning of quality control. As pointed out earlier, QC has already been applied in all of Japan's industries, not only in manufacturing but also in the construction industry and many service industries, and the benefits and potential of its application have been amply demonstrated.

The question is not whether quality control is or is not applicable to a particular company, but whether that company has the will and the ability to apply it. Excuses such as "We are still at the pre-QC stage" simply do not pay. Let us not discuss reasons for not being able to practice quality control, but think positively and work out how we *can* implement it.

Recently, many other countries have realized that Japanese-style QC methods are good, and many companies are applying them, with suitable modification, to their own situations.

1.3 Advances in Quality Assurance

The essence of TQC is quality assurance.

Quality can be assured by various methods, some of which are listed below:

1. By inspection (100% inspection, sampling inspection, check inspection, patrol inspection, or autonomous inspection).
2. By the process (process control, process capability research, and autonomous control).
3. During new-product development.

Particular care must be taken to implement items 2 and 3 above properly in order to guarantee reliability, which is an integral part of quality assurance.

After World War II, quality assurance methods (see also Sections 1.6.1 and 1.6.2 and Chapter 6) progressed as described below*:

(1) Inspection-oriented quality assurance

If a process produces defective products, they must be eliminated through careful inspection, but there are still some companies that do not practice final inspection prior to shipping, even though defectives are present. We would be perfectly justified in accusing such companies of living in the Dark Ages before the advent of quality assurance. Unscrupulous organizations such as these are totally beyond the pale.

---

* See "*Hinshitsu*" ("Quality" *Journal of JSQC*), Vol.10 (1980), No.4, pp.205–213, or Ishikawa: *Nihonteki Hinshitsu Kanri* ("What is Total Quality Control? The Japanese Way," English trans. pub. Prentice-Hall, 1985), Ch. 4, Section 4, pub. JUSE Press (1981).

Historically, quality assurance started with the institution of rigorous inspection, and people who know nothing about quality control still mistakenly believe that it consists of tightening up inspection procedures. Such an approach, however, has many disadvantages, some of which are listed below:

1. The inspection process is never perfect, and the goal of zero defects can never be achieved, even with 100% inspection.
2. Inspection staff are superfluous manpower, which reduces productivity.
3. When there is inspection, production department personnel tend to think that simply getting the products passed is good enough. But *the responsibility for quality assurance actually lies with the producer and the production department.*
4. Inspection department data are often not stratified and are fed back too slowly. Such data are useless for process control and analysis.
5. Statistical sampling inspection cannot guarantee fraction defective values of the order of 0.01% or parts per million.
6. Many items cannot be guaranteed through inspection; complex assemblies and materials cannot all be tested, and destructive testing and reliability tests cannot be performed on all products.
7. Even when defectives and defects are detected, this only results in an increase in scrap, rework, and man-hours spent on adjustment.
8. An increase in production speed means that the inspection process must be automated.
9. Having inspection means that quality control tends to be left up to the inspection department.

Inspection must of course be carried out as long as a process produces defectives, but quality control that relies solely on inspection gives imperfect quality assurance and raises costs.

(2) Process-control-oriented quality assurance

Soon after the promotion of QC was started in Japan in 1949, industry entered a second phase, process-control-oriented quality assurance, in which it tried to produce non-defective products by strictly controlling processes. The slogan "Build quality in during the process" appeared at this time. As quality improved, the numbers of defectives and other problems decreased, and productivity and reliability improved. However, process control cannot by itself give satisfactory quality assurance; however closely a process is controlled, it is impossible to guarantee the quality or reliability (in a wider sense) of a poorly designed product or a product made with ill-chosen materials.

(3) New-product-development-oriented quality assurance (see Section 1.6.2)

Because of the limitations of process-control-oriented quality assurance, in the latter half of the 1950s Japanese industry entered a third phase, in which it tried to implement quality assurance during new-product development. The slogan now became "Build in quality during design and the process." In other words, industry started to build quality into its products by performing careful evaluations at every stage of product development from new-product planning through design to pilot production, and by using the QC approach to investigate reliability in its broad sense. This approach makes it necessary for all employees in all departments to participate in implementing quality control and quality assurance.

This fitted in well with total quality control, which had developed as a result of the different social backgrounds of Japan and the West, and excellent results were achieved. Since these efforts have been continued for so long, many of Japan's products are now the best in the world and are exported at reasonable prices all over the globe.

Of course, even when new-product-development-oriented quality assurance is practiced, process control is still indispensable, and inspection to detect and eliminate defectives and defects must still be carried out.

1.4 WHAT IS QUALITY?

To understand the philosophy of statistical quality control (SQC), it is probably best to split this term into its component parts and clarify each before putting them together and looking at the term as a whole. I will start by discussing the meaning of the word "quality."

In Japan, "quality" is translated as 'hinshitsu,' a word written with two Chinese characters, one meaning "goods" and one meaning "quality." I think this is an excellent rendering. When I went to the U.S. in 1958 to study quality control, I found that even there the interpretation of the word "quality" differed from company to company. For example, quality control in the Bank of America's QC program meant controlling the quality of branches, borrowers, and policy making; while United Airlines, in a service industry, was running an excellent statistical quality-control program based on an interpretation of quality as quality of service. Companies like Bell Systems and General Electric were implementing quality control from the design stage right up to use of the product by the consumer.

Thus, in quality control, the meaning of the word "quality" need not be restricted to quality of product, but can be used for quality in general, including quality of management, and in Japan we are seeing the successful promotion of this wider sense of quality control. However, when starting to promote quality control in Japan, with its scarcity of natural resources and its need to survive through trade, I treated quality as meaning "product quality," and moreover, "the quality that people will buy with satisfaction."

The meaning of quality may also differ from product to product, from general consumer goods and consumer durables to industrial substances and other production materials, but there is in fact very little basic difference, whatever the type of product or industry.

Thus, although the type of quality discussed in this book is mainly that of industrial products ("hard quality"), quality in the service industries ("soft quality") can be regarded as an extension of this. The approaches to quality discussed below can be applied with very little modification to both manufacturing industry and tertiary (i.e., service) industries. At first, the use of the word "hinshitsu" for quality of service as well as for quality of goods seemed strange, but total quality control is now widespread, and many service industries are implementing TQC programs. In Japan today, "hinshitsu kanri" means controlling the quality of both products and services.

1.4.1 Quality to Satisfy the Consumer

Talk of making good-quality products is often misunderstood as making products of the best possible quality. However, when we talk about quality in quality control, we are talking about designing, manufacturing, and selling products of a quality that will actually satisfy the consumer in use. In other words, "good quality" means the best quality that a company can produce with its present production technology and process capability, and that will satisfy the consumers' needs, in terms of factors such as cost and intended use.

Example 1: Which would you prefer to buy, a top-of-the-range camera costing $1,000, or an ordinary camera costing $200 which is perfectly adequate for family snapshots?

Example 2: Which would you buy, a newspaper printed on top-quality paper costing $10, or the same newspaper printed on ordinary newsprint priced at 50 cents?

As the above examples suggest, people will not buy products that are out of their price range, no matter how good the quality (in its narrow sense) might be; conversely, they will not buy a product that does not do its job (such as a camera that only takes blurry pictures), no matter how cheap it is. We buy goods fit for our purposes and our incomes. In today's diversifying and polarizing consumer markets, this is a particularly important consideration at the stages of new-product planning, quality design, new-product development and the selection of research topics, since this is when we decide what products to make and at which market sector they will be aimed.

Some manufacturers and trading companies cling to outdated commercialistic attitudes, acting as if they were still operating under the wartime rationing system, when anything made could be sold. Such organizations, which believe they are doing their job if they somehow manage to fool people into buying their products, have been left behind in the march of civilization and are out of tune with the present democratic age. When viewing our enterprises from the long-term perspective and considering their survival and their use to the community, it is clear that the very least they must do is switch from the old "seller's-market" way of looking at things (the "product-out" approach) to a consumer-oriented, buyer's-market philosophy (the "market-in" approach).

(1) The four aspects of quality

We wish to produce good quality for the consumer; we must therefore decide in advance what quality of product to plan, produce, and sell. To do this, we must consider the following four aspects of quality, and plan, design, and control it comprehensively.

1) Q (quality): quality characteristics in their narrow sense.
   Performance, purity, strength, dimensions, tolerances, appearance, reliability, lifetime, fraction defective, rework fraction, non-adjustment ratio, packing method, etc.
2) C (cost): characteristics related to cost and price (i.e., profit); cost control and profit control.
   Yield, unit cost, losses, productivity, raw materials costs, production costs, fraction defective, defects, overfill, cost price, selling price, profit, etc.
3) D (delivery): characteristics related to quantities and lead times (quantity control).

Production volume, sales volume, changeover losses, inventory, consumption, lead times, changes in production plans, etc. Quality control is impossible without numerical data.

4) S (service): problems arising after products have been shipped; product characteristics requiring follow-up.

   Safety and environmental characteristics, product liability (PL), product liability prevention (PLP), compensation period, warranty period, before-sales and after-sales service, parts interchangeability, spare parts, ease of repair, instruction manuals, inspection and maintenance methods, packing method, etc.

When products are accompanied by good after-sales service, are of reliable quality, and have good compatibility and long lifetimes with little dispersion, the consumer will probably buy them with confidence. Conversely, the consumer will be unsure about buying products with short lifetimes and poor reliability with which something goes wrong a few days or a few months after purchase. The number of complaints against products is also likely to decrease if instruction manuals are written clearly enough for amateur users or children to understand. Customers will not bother to read instruction manuals that are either boastful or so complex that only specialists can decipher them. Is your company receiving complaints arising from misoperation because customers are using the product incorrectly? Do your instruction manuals contain precautions ex-

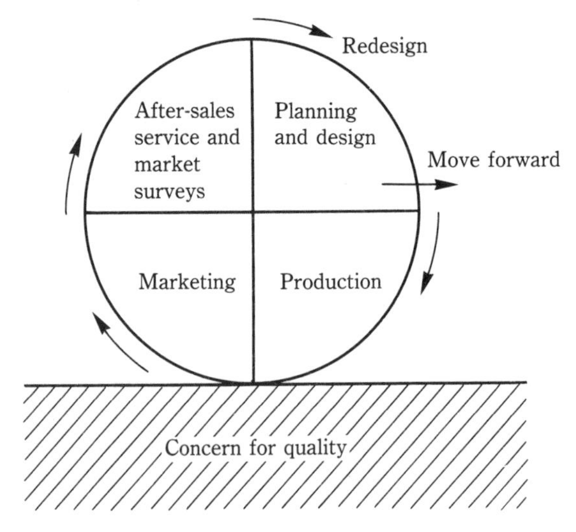

Figure 1.2 The Philosophy of the Control of Quality: The Deming Cycle

plaining unsuitable conditions or methods of use? Are your products being broken or damaged, or are their lifetimes shortened because of poor packaging and transportation methods? Packaging is also an important quality, but are you paying too much attention to your products' visual design and not enough to their real quality?

(2) Complaints

Market research and how complaints are handled are also important. It is easy to forget about a product once it has been sold, but if a company is to make products that customers want, they must find out how customers feel about the products, compared with competitors, after they have bought them.

Complaints often involve money; for example, customers may claim financial compensation or replacement products if a product breaks down or does not meet the terms of a purchase contract. For this reason, sales staff tend to adopt the attitude that dealing with complaints means refunding part of the purchase price or exchanging the faulty product for a new one. Japanese-style quality control, however, is concerned not only with complaints accompanied by claims, but with all types of consumer dissatisfaction. In this book, the term "complaint" is used to cover not only actual claims for compensation but also complaints unaccompanied by claims.

Old-style companies used to try to hush up complaints as much as they could, but companies practicing quality control do exactly the opposite; they see how much information on complaints and dissatisfactions (i.e., latent and actual complaints) they can collect, how many latent complaints and dissatisfactions they can bring into the open, and how well they can listen to their customers. When a company starts practicing quality control, dissatisfaction is usually brought to the surface and the number of complaints increases dramatically (see Section 4.3.4). This is one of the sales department's major QC tasks.

Handling complaints is dealt with in more detail in Section 6.14, but the following two major aspects need to be considered:

(a) Outside action — satisfying the customer. Speed, sincerity, prevention of recurrence.
(b) Internal action.
Recurrence prevention, accounting, disposal of returns.

If priority is given to recurrence prevention (see Section 1.5.3), the quality of the company's products in terms of satisfying consumer requirements will gradually rise. In other words, the process consists of redesigning the quality of products by feeding back and using information obtained from market research and complaints handling, improving process control and inspection methods, and improving quality by smoothly rotating the control cycle (the Deming Cycle). If this is not done, all staff—whether design, technical, workplace, inspection, or other — will become complacent; they will concentrate on quality characteristics that do not really concern users, while they remain ignorant of characteristics that are actually causing users problems. As a result, their products will gradually become unsaleable.

(3) The Deming Cycle

Approaching the four aspects of quality (quality, cost, delivery, and service) comprehensively is particularly important during new-product planning and quality design. As Figure 1.2 shows, the first step is to decide on the quality of the product to be manufactured and set initial technical and other standards specifying how the work is to be apportioned to different parts of the organization, and how it is to be carried out. The product is then produced according to the standards and put on the market. Surveys are then carried out to determine what customers think of the product and what further requirements they may have. The information obtained is then used to revise the quality and standards, and production is continued while quality is constantly improved.

The effectiveness of the quality-control cycle is determined by the weakest step. In this way, based on sound management philosophy and a responsible attitude towards quality, it is possible to keep on producing what customers want, constantly improve, and move forward one step at a time. From one viewpoint, this is the fundamental ideal of quality control. Since this approach was introduced to Japan by Dr. Deming in 1950, it is also known in Japan as the Deming Cycle. Dr. Deming himself, however, said that it was Dr. Shewhart's idea and that the diagram should be called the Shewhart Cycle.

(4) The next process is your customer

Previously, when a problem occurred at a company, people usually tried to cover up or gloss over their own responsibility and blame it on someone else. With this approach, problems can never be solved.

Until now, I have been talking about individual companies, but what I have said applies equally to individual departments within a company and individual sections within a factory. Whenever processes take place within a company, each process is the previous process's consumer or customer, while the previous process is the producer. If the people responsible for each process consider the next process as their customer, listen carefully to its requirements, and are prepared to discuss them sincerely, then problems such as sectionalism will disappear from the company.

For example, at a steel mill, the steelmaking section is the supplier to the rolling section and the customer of the ironmaking section. It is therefore responsible for providing a product of a quality that will satisfy the next process, the rolling section. This means that the steelmaking section must order a statistical investigation to determine the effect of the quality of the steel on the products produced by the rolling section, must visit the rolling section and listen openly to its requirements, and must discuss how those requirements can be satisfied. The steelmaking section is also responsible for clearly explaining its own process to the previous process (the ironmaking and scrap sections), finding out how the pig iron and scrap affect the steel, determining rational quality standards, explaining the results, and making reasonable requests.

Quality requirements passed between processes must of course take into account cost factors and technical conditions.

In many of Japan's factories, it used to be common for those in charge of a process to have no clear idea of the kind of quality they should request from the previous process. Even when they did know, their demands were either too strict or too lax, and the upshot was that the workers on different processes either quarrelled openly or grumbled about each other behind one another's backs. This kind of problem disappears when the people in charge of different processes become able to investigate what happens after they pass on products to the next process, liaise with each other closely, and work as a team. The different processes in a factory begin to cooperate better, barriers are broken down, and work goes more smoothly.

When viewed in this way, the whole series of operations from after-sales service, distribution, and sales and marketing through product warehousing, packaging, production, raw materials storage, design, research and development, and procurement back to suppliers will fall within the province of quality control, and it is everyone's duty to identify and control quality levels accurately within their own area of responsibility.

When we look at the duties of a company's head office and factory management staff, we see that approximately one-third of their work is general staff work, while the remaining two-thirds involve the provision of services to line departments (design, purchasing, production, sales, etc.). Staff departments should therefore act as service departments serving the line departments, which are their "next process," and therefore their customer.

If everyone in the company from the president to the ordinary worker, in every department from sales to purchasing, both on the production line and in staff departments, develops a quality consciousness like that described above, identifies their customers within the company, considers how to satisfy their needs, and takes action to this effect, sectionalism will be broken down. This alone will improve quality, and the foundations of quality control will have been laid.

(5) QC = Business management

As discussed above, quality control in the wide sense is one aspect of business management. After World War II, various management methods were introduced into companies under the name of management science, but they were only adopted piecemeal. When modern quality control was introduced into Japan from about 1948 to 1950, its American interpretation was narrow; when we brought quality control to Japan, we gave it a much wider meaning. In order to make more comprehensive use of the scientific management methods that had been introduced sporadically into Japanese industry (which still had no reliable management base), we took the stand that quality control was a form of business management, and placed priority on the comprehensive implementation of the four approaches described above. Statistical quality control, total quality control, and QC circle activities were promoted as a means of improving the health and character of companies.

There was some opposition at first, but the companies where quality control is going well are those that introduced it in this comprehensive way. And recently, countries all over the world, not only America and Europe, are beginning to adopt Japanese-style TQC, CWQC and QC circle activities. To be effective, control must be comprehensive.

1.4.2 True Quality Characteristics and Substitute Characteristics; Product Research

Example 1: The number of complaints received by Company A were halved after sales staff began asking the following questions when taking orders:

"These are our product specifications. How do you intend to use the product, and for what purpose? Are there any other features you would like to have apart from those in the specifications?"

Example 2: At Company B, a chemical company, some raw material lots were extremely difficult to use, even though all lots gave the same analysis results and were of the same purity.

Example 3: A certain paper manufacturer found that customers often complained that their newsprint rolls kept tearing in their rotary

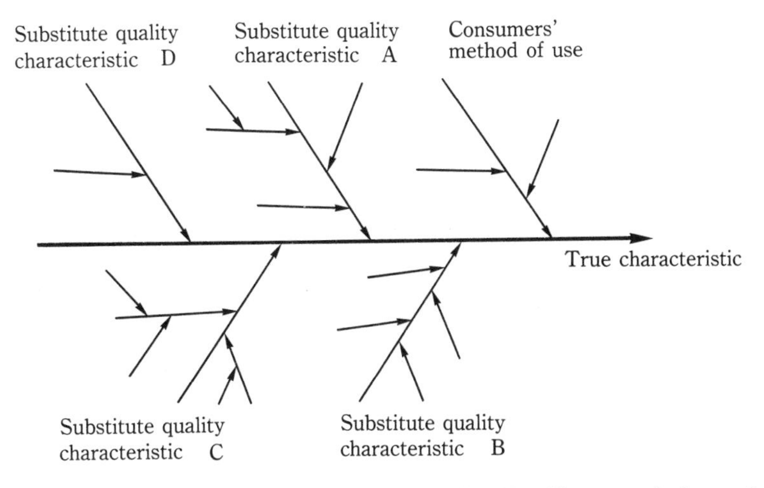

Figure 1.3 The Relation between True Quality Characteristics and Substitute Characteristics: Quality Analysis

presses, even when all the paper's characteristics met JIS standards. Sometimes no such complaints were received even when some of the paper's characteristics were not up to standard.

Example 4: Aren't some products defective even though they meet the manufacturing tolerances, while others are not defective even though they fail to meet the tolerances?

In all of the above examples, the characteristics and values specified in the so-called product standards, raw materials standards, and tolerances are substitute characteristics that do not match the true characteristics actually required by consumers. Their values are also decided with a good deal of guesswork.

The example of the newsprint rolls shows that one of the true quality characteristics required by customers is that the paper should not tear in the presses. Substitute characteristics such as tensile strength are the causal factors behind the true characteristics. When shown on a cause-and-effect diagram, the relationship appears as in Figure 1.3.

True quality characteristics should initially be expressed in the customer's own words, not in the language of engineers.

Example 1: Newsprint rolls: "The paper should not tear during printing, the printing ink should not soak through to the back," etc.

Example 2: Cars: "Good style, easy handling, good acceleration, comfortable ride, good fuel economy, good stability at high speeds, will not break down," etc.

If we do not identify the true characteristics demanded by customers and use technical and statistical methods to determine their relationships with substitute characteristics, we can expect to receive complaints, no matter how well the substitute characteristics are catered for in the design drawings and expressed in product standards, and no matter how strict inspection procedures are. Words such as "performance" or "function" are sometimes used instead of the term "true characteristic," but the term "true characteristic" has a wider meaning than either of these. Many conventional product standards also fail to specify reliability. However, it is often difficult or impossible to carry out inspection based on true characteristics and it is therefore necessary to identify substitute characteristics that are closely related to the true characteristics and significantly affect them, and find out how these are related to factors such as how the products are used. This process is called "quality analysis" or "quality deployment." For this purpose, it is necessary to determine the following:

1. How the customer uses the product and how it should be used.
2. The relationship between the true and substitute characteristics during use.

Here, we will call this quality analysis, or product research in its wider sense. American companies put a lot of effort into product research and are good at quality control even if they have absolutely no knowledge of statistical quality control. Quality control in Japan used to be the opposite: many companies were way behind in product research and were not implementing quality control, even though they had a good knowledge of statistical methods. This meant that quality planning and design were often at the whim of designers or company directors, and inspection was carried out for its own sake, not from the standpoint of the consumer. To improve this state of affairs, more effort must be put into quality analysis and product research, not just into production research. More joint research together with customers is also needed.

1.4.3 Quality Analysis and Product Research

Simply identifying a product's true quality characteristics does not make quality design, process control, inspection, or quality assurance possible. The true quality characteristics must also be successively analyzed and evolved into substitute characteristics expressed in specific engineering language through the use of tools such as the cause-and-effect diagram (see Figure 1.3) and the quality deployment table. This "quality analysis" makes design, process control, and quality assurance practicable. Carrying out quality analysis using cause-and-effect diagrams and quality deployment tables to clarify the quality of substitute characteristics is undoubtedly important, but doing nothing more than drawing up the charts and tables is extremely risky. As with process analysis, everything must be *checked against the facts.* The important things here are product research and the testing of pilot products. If this kind of experimental confirmation is not obtained, some necessary substitute characteristics will be overlooked, while some unnecessary ones will be strictly specified.

Product research usually requires a lot of money and time, but it is an indispensable part of quality control. Companies practicing Japanese-style quality control effectively have achieved many successes by drawing up agreements with users to carry out joint testing of products.

1.4.4 Clarifying Definitions Concerning Quality

Even when true and substitute quality characteristics have been determined through this type of quality analysis, there are still problems with deciding on their meaning, degree of importance, and numerical value. In the strongest terms, we can say that many factories are at present producing products without knowing what they are trying to make.

From this point of view, standards such as the Japanese Industrial Standards, the national standards of other countries, and international standards such as those of the ISO and IEC contain many irrationalities. Thus, although we must use such standards for reference, we should also bear in mind that manufacturing to these standards is inherently fraught with irrationality.

(1) Assurance units

With items such as light bulbs or televisions billed as separate units, the customer is happy if the quality of each unit is satisfactory. However, when it comes to the strength, composition, and other properties of continuous products such as electric cable, yarn, textiles, paper, or steel plate, the composition of chemical products or ores, or the properties of powders and bulk materials, it is necessary to specify the unit quantity on which the quality is based. If this unit quantity (called the "assurance unit") is not fixed, the meaning of the quality figures will be unclear.

For example, simply specifying the electrical resistance of an electric cable does not indicate whether the resistance value is the average for each 100-meter or 10-meter length of cable, or whether the value should hold for every single millimeter of cable. What is the unit length of cable for which the quality is assured?

As another example, what is the meaning of a quality standard that guarantees the calorific value of a certain type of coal as 6,500 calories? Does this mean the average value for each month's shipment, each wagonload, or each sack, or does it assure that each individual lump of coal will have a calorific value of at least 6,500 calories? Such a standard does not indicate what comprises the product lots; in fact, it is not at all clear what it means.

There is also some doubt as to whether the strength of test pieces of the sizes usually used is satisfactory as a substitute characteristic for the strength of steel. In other words, the size of test pieces (assurance unit) should be reconsidered.

This vagueness about the unit quantity for which the quality is assured tends to create problems between suppliers and purchasers, and between official inspectors and manufacturing companies.

(2) Methods of evaluating and quantifying quality

Quality cannot be defined precisely without quantifying it. This means that we must put as much thought and effort as we can into devising methods of measuring it. Particular ingenuity is needed when contriving measurement methods for true quality characteristics, since many of these are expressed in consumers' own words and are therefore difficult to measure; in fact, we often end up having to rely on sensory tests. It is not easy to quantify characteristics such as damage, dirt, color, sound, smell, taste, and texture, which depend on the five senses, or the quality of services, which again depends on human senses; however, advances have been made in physical and chemical measurements, batteries of standard samples have been prepared, and panel evaluations, ranking tests, market surveys, and other forms of sensory testing have progressed and should be studied. In sensory testing, reference samples are no good; boundary samples are needed. Since quality is often judged by lumping all the different quality characteristics together and making a sensory evaluation such as "That's a good car," "That's a good shop," or "She's a fine woman," proper quality analysis is also needed here, and measurement techniques and methods of judging overall value should be investigated.

Even when the problem is not as difficult as this, many companies continue to produce defective products and receive claims against them because upper management and supervisors fail to provide the means of measuring quality, even though a small investment would enable quality to be quantified. Good quantification of quality represents useful knowledge for a company and enables quality control to proceed smoothly. More fundamentally, a company's top management is responsible for indicating the methods and standards to be used for evaluating quality.

The sampling and measurement methods used for this purpose must also be made clear. Quality is often defined by specifying the unit quantity certified by inspection and stating the sampling and measurement methods used. Put the other way round, if the sampling and measurement methods are not specified, it is often impossible to say what quality is being talked about. Once the assurance unit has been decided, the sampling and measurement methods suitable for certifying it will be determined; but if this is not made clear, quality cannot be defined. The question of tolerances is in many ways similar to that of quantification.

Example 1: Does your company have specified methods for evaluating the quality of your products, e.g., cameras, automobiles, or ideas?

Example 2: Since no product can be expected to be completely free of blemishes, are the limits of acceptable surface damage to your products clearly specified?

Example 3: When setting tolerances, is a distinction made between variation within a part (including the tolerance and sampling and measurement errors) and variation from part to part and lot to lot?.

Example 4: Do the design values take into account sampling and measurement errors? What is the relationship between these and the inspection criteria?

(3) Forward-looking qualities and backward-looking qualities

Freedom from defects does not necessarily make a product saleable. Qualities that are positive advantages of the product, i.e., the special features that make it superior to competing products and can be used as sales points—for example, "easy to use," "feels good to use," etc.— are called "forward-looking qualities" (also known as "attractive qualities"). In contrast to this, the absence of flaws or defects is a "backward-looking" (or "must-be" quality). Products are expected as a matter of course to be free of defects and serious disadvantages, and this is why such a condition can be called a must-be quality. The absence of flaws and defects is a necessary but not sufficient condition for the saleability of a product, since products that lack selling points to match consumers' requirements will not sell; this is why we can call forward-looking qualities "attractive qualities." These forward-looking qualities must be clearly set out in new-product planning documents.

Example: A car is a car, whatever its manufacturer. How can the advantages of a particular model be brought out, and what kind of customer is likely to buy it?

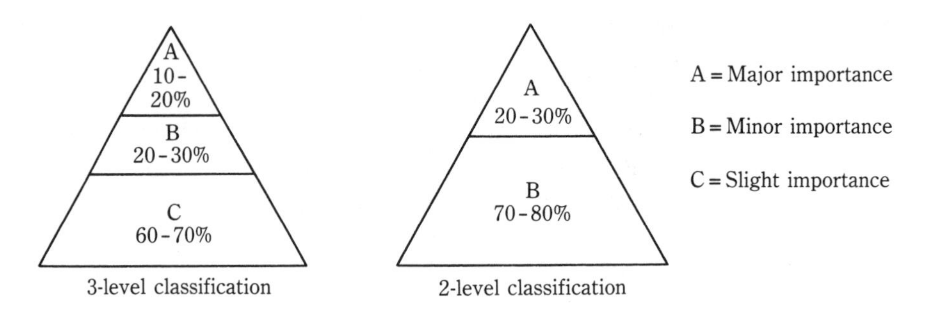

Figure 1.4 Priority Classification of Quality Characteristics

(4) Deciding the priority of quality characteristics; quality weighting

A given product has at least twenty to thirty quality characteristics and may have as many as several hundred. People who do not understand quality control will say that every quality characteristic is important. This approach, however, means that either the price of the product will be extremely high, or it will end up as a product that is neither one thing nor the other and has no distinguishing features. We must therefore classify both positive and negative quality characteristics as being of major importance (A), minor importance (B), or slight importance (C), or at least into two priorities (A and B). We should consider classifying backward-looking quality characteristics (i.e., defects) in even greater detail, e.g., into the following four classes: critical (defects that might endanger life or limb); major (defects that seriously affect performance); minor; and slight. Among forward-looking quality characteristics, the class-A characteristics are the important selling points.

In my experience, a maximum of three classes — two for general products — is sufficient. I usually recommend that the proportion of characteristics assigned to each class follow the Pareto principle (vital few, trivial many), as shown in Figure 1.4. The same procedure can be used for the dimensions of parts and assemblies, with the most important dimensions denoted by the letter A. Class-A defects must be completely eliminated, while class-A forward-looking quality characteristics should be clearly highlighted. Some class-C defects, such as very slight damage to a product's finish, may be allowable. In other words, class-A characteristics must be strictly controlled, while the control of class-C characteristics may be more relaxed. If this distinction is not clearly made, too much effort will be put into controlling class-C characteristics, attention will be drawn away from class-A characteristics, and serious claims may be made against the product.

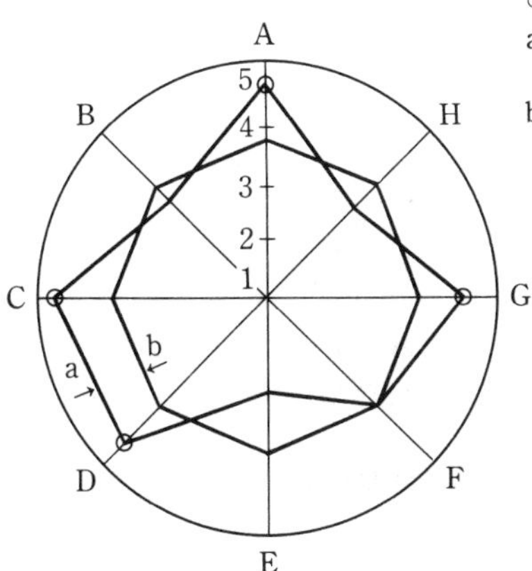

○ : Selling points
a : A product that has drawbacks but also has clear selling points.
b : A product that is neither one thing nor the other and has no distinguishing characteristics, produced by trying to make all of its qualities good.

Figure 1.5 Products with and without Selling Points; a Quality Radar Chart

Also, some quality characteristics, particularly true quality characteristics, may be mutually incompatible (i.e., when characteristic A is improved, characteristic B deteriorates), or trying to improve all the quality characteristics will make the product too expensive. As shown in Figure 1.5 a, when certain characteristics are sacrificed in order to emphasize the selling points, purchasers are happier with the product. As mentioned before, care should be taken not to try to improve every characteristic, since this will result in a product or service that is neither one thing nor the other and has no distinguishing features.

(5) Clarifying the definition of defectives and defects; actualizing latent defectives

As long as the terms "defective" and "defect" are variously defined by different people and different departments, they are not useful. Their definitions should be standardized; the following four points must be taken into account when doing this:

1. The concepts of defectives and defects should be standardized throughout the sales, design, production, and inspection departments. Any differences between terms or use of terms between manufacturers and users should also be clarified.

Example 1: Some glass to be used for windows is slightly scratched. Is this a defect or a defective? What about boundary samples? Would it not be possible to build perfectly satisfactory houses very cheaply using such a second-grade product?

Example 2: A part does not meet the tolerances specified on the drawings but is still usable. Does this make it non-defective?

The terms "defect" and "defective" are widely used in the language of quality control. However, because these can easily give rise to misunderstandings in the legal area of product liability, ISO and American standards draw a distinction between "defect" and "non-conformity," and between "defective" and "non-conforming unit." This distinction is not made in Japanese, and is not incorporated into Japanese industrial standards.

For definitions of defective unit, proportion defective, defect, defect ratio, non-conformity, and non-conformity ratio, see Section 2.3.

2. Go-through rate (non-adjustment rate): Used in the assembly industry, this term means the proportion of products that, once assembled, perform as required without any adjustment or rework. From the QC viewpoint, products that have to be reworked, adjusted, or modified on site are defective. The rework fraction and adjustment fraction should be treateu as part of the fraction defective. Products with a good go-through rate are usually highly reliable and do not break down later on. It is therefore extremely important for assembly industries to carry out thorough quality and process analysis, and then to control their processes in such a way as to raise their go-through rate.
3. Exposing latent defectives: Companies that are not practicing quality control tend to classify only scrapped products as defectives, but as discussed above, reworked or adjusted products, as well as those that are accepted as is, are all in fact defective. There are also hidden defectives. When quality control is implemented, all these latent defectives must be exposed and clearly identified (see Section 4.3.4).
4. As-is acceptance: Products that marginally fail to meet specifications or tolerances are often still accepted for use, without reworking, repair, or blending; this practice is known as "as-is acceptance," and it may involve a variety of raw materials, work-in-hand, or final products that fail raw material, intermediate or final inspections. As-is acceptance is practiced for various reasons, in all companies, even those that are doing their best to practice quality control.

Standards and tolerances are set in numerical terms, but products are sometimes usable even if they deviate slightly from the specified values. There are two reasons for this:

1. Standards and tolerances are often not considered sufficiently carefully, and may be stricter than necessary.
2. It is often illogical or odd for a product to be considered usable up to a certain cutoff value of a characteristic and suddenly to become unusable above this value.

The use of as-is materials often does not affect quality or cost. However, as-is acceptance of materials or products that do not conform to standards does require special action; the product or lot must be carefully followed up to determine whether the special acceptance has resulted in any changes in performance or reliability. The following action should then be taken:

1. If no undesirable effects occur as a result of the as-is acceptance, the standards and tolerances may be relaxed slightly.
2. If some slight undesirable effects are observed, process control and inspection should be tightened up. When as-is acceptance is practiced, it is very important to clarify the following two points:

   (a) Whose permission is required for as-is acceptance?
   (b) Over what range is as-is acceptance allowed?

The special precautions necessitated by a policy of as-is acceptance are explained in more detail in Section 6.14.3.

(6) Statistical quality: think of quality as a statistical distribution

From the consumer's viewpoint, each individual product must be of satisfactory quality. However, when considering the quality of a particular product, both producers and consumers think not so much of the quality of individual units, but rather of the quality of groups of dozens or hundreds. For example, let us consider the lifetime of electric light bulbs, which are produced and sold in the tens of thousands daily. If two types of bulb were available—one that lasted for about 100 to 2,000 hours and another that lasted for about 900 to 1,100 hours—ordinary consumers would probably be reluctant to risk getting a bulb that might only last for 100 hours, and would choose the latter type, with its lesser variation in lifetime and consequent lower risk. A wide variation in the quality of replacement parts would also cause problems for the consumer. In other words, consumers want products of uniform quality in terms of groups of products and statistical distributions. However, since our work is affected by an infinite number of different factors, it is impossible to manufacture products without any variation. What we aim to do, rather, is to control the variation within certain limits; this variation is known as the process capability.

In quality control, we set quality levels for groups of products and we control these levels companywide. On the shop floor, we try to control the process in such a way that we will obtain product lots with the specified statistical distributions.

One of the first steps in quality control is deciding what to make in order to satisfy customer needs in terms of the statistically distributed quality of product lots. Statistical ideas and methods are extremely useful for balancing market research information with existing technical knowledge, factory technical levels, and process capabilities in order to set quality with a reasonable degree of variation. Many existing standards and specifications are either too strict or too lax, perhaps because those who set the standards did not fully understand this concept of statistical distribution.

Statistical quality is not fixed; it always has a range of variation and is a living entity that changes according to technical and economic conditions and advances in process capabilities. Even determined standards and specifications are not constants; they must be continually revised and updated.

(7) Four definitions relating to quality

Many of the quality standards traditionally used in companies are not clearly defined. Considering statistical aspects such as distribution, dispersion, and error, as well as responsibility and authority within the company, the following four types of quality standard should be distinguished:

1. Quality standards applied to processes: these are quality standards for which the production department is responsible.
2. Quality goals applied to research and technology: these "quality targets" are the responsibility of research and technical departments.
3. Assured quality level for consumers: it is the responsibility of the marketing department to determine the level of assured quality.
4. Inspection criteria: the inspection department is responsible for inspection standards.

This categorization is based mainly on a consideration of responsibility and authority within the company.

The quality standards applied to processes can be called quality standards because they should be set by combining company policy with a consideration of the quality levels attainable (in light of factory process capabilities) if work is carried out according to the operating standards (i.e., the quality levels attainable in the well-controlled state). The production department is responsible for controlling the process for products to meet these standards. Within these quality levels, if some of the products fall slightly outside these quality standards, it is not the fault of those who made the product, but rather of management for not providing machinery and equipment capable of meeting these standards. (However, since the people who make the product are responsible for quality assurance, it is the production department's job to detect and eliminate defectives.)

The quality goals assigned to the technical and research departments ("quality targets") are determined in accordance with company policy regarding what the company will make in the future; these targets must take account of surveys of customer needs, along with other factors. Such targets constitute objectives for technical improvement.

In Japan at present, it is the responsibility of the production department, as well as QC circles, to control processes in order to prevent defect recurrence, as well as to improve processes. Evolutionary quality improvements should therefore be the responsibility of technical departments, while constant small quality improvements should be that of production departments.

"Assured quality" levels, or the quality levels assured to consumers, are self-explanatory; they are the quality levels that the marketing department is responsible for communicating to consumers. They might also be called "catalogue Standards." However, some companies take their average quality and misrepresent it as an exaggerated assured quality level, as shown in Figure 1.6. This practice is bound to produce complaints.

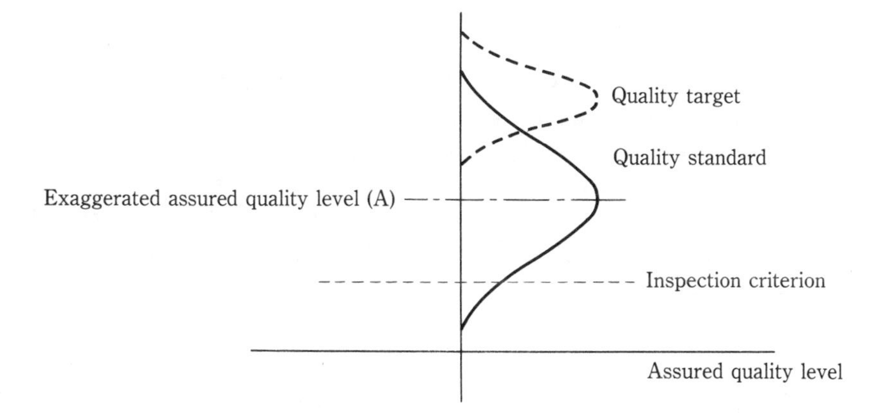

Figure 1.6 The Four Different Types of Quality Level

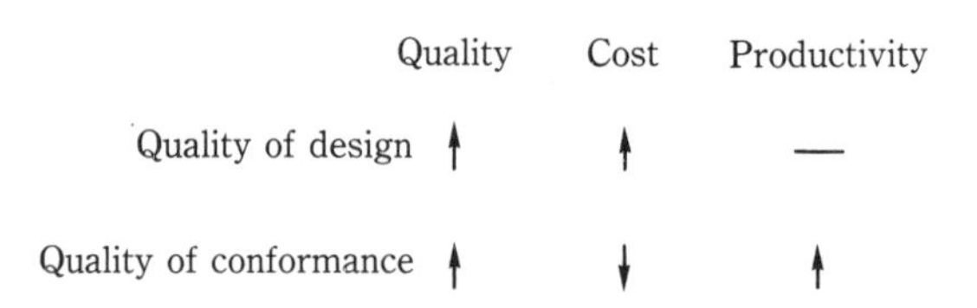

	Quality	Cost	Productivity
Quality of design	↑	↑	—
Quality of conformance	↑	↓	↑

Figure 1.7 The Relationship between Quality, Cost, and Productivity

Inspection standards are also just what their name implies, the values used by inspection departments as criteria for evaluating products. Sampling inspection and even 100% inspection are often accompanied by sampling, measurement, or experimental error; inspection standards must generally be set at a higher level than the assured quality level.

The above four types of quality level are interrelated, but each stands in its own right. Figure 1.6 illustrates the relationships among them.

One of the first steps a company should take when implementing quality control is to review exactly what is meant by its vaguely titled "Quality Standards," "Company Standards," etc. This is particularly important in the electrical and machinery industries, where it is not clear to which of the above four types of quality level design values and drawing tolerances correspond. This varies according to how the design values are used, but design will probably prove impossible if the company's definitions are not clarified.

(8) Quality of design and quality of conformance

The terms "quality of design" and "quality of conformance" were first coined by J. M. Juran. "Quality of design" is the level of quality a company plans to achieve for its product; in general, costs rise as this level is raised. "Quality of conformance" is the difference between the actual quality of a product and its designed quality (i.e., the quality for which the company aims). As quality of conformance is raised, it approaches quality of design; and, if the number of defectives decreases, costs usually come down, while productivity goes up. It is extremely important to distinguish between these two types of quality (see Figure 1.7), and care is needed when setting the quality of design, since various problems will be caused if process capabilities and other factors are not sufficiently taken into account.

1.4.5 What are Good Quality and Good Products?

The various points discussed above must be taken into account when setting quality, but what exactly is a product or service that sells well over a long period? Put simply, it is a product or service that suits consumers' requirements (i.e., needs and desires) and is of fair price and quality. This is, however, an extremely hazy definition.

As far as backward-looking quality is concerned, the basic condition for a successful product is of course either zero defectives and defects or a PPM fraction defective, but a good-quality product is one that receives a good overall evaluation in which all the qualities (including the forward-looking) are appropriately weighted. In a nutshell, then, a good-quality product is one with a good overall balance of features. Various methods of calculation have been proposed to deal with this point, but none of them amounts to anything more than playing with formulas and numbers. In practice, a product will succeed if it satisfies certain basic conditions, if it suits the requirements, hobbies, and interests of the consumers at whom it is aimed, and if it is launched at the right time. The products that succeed are not those that are fairly good all round, but those that excel, according to the Pareto principle, in from one to three particular quality characteristics. The fact that the success rate for new products is as little as 1 to 5% shows us, however, that this is still not an easy problem.

Various other terms are used to describe quality—for example, quality in the various phases of new-product development—but these will not be discussed here.

Finally, I would like to sound a few words of warning in relation to quality:

1. Consumers' quality requirements and conditions of use vary from country to country and time to time. Since consumer requirements are both diversifying and becoming more demanding, companies must constantly gather information, anticipate customers' needs and wants, and strive for improvement. Quality is never good enough.
2. The biggest problem during new-product development is winning over opponents within the company. This is done by furnishing new-product planning documents with data that show that the plans are so good that people will be persuaded to support them.
3. Whenever a completely new product that does not yet exist anywhere in the world is under development, action must be taken to ensure that the new product's growth curve rises quickly.

1.5 WHAT IS CONTROL?

1.5.1 The Old-Fashioned Approach to Control

The biggest problems experienced when quality control was being introduced in Japan were the result of the confusing variety of terms denoting "control." People at offices, factories, and sites were clearly not aware of the idea of control, and the control approach was very rarely built into the organization of such places. Control charts and other statistical tools are only effective when linked with the philosophy of control described here and incorporated into the system of responsibility and authority for control in companies and factories, i.e., with their overall organization.

Expressed simply, control consists of "checking whether or not work is being done according to policies, orders, plans, and standards, and, if it is not, taking action to correct any deviation and prevent it from recurring, then proceeding according to plan."

In the past, all that happened was that orders (or prayers?) such as "Make good products cheaply," "Cut costs," "Improve quality," or "Don't make defective products" were simply passed down without modification in the form of orders, as if through a tunnel, from the company president to the managing director, and thence to head-office department managers, factory managers, section managers, supervisors, foremen, and workers. Moreover, the tunnel down which these orders were passed was often kinked or blocked. If orders are not communicated to everybody, control cannot get off the ground.

I call this type of management, in which orders such as "Do your best" are simply passed down a chain of command and people are exhorted to achieve an objective without being given the means to do so, "management by exhortation," "Yamato-spirit management"* or "management by the whip." Morale is of course important whenever people are being managed, but good control cannot be achieved through morale alone.

Various forms of management were being exercised in Japan in the past, but they did not go well for the following reasons:

1) As with the English terms "control," "management," and "administration," various terms were used to indicate control, and their definitions and shades of meaning differed from person to person. Someone

---

*Translator's note: "Yamato" is the old name for Japan, and "Yamato spirit" means the type of do-or-die approach which, like the Charge of the Light Brigade, is heroic but not very practical.

happened to translate "quality control" into Japanese as "*hinshitsu kan-ri*"; this term stuck, and it was decided to use the word "*kanri*" for "control" in the terminology of Japanese industrial standards.

2) Discussion of control was too abstract and too idealistic.
3) Some people thought that "control" meant "tying people down."
4) Specific methods of gradually achieving control targets were not satisfactorily considered. Orders were simply passed down the chain of command, which was often twisted or broken.
5) Nothing was known about statistically based analysis and control techniques.
6) Only directors and foremen, and others in positions of responsibility, were kept informed, and there was not enough effort to educate everybody.
7) Only complex, detailed methods were used, and control was not exercised from the broad, overall standpoint.
8) Sectionalism was rife, there were barriers between departments, and communication was poor.
9) People tended to be satisfied if they simply got good results; they did not think about methods and processes.
10) Companies considered that "management by exhortation" satisfactorily maintained the status quo. When mistakes were made, people promised that they would not happen again, but no specific actions were taken to prevent recurrence.

This confusing situation is exactly the same in quality control; things will not go well if contradictory orders such as "Reduce costs" and "Raise quality" are issued independently by different departments. Like quality contol, control must be comprehensive; it will not proceed smoothly if exercised sporadically on the basis of sectionalism or power struggles among the different departments of an organization.

1.5.2 Control Methods and Philosophy

The slogan "plan-do-see" has long been associated with scientific management, but it does not suit the Japanese. They learn at school that "see" is "*miru,*" which also means "look," and therefore tend to think that "plan-do-see" means doing something and then just standing and gazing at the result.

We are usually told to "rotate the plan-do-check-act (PDCA) cycle," also known as the Deming Cycle (Figure 1.8), but this is also insufficient. I have had success with splitting the cycle up into the six steps, shown in Figure 1.9. These six steps are as follows:

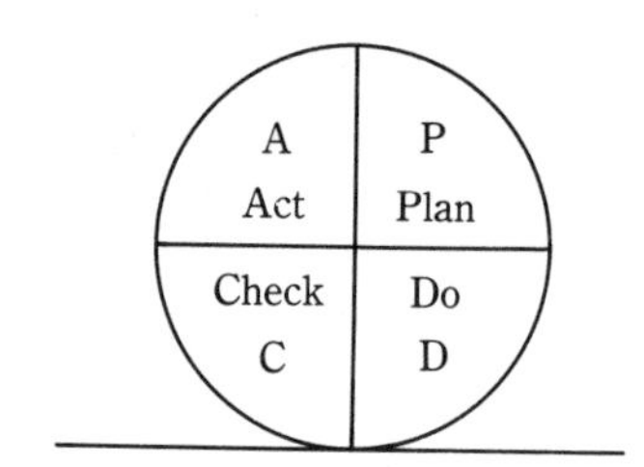

Figure 1.8 The Control Cycle (with 4 Steps)

1) Decide on an objective
2) Decide on the methods to be used for achieving the objective } plan
3) Carry out training and education
4) Do the work } do
5) Check the results } check
6) Take corrective action } action

After these six steps have been followed, a seventh step is to recheck to see whether the corrective action has worked or not. This really is a scientific management procedure using the QC approach.

(1) Decide on objectives and targets←policy←information/surveys

Control is impossible unless objectives and targets are clearly defined, and it

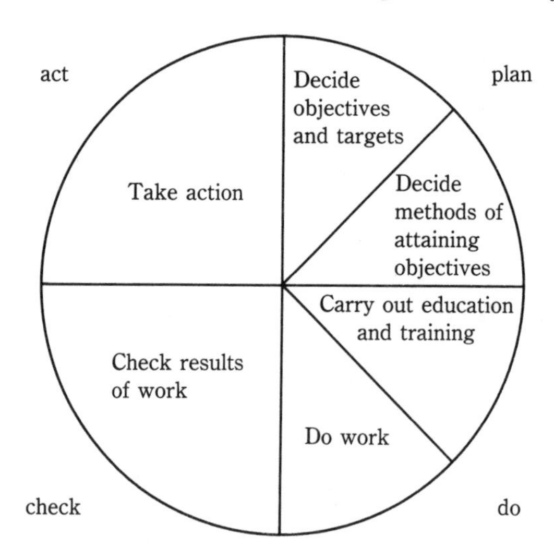

Figure 1.9 The 6 Control Steps

is also impossible if objectives and policies change with every passing whim. For example, we cannot control a design or process without setting quality standards, and we cannot control research and technology without setting quality targets. Some general points to note when setting objectives and targets are:

(a) Objectives are decided as a result of policy. While top-management policy is the most important type of policy, but it must be broken down into subpolicies at the level of the department manager, factory manager, section manager, supervisor, and foreman. These subpolicies must be clearly spelled out so that people at every level know what to do within their range of responsibility and authority. Everyone in a position of responsibility should have a policy, and all these policies, from the company president's down to the foreman's, must be mutually consistent. In many cases, however, a company's president and directors, as well as the people under them, have either no policy at all or extremely abstract ones. Control is impossible without policies, and all leaders need the courage and sense of responsibility to decide on them. Policy clarifies the criteria for setting objectives and taking action.

(b) Policy must have a proper rationale; a policy that boils down to someone saying, "This is what I believe; follow me!", is an ungrounded, or "Yamato-spirit" policy. It may sometimes succeed but it is rather risky and the failure rate is high.

To provide a rationale for policy, we need accurate information from inside and outside the company, including such sources as market research information; survey data on consumers, competitors and foreign markets; in-house technical and research capability data; process capability data; and raw material information. This information should be promptly stratified, collated, and comprehensively analyzed. Many companies either lack the data they need for policy setting, or have the data but fail to get information to the right departments at the right time. In other cases, the company lacks a function for analyzing data comprehensively. Tools such as Pareto charts, frequency distributions, graphs, control charts, and operations research methods are extremely useful for this kind of analysis.

Providing and analyzing this kind of information is the duty of staff departments or subordinates. If possible, a number of different proposals should be prepared and used as the basis for setting specific objectives and targets in accordance with the various policies. Of course, no information is 100% certain. Thus, when information which is 70% or 80% complete has been collected, the rest is up to the courage, decisiveness, and executive abilities of managers, with due regard to the probabilities of failure. If we have collected as much information as we can, we can call our decision "scientific." In other words, it is a question of how quickly we can proceed with caution.

Everyone in a position of responsibility, from the company president down to foremen on the shop floor, should reflect on the kind of data they should collect in order to decide on their own particular policies.

(c) Policy must be decided from an overall standpoint; in other words, policies must not be issued in fragments, and there must be no inconsistency between different policies. In Japanese companies, the head office does not have the function of making comprehensive judgments, and sectionalism is very strong. For this reason, policy often emerges piecemeal from the various departments. I call this kind of company "many-headed." Control is completely impossible with such a system, no matter what policies are issued.

(d) When setting policy, we must decide on our priorities. We should remember the Pareto principle (vital problems are few but trivial ones abound), since a mere listing of ten or even twenty policy items says nothing about their relative importance, leaving the question of priority untouched. Many people simply list the things that occur to them, but this is not policy. In practice, the really important problems are not the sporadic problems, but the chronic problems on which everyone has given up. It is best to narrow down important policy items to two or three, or at the most five, while the rest should be treated as routine management points.

(e) Objectives and targets should be stated clearly and specifically, if possible with concrete deadlines. Abstract policies such as "Good products at a reasonable price" or "Good, cheap and quick" are not very useful by themselves. Of course, this kind of abstract, motivational policy is acceptable as basic company policy, but more specific policies expressed numerically (with methods of measurement, mandatory targets, and desirable targets) should be added; for example, "Based on January to March figures, halve the number of defectives produced from April to September," or "From next March, market 20,000 amateur snap shot cameras per month, at a cost of approximately $150." These will then become control characteristics.

Targets should be classified into two types, mandatory targets and desirable targets.

(f) Policy can also be split into two types, as follows (see Table 1.3):

i) Methodological policy
ii) Objective policy

The former type concerns ways and means of achieving objectives, e.g., "promotion of standardization," "clarification of responsibility and authority," "use of control charts," or "faithful execution of operating standards," while the latter means policy that states specific objectives, e.g., "Halve the fraction defective of product A by December," or "Cut the cost of part B by 20% in six months."

In quality control, methodological policy has tended to predominate. Such policy is needed when quality control is first introduced, but actually to promote quality control and achieve results, objective policy centering on QCDS (quality, cost, delivery, and safety) aimed at more specific practical problems is needed.

(g) As it proceeds down the hierarchy of an organization, policy must be broken down into more and more detail and be made more specific, while remaining consistent. (This process is called "policy deployment" or "goal deployment.")
(h) Policies should not focus on departments or organizations but on objectives and problems. They should allocate responsibility to different teams or related departments.
(i) While objectives and policies must be announced annually or at the end of each accounting period, they must be formulated on the basis of long-term policies and plans.
(j) Objectives and policies should be put in writing and widely distributed.

To summarize, the basic spirit of policies, objectives, and targets should be expressed in words, while specific targets should be expressed in figures. It is insufficient to use either words or figures exclusively.

(2) Deciding on the means of achieving objectives←standardization←technology and administrative techniques

It is not enough simply to state objectives and targets without indicating the means by which they can be achieved. For example, announcing quality objectives or cost objectives without also deciding how these are to be achieved and what each person should do to achieve them, and leaving people to do what they think necessary, is just another form of the "Yamato-spirit management" or "management by the whip." Good work cannot be produced in this way. A company must formulate rules of operation, indicating what should be done to achieve its targets and objectives, and what every employee must do. In other words, the company must prepare standards in the wide sense—operating standards,

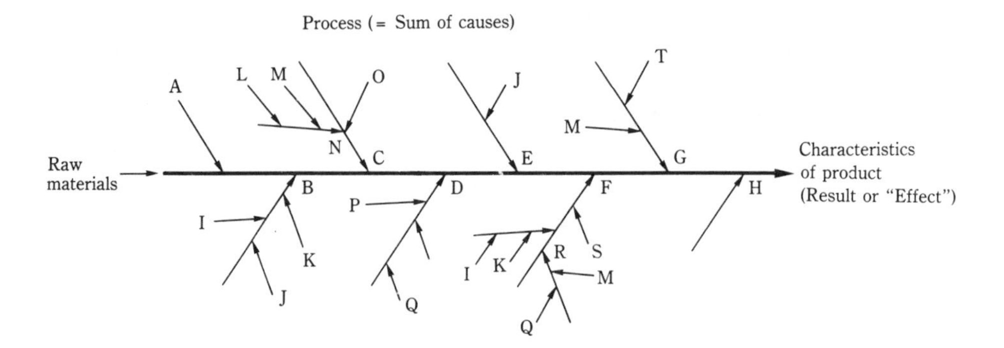

Figure 1.10 Cause-and-Effect Diagram

technical standards, design standards, staff regulations, etc. To clarify this approach and make it more concrete, I proposed the cause-and-effect diagram. Standardization in a narrow sense means "unification," but in this book I will use the word in its wider meaning.

Formulating and continuously improving standards is the work of technical and administrative staff; it is also a QC circle activity. Standards are needed for the general running of a company; they are essential for every type of management and should not be devised solely for quality-control purposes.

Since standardization is treated in more detail in Chapter 5, I restrict my discussion here to a number of problems focusing on process control.

In process control, we first design a process, then prepare QC process charts, carry out process analysis, formulate standards, and revise those standards.

(a) Standards particularly technical and operating standards, should deal with causes. The practice of "preemptive control" is recommended, whereby it is necessary to distinguish between causes and effects (i.e., objectives); the cause-and-effect diagram (see Figure 1.10) can be used for this. When preparing cause-and-effect diagrams, relevant technical, production, design, and inspection staff, together with foremen, line workers, and others, should pool their knowledge using brainstorming and other techniques. Based on this, preemptive standardization—i.e., standardization that anticipates problems as much as possible—should be carried out on the basis of proper technology, experience, statistical tools, etc. Seeing a result and then rushing to do something about it is crisis control. In the case of process control, feed-forward (preemptive control) is more important than feed-back (crisis control).

It should be noted that processes are affected by an infinite number of factors, and those with which we are concerned in quality control are referred to here as "causes" or "key factors." Methods of preparing cause-and-effect diagrams are explained in Section 4.7.4.

(b) How to control the key factors for control of the process must be decided; this involves preparing specific standards relating to the really important causes. There are a limitless number of factors causing dispersion in the targeted characteristics, and it is uneconomical to try to control all of these in achieving our goals. Our work is generally affected by an infinite number of causes, but they all act in accordance with the Pareto principle. We will always achieve results if we choose from among the causes we have identified the really important ones, and then standardize these.

To identify these key factors, we must have our own proper technical knowledge concerning the process, we must carefully observe the actual conditions at the workplace, and we must also be capable of analyzing the process statistically. This means that, from now on, all engineers must master statistical techniques as a matter of common knowledge along with physical, chemical, electrical, and other specific technologies. However, since most problems (in fact, approximately 95% of them) can be solved using the seven QC tools (i.e., check sheets, graphs, histograms, Pareto charts, cause-and-effect diagrams, scatter diagrams, and control charts), everybody in a company from the president down to QC circle members should learn the statistical approach and these tools.

(c) Standardization is carried out in order to delegate authority. While authority must be delegated, not all responsibility can be delegated. It is therefore necessary to standardize what should be done in exceptional or abnormal situations. Such standards could, for example, be called control standards. When an abnormality occurs in the process, the following should have been decided in advance:

— Who should do what (responsibility)
— How far they should go (authority)
— Who they should receive instructions from.

Once authority has been delegated concerning a particular matter, further orders concerning this matter should not be given.

(d) Standards should be formulated to state the objectives (i.e., the characteristics) clearly.

(e) As many as possible of the people involved should have a say in standardization. People tend to observe standards and regulations that they themselves have set.

(f) Human beings naturally make mistakes, and it is wrong to get angry at a subordinate's errors. If a mistake is made, everyone involved should be invited to discuss how such slips could be prevented.

(g) A standard that has not been revised, is a standard that is not being used.

(h) Standards must be properly documented, with a record of all changes. A particular effort should be made systematically to accumulate a body of technology within the company. This will put the company's technology on a solid basis and enable the technology to be upgraded and exported.

(i) All standards must be mutually consistent.

(3) Carry out education and training

All those in charge of others are responsible for training their subordinates.

Even if standards and regulations are properly prepared, simply handing them out will have no effect, as they will probably not be read. Even if they are read, what can be written down is never sufficient and is likely to be misunderstood. Even if the standards are understood, many will be impossible to carry out. In the past, the Japanese, particularly those in bureaucratic positions, tended to set strict standards and regulations but were lax in their observation. The habit of following specific directives was not strong, and the required ambience was lacking; in fact, some people even took a kind of pride in doing as they pleased and deliberately breaking the rules. This suggests the need for training and education. Quality control in particular is a revolution in the philosophy of management, and it needs a complete turnaround in the attitudes of everybody in the company from the president to the workers on the shop floor. Quality control truly begins and ends with education.

Training and education within a company consist of the following three types, each of which is as important as the other two:

(1) Group training
(2) On-the-job training of subordinates by superiors
(3) Letting people teach themselves by boldly giving them full authority for their work

I think of management as being of two types: one based on the view that mankind is fundamentally good, and the other based on the view that mankind is fundamentally evil. According to the latter stance, since people are bad by nature, and since we do not know exactly when they are going to do something bad, they must be closely watched. With this approach, no one can feel relaxed in their work, much time and money are spent on controlling and checking, costs go up, and people forget why they are managing. Conventional centralized management tends to duplicate checking and become a form of management based on the view that mankind is evil.

I believe we should promote the type of management based on the view that mankind is fundamentally good, according to which people are capable of doing excellent work if properly schooled and able to change their way of thinking. Of course, aptitude tests are needed for certain types of work. The ideal type of management is thus management without checking, a situation in which everybody manages him or herself, or self-management. Education generally increases people's span of control, that is, the number of people whom they can manage, and makes them more and more able to delegate authority. Without education, a person cannot control or supervise even one other, and is probably incapable of delegating work. At all events, education is absolutely vital, and no business can make progress, however well-constructed its organization, if the people in the organization are no good. Industry generally should invest more in education. Like an orchestra conductor, one person should be able to manage a hundred others.

Teaching is an excellent way of studying. When people actually try to teach standards to others, they understand them better themselves and appreciate their difficulties, imperfections, and defects. This helps in the rationalization of standards.

Educating their subordinates properly and creating an atmosphere in which standards are obeyed is an important duty of supervisors and managers. Achieving good control by educating people skillfully in this way is the essence of management based on the view of mankind as fundamentally good. It is also necessary to revise operating standards and other standards and regulations to ensure that they are appropriate. At any rate, no good work can be expected if supervisors jealously guard their working secrets and fail to teach them to their subordinates. Managers who are always bawling out their subordinates but never teaching them anything have no right to be called managers.

The basic approach to education and training described above is one of the foundation stones on which QC circle activities were started. Statistical methods are also extremely useful in researching educational methods and evaluating their benefits.

(4) Do the work

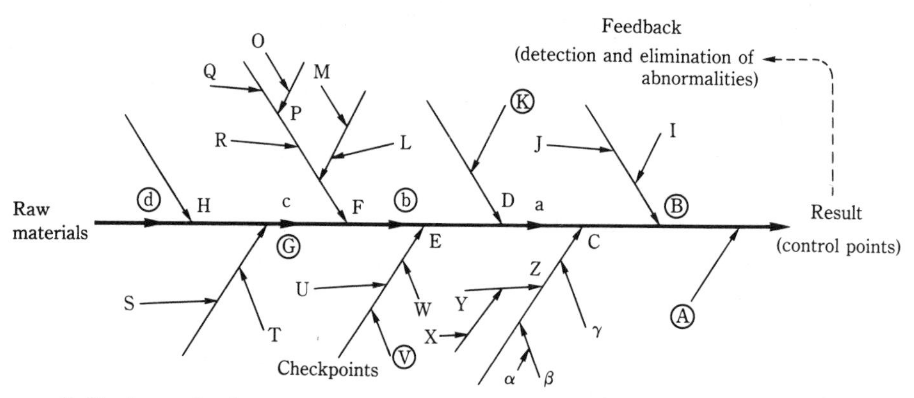

O: Key factors, i.e., important causes which must be checked by walking round the workplace

Figure 1.11 A Model of Control

(5) Check the results

Checking results consists of seeing whether the work is being carried out in line with policies, instructions, and objectives. Issuing an order and teaching people how to carry it out is not the end of a manager's job; he must also check the results and find out whether things have gone well. Orders and standards are of course imperfect, but so are people and equipment. However good people's intentions may be, mistakes and accidents will happen and misunderstandings will arise. When necessary, executives, managers and supervisors must therefore check whether the work is proceeding well and take the necessary action to ensure that it goes smoothly according to policies and directives.

All those in positions of authority must think carefully about where, when, what, and how to check, and then actually carry out these checks. Then, as everyone learns what control is, and becomes familiar with it, they begin to exercise control by themselves, i.e., they become capable of autonomous control.

Managers should leave well alone when work is proceeding smoothly, but step in and take action when things are not going well or there are problems. In other words, control should be based on the exception principle. We must check the work and decide whether or not it is in an exceptional or abnormal state.

When controlling work (i.e., processes), it should be checked by the following two methods, according to the cause-and-effect diagram approach (see Figure 1.11):

(a) Check the causes: Take a walk around the workplace and see whether everything is being carried out according to policies and standards, e.g., working methods, setting up, measurements and so on. This is mainly the job of supervisors at the end of the management chain.

One of the purposes of walking around the workplace is to check whether or not it is operating according to instructions and standards; for example, whether or not raw materials, equipment, measuring instruments, automatic control systems, jigs, and tools are functioning well and work is proceeding smoothly. In the case of process control, this means going around to see whether or not the key factors are being well controlled. In companies with no policies, directives, or standards, it is not clear what should be checked and what should be used as a standard for comparison. Also, if we do not consider what items have priority, make checklists, and use them for checking, we will be doing no more than taking an aimless stroll.

The causes on such checklists are called checkpoints. Checking these is the responsibility of lower-level managers, and it is better for more senior managers not to carry out very detailed checks of causes. People who like to carry out detailed checks even after promotion to department manager or director deserve the title "artisan director" or "artisan department manager." Rather than spending their time checking details, department managers and directors should check the work through its results, as described below, and make time for themselves to think about the future (see Section 5.5.1).

Although industrial processes are affected by an infinite number of causes, the following must be borne in mind:

i) Only a small number of causes can be pinned down in operating and other standards, and even these cannot be controlled 100% correctly.
ii) Because we only have limited time and cannot be everywhere at once, we can only see a small part of the work going on when we walk around the workplace.

It is therefore not enough simply to make rounds of the workplace attending to the priority items, and it would take an extraordinary amount of time and effort to monitor all the key factors perfectly. This is why we need a second method of checking, as described below:

(b) Checking through results: This method consists of checking whether work and processes are proceeding well by examining the results of the work, e.g., changes in quality in its narrow sense, production volumes, delivery times, inventories, amounts of materials and labor required to make a product unit, unit costs, safety, pollution, etc. This means trying to control processes and business operations by observing the results, feeding back the information thus obtained to the process, discovering abnormalities in work, processes, and operations, and eliminating the causes of those abnormalities. To understand the relationship between the causes and the results, we should use the cause-and-effect diagram approach, as shown in Figure 1.10. In checking through results, a number of points may be noted:

i) The method consists of controlling processes and business operations *through* results. It does not mean checking *the results themselves*. For example, if work is controlled through quality, and processes and operations come under control, good things will naturally be produced cheaply. Thus, if we are talking about quality control, "building in quality via the process" is one of its basic tenets. The idea of controlling and checking quality itself leads to old-style, inspection-oriented quality control and will end in failure. Many attempts at cost control also fail, and cost accounting loses much of its meaning, if pride of place is given to the idea of controlling cost itself. The significance of cost control lies in controlling *through* cost. We should be careful to distinguish between controlling through something and controlling the thing itself. Control must not be controlled with inspection. Taking action with regard to individual products or lots based on the results of quality measurements is inspection. Taking action with respect to work, processes, and business operations based on the results of quality measurement is control (i.e., process control).

ii) We must constantly ask ourselves what is the most suitable characteristic of the results of our work through which we can check the work? In TQC, such characteristics are generally called control characteristics or control points (see Section 5.5.1). In the network of control, the company president, department managers, factory managers, section managers, supervisors, foremen, and line workers must each decide on their own control points and ask themselves what they should use to check these in order to fulfill their control responsibilities.

    The items that can be checked are not restricted to quality; they can also include unit cost, production volume, amount of labor and material required to produce a product unit, sales volume, personnel, safety, and other items. If the control net is skillfully spread, control can be carried out easily and with confidence. Some control characteristics are decided according to policy, and others come from the daily routine. Managers from section manager up to company president usually have from twenty to fifty control items, while supervisors and below have from five to twenty.

    If the chosen control characteristics relate to quality, they are called quality characteristics. Readers should note that a distinction is drawn between control characteristics and quality characteristics in this book.

iii) Results are always scattered. People often think that the same thing will be produced if the same raw materials and equipment are used and the same person uses the same method. This is a serious misunderstanding. Some people have no appreciation of statistical dispersion and start screaming if production yields fall off by even a small amount. This type of behavior will probably result in the emergence of false data from the workplace. As stated before, the number of factors affecting the results of our work is limitless, and we cannot control all of them. However closely we stick to the standards, the results of our work (quality, production volume, yield, etc.) will always follow a certain statistical distribution.

Thus, in judging whether or not an abnormality has occurred in the work or process, we must examine how the distribution has changed. In other words, we must make judgments and exercise control on a statistical basis. A chart on which a pair of control limit lines is drawn (called a control chart; see Figure 1.12) can tell us whether or not a distribution has changed with time or whether a result is out of the ordinary. Using these control charts enables us to make judgments most easily and objectively. Graphs without control limits, not drawn up in the form of control charts, are useful, but control charts are easier to use. When we use control charts, if we control a process by taking quality characteristics as our results and plotting them, we are controlling the process through quality.

Many people call control charts "quality-control charts," but control charts are useful whatever we plot, whether production volume, unit cost, yield, or any other characteristic value equivalent to a control characteristic. Since control charts are also employed extensively outside the quality-control field, the term "quality-control chart" is inappropriate.

Control charts facilitate the work of all those with control responsibilities, and control charts and graphs should be seen and used by leaders at every level. However, although control charts tell us of the existence of an abnormality, they do not indicate its causes, and we still have to find those causes and eliminate them. Figure 1.12 is an example of a control chart for a process in the controlled state. If the average yield for your process were 90%, would you scold your subordinates on days when the yield dropped to 83.5% as at points (a) and (b)? The workplace that produced these figures was actually functioning normally.

iv) To control a process by checking through the results, it is essential to clarify the history of product lots and data, i.e., to stratify (see Section

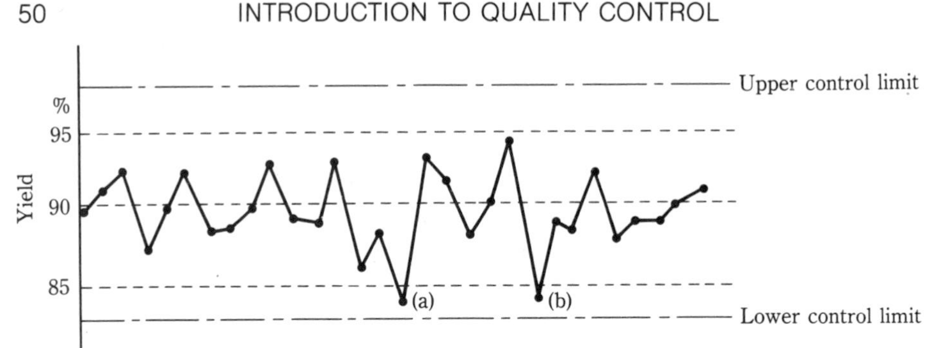

Figure 1.12 A Control Chart

4.3.2) the data carefully (this is also explained in Section 2.1). It is impossible to track down the cause of a product defect if we do not arrange the data so that we know what raw material was used to make the product, what machine was used to make it, who made it, and when it was made. Good control and analysis are impossible without skillful stratification. Conventional inspection data often cannot be used for control because the history of the data is not known.

v) Information must be fed back as quickly as possible, accurately, at the appropriate time, to the appropriate person. For example, to use inspection data for control purposes, inspection results should be stratified by lot and should be communicated directly to the workplace.

The various checking methods discussed above should be considered by all managers and supervisors with regard to their own areas of responsibility and authority. It is of course good to discuss with quality-control specialists whether or not their ideas are sound, but it is also important for non-specialists to think for themselves.

(6) Take action

Just checking and leaving things at that will not do any good. If the work is not going well or there is some abnormality, something must be done about it. The causes must be found and eliminated from the process to ensure that the process or work proceeds smoothly. When doing this, our main aim is not to remove the symptom, but the causes of the symptom, i.e., the root causes.

To perform work according to standards, we need to have standards, especially operation standards. It is the job of managers to train people to observe the standards and to ensure that they do so properly. However, it is not always their subordinates' fault when they fail to follow the standards or produce odd results. The standards may be incomplete, an atmosphere of observing standards may not have been established, or the boundaries of responsibility and authority may not be clear. The reason for not following standards is usually one of the following:

i) The worker responsible is careless or lacks awareness that he should work according to the standards.
ii) Insufficient training and education in standards have been given, or the worker misunderstands them.
iii) The standards are inadequate and impossible to follow. The specified methods may be difficult or prone to error, or require an extremely high skill level.

The first of these possibilities is the responsibility of operators and other front-line personnel, but the second two require action by upper-level managers. When shop-floor workers make a mistake, they usually only bear about one-fifth to one-quarter of the responsibility; the other three-quarters to four-fifths are management's responsibility.

This means that when work is not being done according to standard, one of the actions (a) to (e) listed below should be taken. Yelling and scolding are neither action nor good management.

(a) Try having the workers do the job according to the standard and look carefully at what is happening.
(b) Retrain the workers. It is probably also necessary to reflect on whether the workers' instruction and training were poor or sketchy, and to retrain them. It is wrong to think that they must be able to do something just because they have been taught to. In many cases, people do not do as they are taught because the teaching is poor. If you teach something and it is not understood, then your teaching was inadequate. However, when workers fail to perform well no matter how many times they are taught, or when they persist in making careless mistakes, we must take one of the following actions:
(c) Institute foolproofing measures or change the workers. Take account of workers' aptitude for the task in hand.
(d) Revise the standards. Some of Japan's standards are so bad that one wonders how on earth any work can be done with them. Work often cannot be done according to standards because the standards are poorly set. If so, the standards must be revised. This is done by utilizing QC circle activities, forming teams including front-line workers and listening to their opinions, collecting and analyzing data, performing careful observations, and carrying out factory experiments. In Japanese-style

total quality control, great importance is attached to this kind of preventive action.

(e) Modify the objectives and targets. Standards are not the only things that need revision; objectives and targets are also sometimes wrong. In this case, we should collect sufficient data, reconsider whether the objectives and targets are correct or not, and revise them if necessary.

Items (a) to (d) above are mainly the responsibility of shop floor foremen.

The actions detailed in (a) to (e) above can also be classified from a different standpoint:

(A) Take immediate action to ensure that people do as they are told.
(B) Take action to ensure that the same mistakes do not occur in the future (this is called "recurrence prevention").

Item (A) corresponds to action (a), while item (B) corresponds to actions (b) to (e).

In TQC, particular importance is given to item (B), recurrence prevention. Without it, we cannot claim to have established a proper system of control, and we will make no progress. Without (A), the process cannot remain in the controlled state, while without (B), no advances can be introduced into the process. (A) is of course the responsibility of the production line, but (B) is the responsibility of QC circles, staff, and managers.

In taking these actions, removing the causes of abnormalities must not be confused with regulation and adjustment. When an abnormality is discovered in the output of a process, its causes must be found and eliminated. Trying to achieve good results through stop-gap measures without looking for the root causes of the abnormality is regulation and adjustment.

Example: In a drying process, some of the products had an abnormally high moisture content. If the cause of this is a high moisture content in the raw material, the cause of the abnormality, moisture in the raw material, must be eliminated. A makeshift expedient, such as raising the drying temperature without doing anything about the moisture content in the raw material, is adjustment. If the cause of the abnormality is not removed, the process will not readily stabilize.

When the causes of abnormalities are often unclear, either the upper-level managers are not doing what they should do, or they easily lose their tempers, or they are telling their workers to do the impossible, or the concept of control has not filtered right down to the end of the chain of command, the control system is inadequate, the technology has not been properly established, or the boundaries of responsibility and authority are not clearly defined. To eliminate the causes of abnormalities, everybody concerned must understand the philosophy of control described above and work together to ferret out the root causes and find methods of dealing with them.

(7) Check the results of the action taken

Those taking action, together with their superiors, are responsible for checking its effectiveness. Just taking action and leaving it at that is not good enough; we can only be said to have discharged our responsibilities when we have checked whether the action was effective. Control charts are also useful for this.

The above is the basic philosophy of control, and control only takes place when the loops shown in Figure 1.13 are followed. If this is done with respect to clearly defined quality objectives, it is quality control.

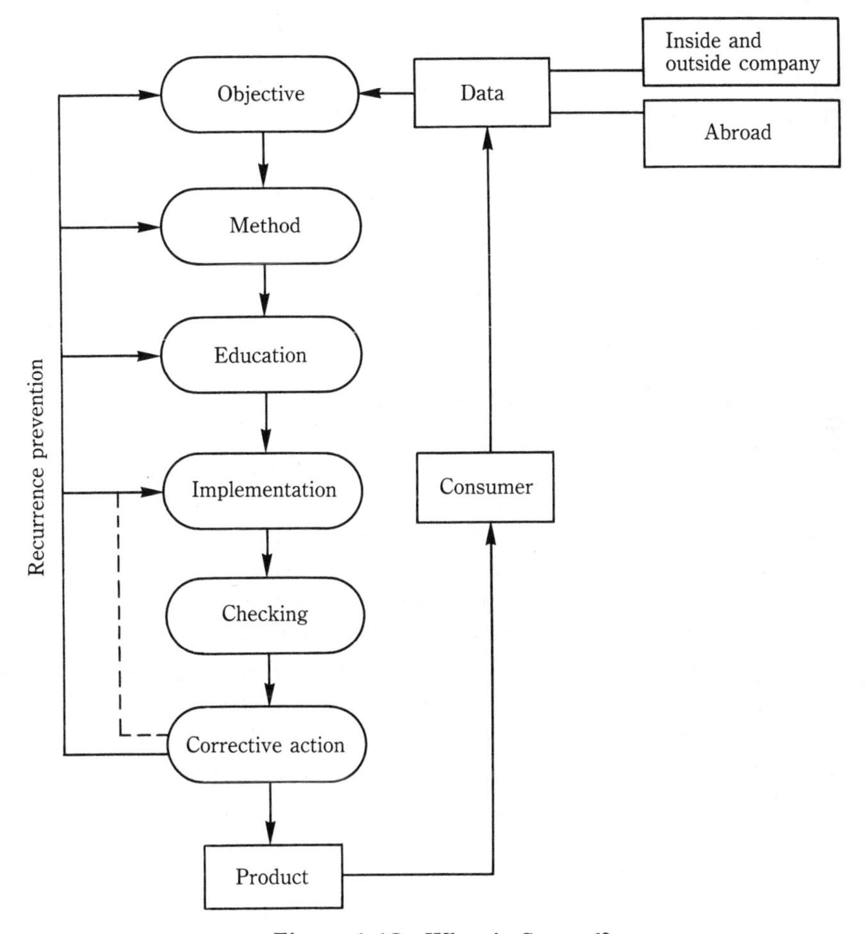

Figure 1.13 What is Control?

Also, control can be exercised effectively if statistical methods are used skillfully at each of the above stages. This is statistical control. When exercised with respect to quality, it is statistical quality control (SQC). As can be seen from the above discussion, our aim is to control quality, and we try to utilize statistical methods from every possible aspect as the means to achieve this.

The use of SQC and TQC benefits consumers, company employees, and shareholders, and naturally makes it possible for company profits to be shared among these three groups.

1.5.3 Action for Recurrence Prevention ("Permanent Fix")

The phrase "recurrence prevention" rolls easily off the tongue, but people are often slow to put it into practice. In TQC, much is made of recurrence prevention in the areas of control and quality assurance (i.e., new-product development problems and customer complaints). Recurrence prevention means preventing abnormalities from recurring as part of process control (this is also known as "permanent fix") and preventing the recurrence of problems and complaints arising during new-product development.

Conventional recurrence prevention measures are of the following three types:

(1) Eliminating the symptom (poor)
(2) Eliminating a causal factor (fair)
(3) Eliminating the root cause (good)

Of these, (1) is a stop-gap measure, not a recurrence prevention measure. (2) is a recurrence prevention measure of sorts, but it still leaves the possibility of recurrence. In addition to implementing the second type of measures, type (3) measures for eliminating root causes must be extended to all areas of the organization, even as far as reforming the management system and revising important standards.

In exercising control as part of TQC, we give priority to preventing recurrence of the causes of abnormalities. Since this means that the cause does not occur again, the work or process improves little by little, as shown in Figure 1.14. Thus, although the type of control we are talking about in TQC may be passive, it produces a constant gradual improvement through preventing the recurrence of the causes of abnormalities every time they are detected, and is therefore not simply maintaining the status quo.

All this is easy to say, but in practice, people often do not make enough effort to come to grips with the true, basic causes of problems. They make do

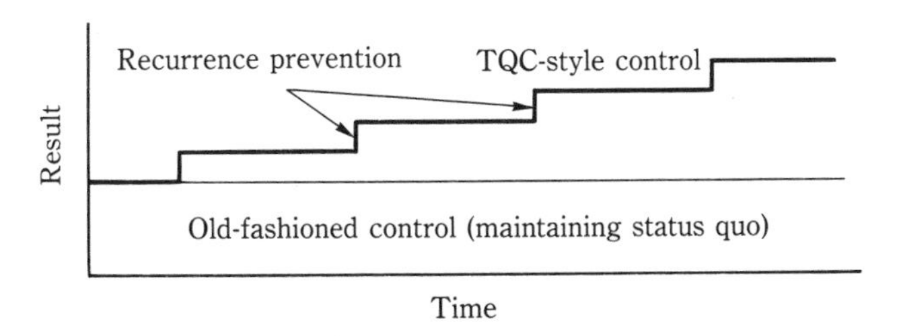

Figure 1.14 TQC-Style Control and Old-Fashioned Control

with firefighting measures and adjustments, forgetting the danger once it is past. It is essential for managers, supervisors, and engineers, as well as those on the shop floor, to pursue effective recurrence prevention measures tenaciously.

1.6 Controlling Quality

The aim of quality control is to assure quality by controlling it. Since this is discussed in detail in Chapters 4 to 6, I will only mention the main points briefly here. As stated in Section 1.3 concerning advances in quality assurance methods, controlling quality means implementing quality assurance using the TQC approach by building quality (including reliability) into a product during its development stage (which starts from the planning of the new product), then carrying out properly executed process control and, if necessary, performing inspection.

When quality control is being implemented for the first time, this procedure should be reversed: initially, rigorous inspections should be conducted to avoid inconveniencing consumers, then tight process control should be instituted, and finally, a quality assurance system should be built up through the new-product development stage.

When implementing quality control, we need to manage the five Ms (see Figure 1.15): men, materials, machines, methods, and measurements. Statistical quality control means using statistical methods to do this in every situation.

1.6.1 The Basis of Quality Control and Quality Assurance

Since the essence of quality control is quality assurance, I would like to mention some basic factors involved in it.

1) Consumer orientation.
2) The "quality first" approach.

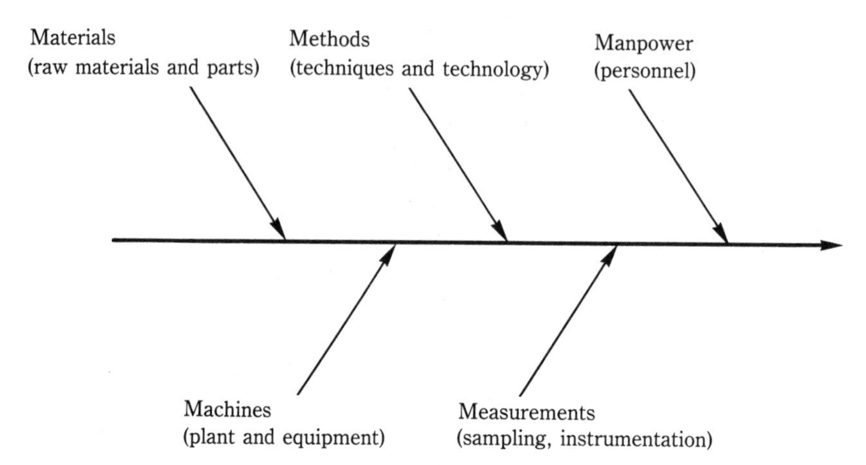

Figure 1.15 For Creating Products and Quality — The 5 Ms

3) Quality is everyone's business—from top management down.
4) Continual improvement of quality by rotation of the PDCA cycle (see Figure 1.2).
5) Quality assurance is the responsibility of the producer (the seller, i.e., the production department or workplace), not of the purchaser or the inspection department.
6) Quality should be extended from the hardware (i.e., the product) to the software (i.e., services, work, personnel, departments, management, corporations, groups, society and the environment).

1.6.2 The Quality Assurance System

As long as a corporation is producing and selling products or services, it must plan and design quality in the sense of what it should produce, and design the processes needed to produce that quality. We are now in an age of international competition in the development of new products, that are based on a balance of customer demands and the various capabilities, especially process capabilities, of the corporate groups (including affiliated companies) producing them. Customer demand and group capabilities must therefore both be considered; this is what is meant by group-wide quality control (GWQC). Through it, a company's capabilities can make rapid advances.

(1) The classification and definition of new products

Since a quality assurance system begins with new-product development, we must start by precisely defining and classifying what we mean by new products.

(A) Products completely new to the world } "front-running products"
(A′) Products completely new to Japan }
(B) Products new to one's own company but already produced by other companies ("rear-guard products")
(B′) Products resembling existing products
(C) Full model change of existing products
(D) Minor model change of existing products
(E) Specially commissioned products

How far down this list can we go and still call the products new? Can products of type (D) in particular be called new? When the customer requires a slight change in the specification of a product, as in order production (item E), the product probably cannot be called new. The distinctions between these types of product are fine ones, but companies must try to make them clear.

In new-product sales and profit policy setting and control, we must specify the number of years for which the product will be called new, and control the new-product sales and profit ratios during this period.

To clarify the criteria by which we can deem a new product a success, we must formulate new-product development control standards that clearly define each of the development steps described in (2), as well as the terminology used in them.

The development procedure becomes simpler as we proceed down the above list of new products from (A) to (E); the procedure for completely new products, (A), contains all the steps while the procedures for the others can be obtained by omitting some of these steps. Order production and high-variety, small-lot production can be thought of in roughly the same way.

There are various methods of classifying new products. One method, which classifies them according to the presence or absence of technology and sales routes, is shown in Table 1.2. In this case, the amount of care needed in considering the new product increases in the order (D) to (C) to (E) to (A).

Table 1.2 A Method of Classifying New Products

Technology / Sales route	Yes	No
Yes	D	C
No	B	A

(2) Quality assurance system

A quality assurance system is shown in Figure 1.16.

(a) This system is extremely simple, but every company must prepare such an organization chart showing what each department should do at each step, what should be done by what committee or other body, and who should decide to proceed to the next step.

(b) The purpose of quality assurance system chart is to unify the thinking of the whole company; it is simply a general outline, and preparing it does nothing for quality assurance in itself.

Starting with top management's quality policy, it is necessary to decide who should do what at every step (in particular, what investigations should be carried out and what kind of tests should be performed under what conditions in order to implement quality assurance) through data analysis, quality analysis, and experiment, all using the QC approach. The number of items tested may range from about 300 for simple producrs to 2,000–10,000 for complex products.

(c) This process is usually broadly classified into the seven steps shown in Figure 1.17. As shown in this figure, subcenters (SC) are usually established at each step to promote new-product development and rotate the PDCA cycle. Directed by a companywide center, these subcenters rotate the new-product development PDCA cycle to expedite new-product development and implement recurrence prevention.

(d) It is extremely important to evaluate and appraise factors such as quality, unit cost, and lead times at each step in order to decide whether or not to proceed to the next step, e.g., from the planning stage to the design and prototype fabrication stage, or from the design and prototype fabrication stage to the pilot production stage.

(e) Evaluations and tests on quality assurance should be carried out at as early a stage as possible; people from departments such as sales, research and development, design, production technology, production, purchasing and subcontracting, and after-sales servicing should take part in this at the new-product planning and first-prototype fabrication stages in order to sift out potential problems.

The important things to be done at each step are listed below.

(3) The planning stage (Step 1)

This stage focuses on the preparation of new-product plans, and looks at factors including what consumers are being targeted; sales price and unit cost; sales

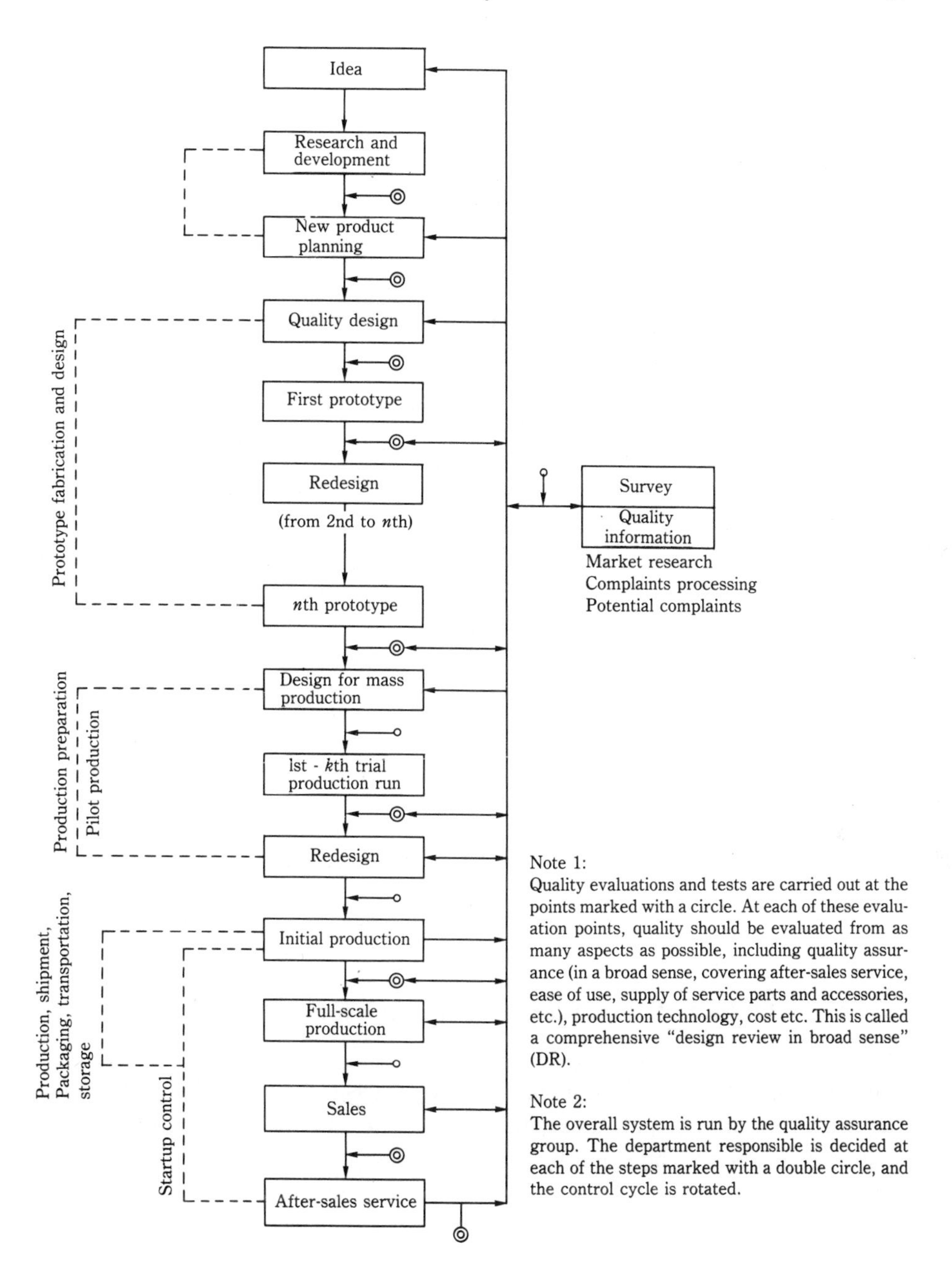

Figure 1.16 A Quality Assurance System

Figure 1.17 PDCA for the Quality Assurance System

volumes (monthly, total, life-cycle); quality (expressed in customers' own words, and properly ranked); and sales schedule. The items to be investigated include market information (consumer needs, dissatisfaction, complaints); technical information (existing technology, process capabilities, production capacities, research, design, technical capacity, presence or absence of these); personnel resources; financial capacity; existence of sales routes, and sales and service capacity; and capacity for materials procurement and subcontracting. These investigations should be carried out not just from the domestic viewpoint but with an international approach, to obtain information to support the new-product plans.

This will help persuade people within the company, particularly top management, that the new product will sell.

Some companies' corporate cultures enable them to succeed with new products, while others do not.

(4) The design and prototype fabrication stage (Step 2)

This step revolves around new-product development plans; quality targets (quality analysis, substitute quality characteristics in engineering terms, and ranking); unit cost targets; process design; QC Process Charts 1; setting of items, methods, and conditions for quality assurance tests; and control of design, research, and prototype fabrication.

For this, the following must be surveyed, researched, developed, and deliberated: manufacturing and production research; production technology; various types of process capability; product research; trials; investigation of methods of use; investigation of methods of evaluating and testing new products; practical tests; joint experiments with users; reliability tests; and serviceability considerations.

This step will also involve the selection of substitute quality characteristics, including review of testing and inspection methods; preparation of QC Process Charts 1; design review; visual design; packaging; test marketing; marketing techniques; distribution organization; and the determination of design standards and design technology standards.

Design and prototype fabrication should be repeated until the above conditions can be satisfied to a reasonable extent. At least three first prototypes should be fabricated, and examined from every aspect (sales and service, performance, production technology, manufacturability, etc.) to identify any unsatisfactory points.

(5) The pilot production stage (Step 3)

At this stage comes the preparation of QC Process Charts 2 and various standards, e.g., final product, intermediate product and raw materials specifications, technical standards, operating standards, process control standards, equipment standards, equipment control standards, maintenance and control standards for jigs, tools, molds, and dies, measuring instrument and measurement control standards, packing standards, transportation standards, and inspection standards for raw materials acceptance inspections and intermediate, final, and pre-shipment inspections.

For the above, it is necessary to decide the following items before the commencement of pilot production, or at the very latest, by its completion:

- Cost control methods (e.g., standard amount of materials and labor for a product unit, standard unit cost).
- Quantity control methods (e.g., production volume, inventory, sales volume).
- Marketing methods, establishment of distribution organization (including after-sales servicing).
- Preparation of sales manuals: catalogues, instruction booklets, spare parts lists, service manuals, claim processing procedures, etc.
- Education and training: production personnel, sales and service staff, subcontractors.
- Test marketing.

The following items should also be checked to lead in to actual production:
- What is the purpose of carrying out pilot production?
- What is the dispersion in the product during pilot production like?
- Are the parts satisfactorily interchangeable?
- What is the value of the go-through rate?
- What percentage of jigs and tools, molds and dies, measuring instruments and inspection techniques for mass production are used?

(6) The purchasing and subcontracting stage (Step 4)
For a discussion of this step, see Section 1.6.3.

(7) The production stage (Step 5)
Whether or not actual production proceeds smoothly is a vital question that affects the survival of the company, but whether or not a new product starts off on the right foot depends on how well the source control consisting of Steps 1 to 4 is done. The production department and the workplace are of course responsible for process control and improvement in Step 5. For this to be done properly, the previous items must be clearly decided, education and training must be given, control and improvement must be exercised through QC team and QC circle activities, and quality assurance must be perfected with the cooperation of the inspection department (see Sections 1.5.2 and Chapter 4).

(8) The marketing stage (Step 6)
However many products or services are produced, their production is pointless if customers are unhappy with them after purchase. To achieve customer satisfaction, we must sell products and services that meet customers' requirements, and we must provide before-sales service. For this purpose, marketing-related standards such as those mentioned in Step 3 must be prepared and continually

improved. Some particularly important guidelines for sales staff and others concerned with marketing are:

(a) Have a good understanding of the philosophy behind TQC, QC and QC circle activities.
(b) Clearly understand customers' requirements, needs, wants, methods of using the product, latent complaints, etc.
(c) Have an excellent grasp of product and after-sales servicing knowledge and technology.
(d) Don't try to sell goods and services by cutting their prices; sell them by virtue of their quality. Control profit margins on sales.
(e) Control sales volumes not in terms of total sales values, but by separate figures for each product for items such as sales volumes, inventories, stocks of defectives, off-the-shelf ratios, delivery ratios, ratios of out-of-stock products, inventory ratios, etc.

(9) The after-sales servicing and survey stage (Step 7)

Selling a product and forgetting about it are a long way from the QC approach. However hard we work at assuring the quality of our products, it is still not enough unless we also provide after-sales servicing (including the supply of consumables), offer regular servicing and repairs, deal with customer complaints, and anticipate customer dissatisfactions and their hopes, needs, and wants for the future. It is also important to control and improve the quality assurance system, including new-product planning, design changes to existing products, the after-sales servicing organization, etc. However, providing these on a nationwide and worldwide basis requires effort, research, and experience over a long period.

I do not care very much for the term "*hanbai-bu*" ("sales department"); it often suggests to people in such a department that their job is selling only (Step 6), and leads them to neglect to carry out Step 7 properly. The term "*eigyō-bu*" ("marketing and sales department"), on the other hand, brings Steps 6 and 7 together. In Europe and America, some of the activities of Step 7 come under the heading of "marketing," but, in Japan, I would prefer to see them all lumped together under the title of '*eigyō*' ('marketing and sales').

(10) Summary: promoting the establishment of a quality assurance system

Constantly rotating the quality PDCA cycle in this way is one characteristic of TQC.

In each of the seven steps described above, the appropriate subcenter ("SC" in Figure 1.17) acts as a focus for the promotion of new-product development schedule progress, quality assurance, and cost control for that particular step, while new-product development is promoted companywide based on information supplied by all the subcenters. Some points to note when this is done are:

(a) The decision of when to move from one step to the next is an important one (important decisions are shown by the symbol ◎ in Figure 1.16). For example, in deciding whether it is all right to proceed from Step 2 (prototype fabrication and design) to Step 3 (pilot production), the quality assurance, unit cost, and ease of fabrication (i.e., productivity) aspects must be properly reviewed. This review is then submitted to a new-product committee chaired by a top manager, and a decision is made as to whether the prototype should be remade or whether to proceed to pilot production and preparation for actual production.
(b) When this is done, in principle, all of the different jobs should have reached the same stage of progress and should be handed over to the next step simultaneously. In practice, however, not all of the problems will have been solved by the same time. Incomplete areas must be clearly identified as such when handing over to the next step.
(c) At each step, the work must be compared with the new-product plans, which are the basis for the whole operation, to ensure that they are satisfied. Naturally, each time this is done, the new-product plans must be rechecked to make sure that they are still satisfactory.
(d) As I have stated before, the purposes of the various tests performed must be reviewed at each step, and their methods and conditions in particular should be revised and augmented in the light of failures and complaints. This will contribute to the company's fund of technology and will become an important source of future know-how. Conscientiously fleshing out the company's knowledge and building up experience in this way will enable it to produce good new products quickly and with satisfactory quality assurance.
(e) The matters described above should be discussed at new-product conferences and in function-specific quality assurance committee meetings, and the system should be improved while ensuring that the responsibility and authority of each department is always made clear.

1.6.3 Control of Raw Materials and Subcontracting (Materials)

On average, Japanese manufacturing industry purchases materials equivalent to 70% of the total manufacturing costs of its products from outside suppliers. This means that quality assurance, cost reductions, and delivery-date control will be impossible if raw materials and parts procurement and subcontracting do not go smoothly.

Good, reliable products cannot be made with poor raw materials and parts. However, technology means making good products with raw materials of as low a quality as possible.

Some important factors concerning raw materials and subcontracting control are as follows:

1) The Ten QC principles for Vendees and Vendors (see Section 7.5).
2) Basic long-term subcontracting and purchasing policy.
3) Raw materials and parts specifications, acceptance inspection standards, stock control standards (the just-in-time system, control by lot, stratification, and leveling).
4) Distinguishing between make and buy items.
5) Selection and development of suppliers, development of specialist manufacturers, TQC education.
6) Contracts and contract documents (bonuses and penalties).
7) Joint experiments with suppliers.
8) Establishment of a group-wide quality assurance system and an inspection-free procurement system.
9) Ordering methods and stock control, stratification of ordering, reduction of lead times, the just-in-time system.

1.6.4 Equipment Control (Machines)

Good products cannot be produced without designing, installing, maintaining, and controlling equipment, machinery, apparatus, dies, molds, jigs, tools, etc., in conjunction with process design. As machinery and electrical equipment factories become more and more automated, they become more and more like process industries; we should therefore study the latter. Recently, mass production has become virtually nonexistent, and most manufacturing is of the high-variety, small-lot type. This means that, while we promote standardization, we must also make our equipment more flexible.

The following are some important points in implementing equipment control:

1) Design, selection, and installation of equipment.
2) Standards for controlling use of equipment.
3) Process and machinery capability studies: investigation and improvement, dynamic and static, statistical, line balancing.
4) Advances in equipment control methods:
   i) Repair if broken down.
   ii) Maintenance of equipment so that it does not break down. Preventive maintenance (PM), lifetime and inspection, replacement intervals, lifetime (reliability) distributions, lifetime data and equipment history, recurrence prevention.
   iii) Control to maintain process capability.
5) Equipment replacement: replacement on the basis of depreciation, replacement because of outdated technology.

1.6.5 Working Methods and Standardization (Methods)

The control of working methods was touched on in Section 1.5.2 (2). For a discussion of standardization—e.g., work standards, technical standards, etc., —please see Section 5.4.

1.6.6 Measurement Control (Measurement)

If measurement and testing are not performed correctly, accurate data will not be obtained. Some important points to note in measurement control are as follows:

1) Error theory and error control.
2) Selection and control of measuring instruments, jigs and tools, gages, and analytical methods.
3) Control of sampling and measurement methods.
4) Checking, inspecting, and recalibrating measurement instruments amount to no more than fixing instruments that have gone wrong; this is inspection, not instrument control. Instrument control means using instruments in such a way that they do not go wrong and no defects are found when they are tested.
5) Measurement control consists of ensuring that data obtained can be relied on within a certain error range.

1.6.7 Personnel (Man) and Education

Quality is planned, designed, manufactured, and marketed by people, and goods and services are bought and used by people. While automation, robotization, computerization, and office mechanization may become more widespread, they are still used by people. We have long been told that "a company is its people," and the reason why Japanese-style TQC is succeeding so well is that it respects humanity and makes human relations go smoothly, while enabling every individual to exercise his or her full capabilities through QC circle activities and the involvement of all departments and employees.

People have various desires, such as wanting to lead happy lives, acquire skills, make friends, be loved, feel pride, and wield influence. Happiness takes many forms, such as financial satisfaction, job satisfaction, and the satisfaction of personal growth and interpersonal acceptance. Total quality control must be implemented in such a way that these desires are satisfied and this happiness is provided.

Also, as long as a workplace is staffed by people, it must be a place where humanity is respected. Philosophers doubtless have much to say about what constitutes humanity, but, as an engineer, I have promoted total quality control and QC circle activities based on the simple belief that humans differ from animals and machines in the following two respects:

First, people work autonomously, of their own free will, spontaneously, under their own motivation. Working according to orders and instructions from above is no different from being a machine, and people will only work grudgingly under such conditions. Using people like machines under the old Taylor system, as is sometimes done in Europe and America, takes all the interest out of work and turns it into something like the industrial world depicted in Charlie Chaplin's film *Modern Times*. Boring people and making them work reluctantly can never produce good products or services.

Second, people think and use their heads while working. If people keep thinking and questioning while they work, good ideas will emerge and they will produce a host of excellent suggestions. This will foster creativity and facilitate the development of new products and new technologies. In companies actively implementing total quality control and QC circle activities, the number of good suggestions increases rapidly and can be expected to reach from 12 per person per year (1 per month) to 50 per year (1 per week), with more than 60% to 70% being taken up.

Finding a management posture that enables people to express their human qualities and motivates each individual in this way is one of the most important philosophies of total quality control and QC circle activities. However, simply adopting this kind of management posture will not in itself improve people and, if people do not improve, good products and services cannot be produced.

This means that the whole workforce must be educated and trained, from the president down to production-line workers, sales personnel and part-timers, as well as the staff of associated companies such as subcontractors and distribution organizations (see Section 1.5.2(3)).

Because a company's workforce is constantly changing and new people are always joining, I say that "QC begins with education and ends with education." Since quality control must be continued for as long as a company continues to sell goods and services, QC training and education must also be carried out without interruption, through good times and bad.

The following points should be noted in this connection:

1) Both education and training are needed: many European and American companies ignore the former.
2) Education and training methods include:
   i) Group education.
   ii) Education and training of subordinates by superiors.
   iii) Delegation of authority.
   iv) Mutual development:
      – In-house: committees, discussion meetings, briefing sessions.
      – Outside: QC conventions, QC circle conventions, QC circle exchange meetings, seminars, etc.
   v) Self-development, self-study.
3) Long-term plans for education and personnel appointing should take into account: reshuffling of organizations, personnel evaluation for educational purposes, job rotation, and development of multi-skilled workers.
4) Organization involves responsibility and authority, delegation of responsibility, reporting and checking, line personnel and staff, service staff and general staff, individual wishes, job rotation, personnel selection, grades and salaries, duties and status, aptitude testing, fair appointing, and remuneration.
5) As far as personnel and character evaluation are concerned, test scores, entrance examination results, graduation results and company entrance examination results cannot be relied on to any great extent. People change rapidly as a result of their own efforts, the behavior of their superiors, and education and training. Evaluation must make use of records of service and self-evaluations, and look closely at originality

and ingenuity, suggestion schemes, initiative and positive attitude, leadership.

6) Personnel management should exercise people's capacities.

1.7 Quality and Process Improvement

1.7.1 Philosophy and Basic Conditions of Control and Improvement

Control leans more towards making maximal use of existing capabilities and producing a gradual improvement by introducing various recurrence prevention measures while maintaining current standards. It does not consist merely of maintaining the status quo. Improvement, on the other hand, means taking positive steps to improve existing capabilities. At first glance, control and improvement therefore appear to be different jobs. The problem in the U.S. and Europe is that people tend to think of control and improvement as separate jobs with different people responsible for each. In fact, they are related in a particular way.

When we try to exercise control, improvement happens naturally; when we try to effect improvement, we naturally understand the importance of control. In other words, control and improvement are like the two wheels of a bicycle; if either does not rotate properly, the bicycle will not move forward smoothly.

While improvement consists of actively seeking out problems and dealing with them, it can be split into two different types: improvement of one's immediate surroundings and full-scale, priority-based improvement. The former consists of individuals in each workplace actively hunting out problems in their immediate surroundings and dealing with them one by one. This is the type of improvement promoted through the use of QC circle activities, suggestion schemes, and other schemes for promoting originality and ingenuity in the workplace. Since the equivalent concept does not exist in English, some English-speaking engineers and managers use the Japanese word "*kaizen*" without translation to describe it. It could also be termed "continuous improvement."

In contrast to this, there is also full-scale, priority-based, breakthrough-type improvement. This is the type of improvement in which a company sets priorities and attempts to bring about an improvement through technical innovation; it requires investment in research, development, and equipment. This type of improvement is carried out by project teams, task forces, QC teams, or functional organizations. However, once priorities have been set and everyone's knowledge has been pooled, problems thought to require this type of improve-

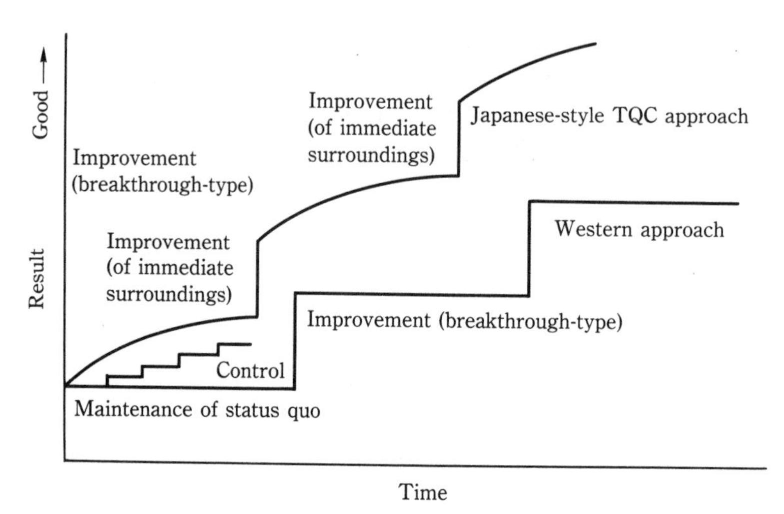

Figure 1.18 The Philosophy of Control and Improvement (see Fig. 1.14)

ment are surprisingly often found to be continuous-improvement problems. The above relationships among maintaining the status quo, control, continuous improvement, and priority-based improvement and breakthrough are depicted in Figure 1.18.

i) Improvement of immediate surroundings
ii) Priority-based improvement
} Improvement

These are covered in more detail in Chapter 4, but the following three basic conditions are the minimum requirement for effecting improvement:

1) Top-management leadership and support concerning innovation and the pioneering spirit, together with an indication of specific policies and goals; the creation of a system and atmosphere in which failure does not provoke fear or anger.
2) Turning the company into an organization permeated by an innovative mood and pioneering spirit. Some companies are continually developing new products and new technologies, while others find this difficult. The bureaucratic type of company in which "the protruding nail is hammered down" — i.e., companies in which initiative is squashed and people believe it better to do nothing than to risk failure — are no good. An atmosphere must be created in which all employees are aware of

problems on their own initiative, and in which nobody, including department and section managers, becomes fearful or angry at failure. It is perfectly satisfactory if 5% of new ideas succeed; over 95% will fail as a matter of course. I like the story of Columbus and the egg* and the words "invention and execution."

3) A company organization that is able to respond promptly to outside stimuli. This means a company organization capable of quick action in response to changes in the international business environment (e.g., slumps and booms, trade friction, etc.), moves of other companies in the same or different industries not only at home but also abroad, audits by outside consultants, Deming Prize examinations, or audits by the Deming Prize committee. Responding to change rather than taking the initiative is a passive approach, but human beings do have their weaknesses and they tend not to move unless subjected to some outside stimulus. However, it would be even worse if a company did not seek stimulation, was insensitive to it, or failed to take action even when it was aware of the need or opportunity for it.

To sum up, it is vital for everyone throughout the company to become problem-conscious and to think constantly about possible breakthroughs and continuous improvement. The company should encourage and utilize this skillfully, and should advance through tireless improvement. When people think they have no problems, they stop moving forward and start moving backward. However, these are all human problems that depend on people's attitudes and ways of thinking.

1.7.2 The Steps in Improvement

The following are the steps that must be taken in order to effect improvement:

1) Carry out research and analysis to identify the status quo and reveal problems.
2) Decide which problems to tackle, and set targets.
3) Fix the structures and responsibilities of improvement organizations (QC teams and QC circles); formulate activity plans.

---

*Translator's Note: When people devalued Columbus' discovery of America by saying that anyone could have done it, he challenged them to stand an egg on end. After they had tried and given up, he showed them how by smacking one down onto a table so that it stood up on its broken end.

4) Identify the status quo.
5) Carry out process analysis.
6) Prepare action plans.
7) Take action.
8) Check the results.
9) Carry out recurrence prevention, standardization, and permanent fix.
10) Establish control.
11) Identify remaining problems and review progress.
12) Prepare plans for the future.

As well as an improvement procedure, this could also be called a procedure for discovering and solving problems.

1.7.3 Investigation and Analysis for Revealing Problems

When the real problem has been discovered, it is already halfway to being solved. If control is not properly exercised, the problem areas cannot be discovered and people simply run about in confusion. The key points in investigation and analysis with the aim of detecting problems are as follows:

1) Both staff and production-line workers are responsible for this kind of investigation. However, managers are responsible for discovering problems and making decisions. All company employees are responsible for investigation, and all should be prepared to act as investigators. Everybody should be problem-conscious and actively point out problems without being told to do so.
2) The real situation must be accurately identified. This may, for example, mean taking a careful look at the workplace and finding out the true capabilities of processes.
3) Identifying the status quo, setting policy, and discovering problems require data. These kinds of data (stratified data, frequency distributions, Pareto charts, graphs and control charts) are often only rarely available.
4) Everybody's knowledge can be pooled (ask the opinion of everybody involved, employ suggestion schemes, hold brainstorming sessions).
5) When there is a clear profit plan, people should be given deadlines for discovering problems which, when solved, will result in more than a certain minimum cost saving.
6) There must be a department responsible for accumulating research data, analyzing it, and detecting problems from the overall standpoint.

However, all departments should submit data, and managers should make the decisions.

7) There must be a means to ensure that information is not distorted or biased, that the information network is intact, and that information is accurate.

1.7.4 Deciding Problems, Targets, and Deadlines

1) Methods of determining which problems should be tackled and how to evaluate the results should be decided in advance. The authority for doing this will depend on the particular company, but in principle, such authority should reside with managers. This should be publicly announced.
2) Staff should take up a number of big improvement problems, estimate the costs of tackling them and the possible benefits (monetary and otherwise), and prepare plans about which managers are to make decisions in line with company policy. In doing this, they should solicit and consider the opinions of as many people as possible. Cost-accounting data and Pareto charts should be used.
3) One must distinguish between chronic and sporadic problems. Sporadic problems do not warrant much attention; the most important problems economically are the chronic ones on which everybody has given up.
4) The biggest problems should be tackled through a companywide cooperative effort. For this purpose, it is better to give each department a role in tackling the problem rather than have individual sections take up improvement topics.
5) Improvement targets and deadlines in the areas of personnel, quality, cost, quantity, etc., should be indicated as specifically as possible, using figures.
6) As far as possible, budgets must be set for improvement costs (including the costs of research and investigation as well as action).
7) It is obviously necessary to discuss the likelihood of solution when deciding which problems to tackle; however, if too much attention is paid to this, there is a danger that the biggest problems will be abandoned and too much time will be spent on trifles. Never abandon hope for a solution.
8) Restrict the number of important improvement problems according to the Pareto principle. If there are too many important problems, they lose their importance.

9) Decide in advance how results will be checked and evaluated.

The above are some points to be aware of when deciding what problems to tackle. The important thing here is to think how problems *can* be solved, rather than why they cannot. Further details on problem analysis and improvement can be found in Chapter 4.

1.8 Statistical Quality Control, Total Quality Control and Technology

The words *"gijutsu"* (technology) and *"gijutsusha"* (technician/engineer) have been used very loosely in Japan. "Gijutsusha" has too wide a meaning, and we should really classify these people more accurately, as follows:

1. "*Kagakusha*" (scientists)
2. "*Gijutsusha*" (engineers)
3. "*Ginōsha*" (technicians)

Scientists are people who patiently study basic science, while technicians are like master-craftsmen, good at running processes or assembling products. The majority of the so-called "gijutsusha" in factories used to be of this type. Engineers are people who are able to apply science skillfully and economically, and are good at developing new products and new technology. At present, many of the so-called "gijutsusha" in Japan are in fact not engineers but scientists or technicians. We have few true engineers; this has prevented our industry from making real technological progress and earned our technology the nickname "copycat technology."

Expressed simply, the relationship between statistical quality control, total quality control, and research and technology is as follows: Good quality control is impossible without proper technology. The motive power behind the search for causes is research, technology, and technical skill (i.e., experience and training). However, technology improves rapidly through the use of the S part of SQC (statistical methods), i.e., through carrying out quality analysis and process analysis using the statistical tools of the QC approach. We must use proper technology, statistical techniques, and control techniques as tools to control quality and promote effective TQC.

There is no end to the number of different terms used for various types of technology: e.g., product engineering, design engineering, process engineering, production engineering, industrial engineering (IE), sales engineering, service engineering, etc. Here, I would like to categorize research and technology into three different types from the QC point of view:

1) Manufacturing research and technology (research for producing products and services)

This consists of research and technology for the design and prototype fabrication of products and services, process design, production technology, preparation for production, process control systems, process analysis, dies, molds, jigs and tools, process automation, computerization, etc; in other words, research and technology for producing products and services. This type of research and technology is being carried out quite enthusiastically in Japan, and is continuing to make big advances through process analysis.

2) Product research and technology (research into the use of products and services)

This consists of product and service planning, quality evaluation methods and conditions, methods and conditions of use, quality analysis (quality deployment), true and substitute quality characteristics, analysis of complaints and dissatisfactions, joint research with customers, testing and experimental methods and conditions (including reliability tests), inspection methods, development of new applications, etc. Nobody really knows the quality of a product or service until it is used. Quality assurance and reliability tests in particular must be carried out during new-product development. This type of product research and technology is extremely important in quality control; although this point has been studied for many years, it is still not satisfactorily grasped, and not enough of this type of technology is being accumulated.

3) Service (marketing) research and technology

This type of research and technology deals with consumers' purposes of use, needs, wants, and requirements, explanation and guidance in methods of use, after-sales servicing and repairs and their quality assurance, before-sales servicing, anticipation of consumer needs and wants, collection and analysis of market data, and materials and manuals relating to the above. This type of technology has improved considerably, but sales personnel still do not have a sufficient grasp of it and are not trying to improve it. In some extreme cases, they do not even have sufficient knowledge of the products they are selling.

The implementation of statistical quality control and total quality control has helped many Japanese products become the best in the world; these products are now capable of being exported all over the globe. At the same time, good progress has been made in the three types of technology described above. However, people who know little about statistical quality control and total quality control mistakenly believe that implementing quality control will stifle creativity and halt technical progress. In fact, technology makes progress as a result of statistical quality control and total quality control, and the export of Japanese technology has recently increased dramatically. My hopes at the time quality control was started in Japan (see Section 6.1) are gradually being fulfilled. However, there is still much room for improvement in technology, and the pace of progress is still rapid. Because of this, it is important to improve technology further while implementing total quality control.

1.9 The Ends and Means of Business Management

A distinction should be drawn between the ends and means of business management. I think of business management in the following way (see Table 1.3):

As long as we are living in human society, the ultimate aim of business management must be human happiness. In a narrow sense, this means everybody concerned with the company—i.e., all employees (including top management), together with the company's customers and shareholders. In a wider sense, it should also include everybody in associated companies and society in general. To achieve these aims, as was also explained in Section 1.4.1, we must control quality (Q), profit, cost, and price (C), quantities and delivery dates (D), and environmental quality and safety (S) as our secondary objectives. I call the control of quality, cost, delivery, and safety (QCDS) "objective control."

There are many means and methods for achieving these objectives (see the left-hand column in Table 1.3). We use these means and methods to try to achieve our primary and secondary aims. However, people often tend to become entranced with the methods and forget their goals. For example, they end up regarding mathematics, statistical tools, standardization, or computerization as the goal, and let the method dictate their behavior. This is a confusion of the ends with the means. We must be careful to remember our goals and not fall under the spell of the methods of attaining them.

What we should be doing is setting clearly defined goals — e.g., improving quality — and using a variety of means to attain them. This is quality control in its narrow sense. In other words, we must be prepared to use all available

Table 1.3 Management Ends and Means

End / Means	People			
	Quality (Q)	Profit, cost, and price (C)	Volume and delivery date (D)	Society and safety (S)
Physics Chemistry Electricity Mechanics Mathematics ⋮ Research and development Surveys and market research Product technology Design Production technology Standardization Industrial engineering Materials control Supplier management Equipment control Instrument control Control of jigs and tools Automatic control Computers Information management Statistical tools Operations research Inspection Education ⋮				

tools in order to control quality skillfully, and we must continually consider their ability to be integrated with other tools and types of control. As can be seen from Table 1.3, quality control is closely related to all other types of control, and it cannot operate in a vacuum. Other types of control must also be promoted in parallel with quality control.

1.10 QC Circle Activities

We officially started QC circle activities in Japan in 1962. Because these activities suit human nature, they were a tremendous success, and now more than fifty countries around the world have begun to imitate them. However, because of this, many people mistakenly believe that QC circle activities are the same thing as total quality control, and that QC activities and campaigns mean QC circle activities. Since QC circle activities are not the main point of this book, I cannot give much space to them here, but I would encourage readers to study them further in the references given below*.

(1) What are QC circle activities?

QC circles are small groups of people from the same workplace who carry out quality-control activities voluntarily. These small groups carry out self-development and mutual development as part of companywide quality control (TQC) activities and use QC tools to control and improve their workplaces continuously, with everybody taking part.

(2) The basic philosophy of QC circle activities

The basic philosophy of QC circle activities carried out as part of companywide quality control activities is:

1) To contribute to the improvement and development of the corporate culture.
2) To create cheerful workplaces that make life worthwhile and where humanity is respected.

---

* *QC Sākuru Kōryō* (General Principles of the QC Circle) (English translation available) and *QC Sākuru Katsudō Un'ei no Kihon* (Basic Principles for Running QC Circle Activities)(in Japanese); ed. QC Circle Headquarters, pub. JUSE, sold by JUSE Press. Also, *Nihonteki Hinshitsu Kanri*, by Kaoru Ishikawa, pub. JUSE Press, Chapter 8 (English translation; What is Total Quality Control? The Japanese Way, translated by David Lu, pub. Prentice-Hall, ISBN 0-13-952433-9); and many other works published by JUSE.

3) To exercise people's capabilities and bring out their limitless potential.

(3) The relation between TQC and QC circle activities

QC circle activities were begun after we had started TQC to ensure that quality control was properly implemented in front-line workplaces. They are part of TQC, not the whole of it. In the manufacturing industries, the importance of QC circle activities relative to TQC as a whole is from one-quarter to one-fifth; other QC activities, such as quality control in new-product development and group-wide quality control, are more important. In the service industries, since people at the end of the management chain have far more opportunities to come into contact with customers, the relative importance of QC circle activities is somewhat greater, approximately one-third.

The relationship between TQC and QC circle activities can be modelled as shown in Figure 1.19.

(4) Misunderstandings and items to note

QC circle activities are subject to misunderstandings of the type described in Sections 1.1.2 and 1.1.3. The following points, which are easily misunderstood, should be checked:

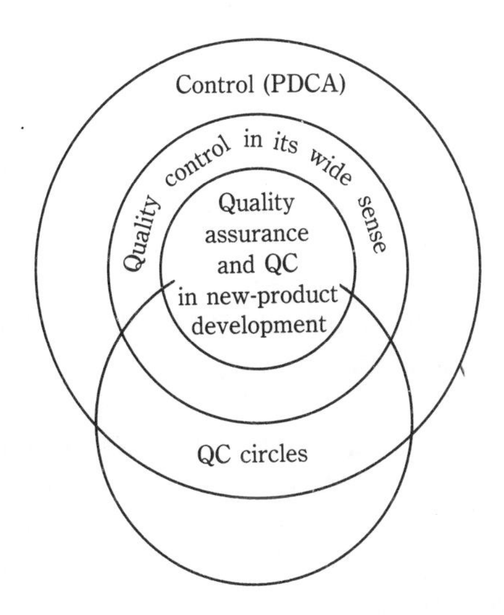

Figure 1.19 The Relationship between TQC and QC Circle Activities

1) QC circle activities are part of TQC and should not be separated from it.
2) In Japan, QC circle activities were started after TQC had been introduced. This is the correct order in principle, but in medium-sized and small businesses and the service sector, it is also permissible to start with QC circle activities. However, if TQC is not introduced within two to three years, the QC circle activities will fail.
3) QC circle activities are voluntary activities which suit human nature. The circles' PTA ("parent-teacher association" consisting of top managers, middle managers, and staff) should therefore not exert any pressure to achieve quick results. Progress should be gradual.
4) Because they are voluntary, this does not mean that their PTA, especially top and middle managers, should simply ignore them. The successes and failures of QC circle activities reflect the attitudes of top and middle management to QC circle activities and TQC. The PTA's attitudes and the measures they take to stimulate QC circle activities are important.
5) QC circle activities are not just a motivation campaign. Training and education in the QC tools and other methods should be given, and the activities should be continued permanently on a scientific basis.
6) QC circle activities and QC term activities are separate activities (see Section 4.5.2).

The following misunderstandings and confusions about QC circle activities are found both in Japan and elsewhere:

(i) QC circle activities are the reason why Japan's products are so good.
(ii) The Japanese can carry out QC circle activities because the quality of their workforce is so high.

The above two statements are not actually wrong, but they are obviously not the whole story; in addition:

(iii) Some people think of QC circle activities as a method of labor management. Companies that introduce them on this basis fail, even in Japan.
(iv) Some companies call QC circles "Q Circles"; this becomes QC when abbreviated and is confused with the QC meaning quality control.
(v) Some companies misunderstand the meaning of "voluntary." They think that a QC circle is formed by collecting together a group of volunteers, and make no effort to have people from the same workplace participate voluntarily.

1.11 The Introduction and Promotion of TQC

As I said at the beginning, this book is intended for people implementing TQC. I will therefore keep my remarks on its introduction and promotion brief; please refer to other works for the management aspects.

(1) The aims of introducing TQC

Different companies have different aims in introducing TQC, and they may have more than one. The following are the most common:

(a) Improving the corporate culture.
(b) Unifying the company's strengths and establishing a cooperative organization with total employee involvement.
(c) Establishing a system for promoting quality and obtaining the trust of consumers and customers.
(d) Aiming at world-beating quality and developing new products for this purpose.
(e) Securing profits and establishing a management system able to withstand low growth and change.
(f) Creating respect for humanity, encouraging personnel development, providing employee satisfaction and cheerful workplaces, handing over to the younger generation.

Some common motives and incentives for introducing TQC are: marking a change of company president; preparing to hand over management to a younger generation of managers; celebrating a certain number of decades since the founding of the company; providing measures to cope with changes in the outside situation (such as trade or capital liberalization, trade friction, oil crises, or currency appreciation); providing countermeasures against recession; or loss of ground against competitors. Since quality control should really be an ongoing, permanent activity, it should be started when a company is in profit. It is a shame that so many organizations only start it when they are feeling the pinch and are clutching at straws to save themselves.

(2) Things to be done when introducing TQC

The following things must be done when TQC is introduced:

(a) Top management must correctly understand the essence of TQC and QC circles and harmonize people's thinking.

(b) The company president must announce the introduction of TQC.
(c) The company should learn from others who have already introduced TQC, invite outside lecturers, and hold lecture meetings and seminars.
(d) A TQC promotion department should be set up as part of the company president's staff to consider methods of promoting TQC.
(e) Separate TQC education programs should be set up for each level of the company hierarchy: top management, department and section managers, staff, and general workers.
(f) A TQC promotion plan should be formulated and put into effect.
(g) The company president must carry out QC audits.

1.12 Methods of Promoting TQC in Individual Departments

Basically, the directors and department managers responsible for each department should voluntarily take the lead in the promotion of TQC by individual departments. If necessary, each department should appoint its own TQC promotion staff to decide policy and promote TQC according to the procedure given below, while educating and training all the department members:

(1) Quality control
Before taking action, each department should consider what it should do as a department to control the quality of its company's products or services. The duties of line departments (such as new-product planning, research and development, design, prototype fabrication and production preparation, purchasing and subcontracting, manufacturing, inspection, sales, and after-sales servicing) will probably be clear, since these departments are directly concerned with quality assurance. Non-line staff departments (e.g. personnel, general affairs, accounting, technical, market research, and warehousing) should consider what kinds of services and cooperation they should offer in order to assist with quality control.

(2) Quality control in its wide sense
Each department should ponder what is meant by good quality in relation to its work, then clearly define and control it.

(3) Control

The control duties of each department were discussed in Section 1.5.2. Control should be carried out according to the basic philosophy of control.

(4) Statistical control

Each department should analyze, control, and continuously improve its work statistically. In other words, each department should consider how to use control charts and other statistical tools. These should be used wherever possible.

Total quality control should be promoted in the above order. For example, when asked what quality control in the personnel department is, many people immediately think that it means drawing control charts; this is a misunderstanding of quality control. Selecting a tool and looking around for a place to use it is not generally a very effective way of working. Likewise, it is no good studying statistical methods, control charts, and/or methods and then looking around for somewhere to use them. It is much more effective to clarify objectives (quality, profit, delivery dates, etc.) and problem areas (see Section 1.9), and then consider what methods could be used to tackle them. This is particularly important in QC carried out by non-line departments.

1.13 Quality Diagnosis and TQC Diagnosis

When quality control is introduced or promoted, it is necessary to diagnose its condition, check whether it is being promoted well or badly in a variety of senses, see what problems exist, and reflect on progress. I will discuss the diagnosis need briefly here. This diagnosis can take the form of quality audits or quality-control audits (for a detailed discussion of these, see Sections 7.10 and 7.11).

(1) Quality diagnosis

A quality diagnosis consists of sampling goods or services from within the company or marketplace and carrying out various tests to check their quality, in order to determine whether or not customers are satisfied with them. This type of diagnosis is carried out with the aim of correcting shortcomings or defects and improving selling points. In other words, it is a type of diagnosis designed to improve quality by rotating the PDCA cycle with respect to both hard quality (the quality of goods) and soft quality (the quality of services).

(2) QC and TQC audits

The quality control diagnosis differs from the quality diagnosis, which checks quality itself, in that the former assesses the process by which quality is built into the product or service. In other words, it examines and advises on the quality-control methods and quality assurance system of the company as a whole, and sometimes also on those of the company's suppliers and distributors. The TQC diagnosis goes a stage further, covering a slightly wider range than the QC diagnosis and examining and advising on the overall management of the company with emphasis on quality. It naturally includes quality assurance, but also covers companywide quality control, policy management, functional management, new-product development, research and development, the management of suppliers and distributors, QC circle activities, etc. Since quality control means companywide quality control in an organization implementing TQC, all the internal quality diagnosis in such an organization can be called TQC diagnosis.

This type of diagnosis covers the methods used to promote quality control, build quality into the product or service via the process, manage subcontracting, deal with complaints, and promote quality assurance at the new-product development stage; in other words, it is designed to check whether the company's quality control systems and the way they are being operated are satisfactory, and to take action to eliminate problems and prevent their recurrence. In brief, it examines the process of implementing and promoting quality control to check whether it is satisfactory, and rotates the PDCA cycle. This type of diagnosis may be carried out be someone from inside or outside the company. The presidential diagnosis is a particular example of the former. The latter type may be carried out by purchasers to check whether the company's quality and reliability assurance are satisfactory.

Presidential QC diagnosis should not be carried out on the premise that everything is bad, using top-management muscle to expose malpractice and reveal shortcomings. Like a doctor who examines a partient in order to diagnose an illness and commence treatment promptly so that the illness gets no worse, the presidential QC diagnosis aims for action. Its purpose is to enlist everyone's cooperation to pinpoint weaknesses and systematically improve the situation. This means that CEOs must never become angry, even when their company's flaws and shameful weaknesses are exposed, and those being diagnosed must also describe their shortcomings clearly and honestly, exactly as patients would explain their symptoms to a doctor.

1.14 The Role of Executives in TQC

The role of executives in total quality control, particularly those in top management, is extremely important. The leadership and attitude of the chief executive officer and his or her deputy can be said to govern the success or failure of TQC, QC and QC circle activities. Executives must therefore do the following:

(1) Study quality control, companywide quality control, and QC circle activities; investigate how these are actually put into practice; and acquire a clear understanding of their fundamentals.
(2) Consider the culture of their company; decide the standpoint from which companywide quality control should be adopted; clarify their policy regarding the introduction of TQC; and announce its introduction.
(3) Take the lead in promoting quality, quality control, and total quality control; for this purpose, TQC promotion organizations (including QC circle activities) should be set up as part of the president's staff, and promotion plans formulated.
(4) Carry out the education needed for implementing quality control; and prepare long-term staff appointment and organizational plans closely linked to this.
(5) Collect information on quality and quality control and establish specific priority policies concerning quality; also, issue a basic "quality-first" policy and decide specific long-term quality targets from an international standpoint.
(6) Set up a quality assurance system.
(7) Check whether quality control, total quality control, and QC circle activities are proceeding according to policy and plans; take corrective action if necessary (through priority management, daily management, and presidential diagnosis).
(8) If necessary, establish control systems for individual functions.

The above applies correspondingly to department and section managers and those below them.

Some Quality Control Maxims

(1) The relation between quality control and total quality control

- Quality control means doing what ought to be done in all industries. Japan and other countries have already proved that such an approach can produce excellent results (see Section 1.2).
- The basic principles of quality control are exactly the same in any industry.
- As long as a company is selling products and services, it must never stop controlling quality.
- Quality control is applicable to any kind of enterprise; in fact, it *must* be applied in every enterprise.
- Implementing quality control will benefit not only consumers, but also a company's employees (including top management) and shareholders, in addition to society as a whole.
- Modern quality control is a revolution in management thinking.
- Quality control will not progress if top management's policy is unclear.
- Any enterprise that does not practice quality control will not be around for long (Section 1.2).
- The further civilization advances and the more modernized manufacturing becomes, the more important quality control becomes.
- Promote trade liberalization through quality control!
  (see the discussion of trade liberalization in 1960, Section 1.2.)
- How can we make a horse drink when it doesn't want to?
- If you will not try a food through prejudice, you will never know its taste or receive a single scrap of nourishment from it. The longer quality control is chewed, the tastier and more nourishing it becomes.
- Implementing quality control requires the constant education of everyone from the company president to workers on the line.
- Quality control fails when everyone misunderstands it, and succeeds when everyone understands it correctly.
- Quality control is the business of every employee and every department. If all employees and all departments work together, it is bound to succeed.
- Quality control is a group effort that cannot be carried out by individuals. It must be done through a system of teamwork and cooperation.
- Quality control should continue from the planning of a new product to its arrival in the customer's hands.
- Quality control should extend from sales through to subcontractors, suppliers, and dealers (GWQC).
- Quality control should evolve from companywide quality control (CWQC) or total quality control (TQC) to groupwide quality control (GWQC).
- Total quality control is not a fast-acting drug like penicillin, but a slow-

acting herbal remedy that will gradually improve a company's constitution if taken over a long period (Section 1.2).

- If quality control does not produce effects, it is not quality control. Super-profitable quality control is the goal (Section 4.13).
- Quality control must be rapacious (Section 4.13).
- Were we really that bad? (see "presidential diagnosis" in Section 7.11).
- The next process is your customer (Sections 1.4.1, 1.6.1 and 6.2).

- Quality control can only be implemented and effects obtained if the company president or his deputy really understands it and takes the lead in promoting it.
- Top management is responsible for setting methods and standards for evaluating quality (Section 1.4.4).
- Quality control will not progress unless department and section managers are won over.

(2) Education/personnel/organization

- Quality control begins and ends with education (Sections 1.5.2 and 1.6.7).
- Education must continue as long as the company survives (Section 7.3).
- If students fail to understand, it is because the teaching methods are inadequate (Section 1.5.2).
- In implementing quality control, all personnel must undergo total immersion.
- As civilization advances, our mental age decreases (Section 1.6.7).
- Managers and engineers who cannot handle subordinates are still only fledglings. They can also only be said to have come of age when they are able to deal confidently with their superiors and people in other departments, i.e., when they are able to work in the way they want (Section 5.5.1).
- People show their true capacity when their abilities are properly used and they are given responsibility (Section 1.6.7).
- Rather than expressing your own opinion, listen to what others have to say (Section 4.7.2).
- Fostering people who can be trusted is essential for management. Management should be based on the belief that human nature is fundamentally good.
- Criticizing failure and neglecting to praise success is a bureaucratic approach that discourages personal growth and blocks ideas for new products and new technology. Failure is the seed of success (Section 1.6.7).
- Workers on the assembly line know the situation best, but their judgment is often one-sided.

- When a problem arises in a workplace, that workplace is one-fifth to one-third responsible. Other workplaces are four-fifths to two-thirds responsible (Section 1.13).
- Managers! Take responsibility for problems and do not blame your subordinates! (Section 1.6.7.)
- Management based on the belief that human nature is fundamentally bad is costly, makes everyone unhappy, and duplicates control.
- Thinking up reasons for being unable to do something is a waste of time; think positively about how something could be done (Sections 1.2 and 1.7.4).
- Never say you are too busy to do TQC. If you practice TQC, you will have time on your hands.
- Only babies are not responsible for themselves (Section 1.6.7).
- To implement quality control, the organization must be rationalized.
- Organization is the clarification of responsibility and authority; it does not always mean setting up a hierarchy of sections, subsections, etc. Authority can be delegated, but not responsibility (Section 1.5.2).
- If quality control is implemented, the duties of line workers and staff become clear, technical departments are set up, real technology is established, and technology export becomes possible.
- An engineer must be an economist (Section 1.1.4).
- Researchers, engineers, and designers! Be humble! (Sections 4.7.1 and 7.4.)
- Things improve when you do the opposite of what engineers say you should (Section 4.13).
- Unfounded confidence obstructs progress (Section 4.7.1).
- Those who try to make a name for themselves by stealing a march on others cause nothing but harm (Section 4.7.1).

(3) Consumers

- Consumers provide us with work.
- When making products, put yourself in the purchaser's shoes; shift from a seller's market to a buyer's market.
- The customer may be king, but many kings are blind; sales staff are obliged to educate them properly. ("Lack of product knowledge"; Section 7.7.)
- Consumers are not guinea pigs (Section 6.3).
- Cakes taste good to those who bake them, but not always to those who buy them.
- Swallowing one's grievances is not a virtue.
- Buying cheap can cost dear.
- The first fruits of the season are always expensive.
- Never buy new products (Section 6.3).

- It is the woman who develops Japanese quality control.

(4) Quality and quality assurance

- Improve quality continuously by rotating the PDCA cycle (Section 1.6.1).
- Rational quality design is the first step in quality control.
- Finding out what the consumer wants is the first step in achieving quality.
- Identifying what to have the consumer buy is the first step in quality control.
- A company that carries out 100% inspection is a company that makes defective products.
- Inspection-oriented quality control is old-fashioned quality control.
- Build quality in during the process (Sections 1.3 and 1.5.2).
- Quality is not created through inspection; it is built in through design and the process (Sections 1.3, 1.5.2, and 6.7).
- Quality control that does not assure quality is not quality control (Section 6.15).
- Quality assurance is the purpose and essence of TQC (Sections 1.3, 6.1, and 6.15).
- Quality assurance is the responsibility of the producer (seller, production department, workplace, etc.), not of the purchaser or inspection department (Sections 1.6.1 and 6.1).
- Quality cannot be defined apart from price.
- When quality control is initiated, defects and complaints will multiply (Section 1.4.4).
- If bosses get angry when defectives are produced, defectives get hidden (Section 1.4.4).
- Trust takes years to build but is lost in a day (Section 6.4).
- How many years' worth of spare parts does your company stock for after-sales service? (Section 1.6.2.)
- Lifetime supply! (Sections 1.6.2, 6.1, and 6.4.)

(5) Design and new-product development

- Some companies are constituted so as to succeed in introducing new products; others are not (Section 1.6.2).
- A company's TQC has come of age if the company can develop new products and start full-scale production on schedule, if planned go-through rates and production volumes are achieved rapidly and smoothly, and if sales increase steadily, with no consumer complaints or dissatisfaction (Section 1.6.2).
- A company's quality control has come of age if its new products always succeed and consumers feel happy and confident buying them.

- Be first with new products; a new product that does not lead the field is just a copy.
- Design products from the standpoint of the user (Section 7.4).
- Never say, "I didn't think the product would be used that way" (Section 7.4).
- Carefully check the conditions under which your products may be used and bear them in mind when designing (Section 7.4).
- Apply quality control to the design process by treating it as a high-variety, low-volume production process for manufacturing products called drawings (Section 7.4).
- Promote design standardization and the use of standard parts (Section 7.4).
- Make drawings that allow products to be produced without any adjustment (Section 7.4).
- Destroy the self-satisfied attitude of some designers that they are artists, and that their work is therefore above criticism or suggestions from others.
- A good design cannot be produced without knowing how the product is to be made (Section 4.7.5).
- Design is not true design unless it considers the method of manufacture (Section 7.4).
- Making drawings gives rise to errors and increases the variety of parts. Reduce the number of man-hours spent on design by 80% (this also applies to software preparation; Section 7.4).
- Tolerances and safety factors should be determined statistically (Section 7.4).
- Do the pilot products match the drawings? (Section 1.6.2.)
- Design is not design unless it considers costs (Section 7.4).
- Select a worse material over a better one if it gives the same performance and reliability (Value Analysis; Section 7.4).
- The secret of success in new-product development is to remove the dross quickly (Section 1.6.2).

(6) Standardization

- Unnecessary or vague standards lead to standardization for form's sake (Section 7.3).
- Standards that do not produce results are "paper standards" only; standards must be effective.
- A standard that has not been revised for six months since it was first prepared, is a standard that is not being used (Sections 1.5.2, 5.4.3, 5.4.6, and 7.3).

- When standards are not being revised, technical progress has halted (Sections 5.4.3 and 5.4.6).
- Standardization is not just for quality control. Standards are devised to ensure effective management and to make work rewarding for all.
- A company that claims that it cannot standardize and must rely on experience is a company without technology.
- Standardization allows authority to be delegated. This, in turn, allows managers time to consider future plans and policies, which is their most important responsibility.
- Quality control brings out the best in people. When a company implements it, deception disappears.
- Standardization is the job of engineers. Engineers must be practical.
- Technology must be standardized and a body of technology for the company must be systematically built up (Sections 1.5.2 and 5.4.3).
- When standards are being drawn up, input should be elicited from as many of the people affected by them as possible. It is human to observe self-imposed standards and regulations (Section 1.5.2).
- The purpose of standardizing is to delegate authority (Section 1.5.2).
- When planning the construction of a new factory, start the work of QC planning and standardization at the same time.
- Always doubt the validity of product standards, materials standards and tolerances, and never trust measuring instruments or chemical analyses.
- Products cannot be produced without knowledge of what kind of product one is trying to make (Section 1.4.4).
- Are you satisfied if your products meet the standards? (Section 1.4.2.)
- Are consumers complaining about products, even though they meet the standards? (Section 1.4.2.)
- Are you receiving complaints about points not covered by the standards?
- Work standards and control charts are two sides of the same coin.

(7) Control and process control

- The only way to clarify what is actually happening in the workplace is through process control. It will enable optimum process performance, will establish technology, and will allow for process and design improvement (Section 1.5.2).
- A process can only achieve its optimum performance when it is in the controlled state.
- Major improvement can only be obtained when adequate control is carried out (Section 4.1).

- Companies, factories, and processes over which no control is exercised are definitely not in the controlled state.
- Control must be comprehensive (QCDS; Sections 1.4.2 and 1.5.1).
- Follow the PDCA cycle to improve the quality of every type of work.
- Control the processes in every type of work.
- Follow the PDCA cycle in every type of work.
- Trying to exercise control will naturally produce improvements, while trying to produce improvements will naturally demonstrate the importance of control (Section 1.7.1).
- Do not confuse inspection with control (Sections 1.5.2, 5.2, and 5.3.1).
- Control and improvement are two wheels of the same cart (Section 1.7.1).
- Understand the difference between control and improvement (Section 5.2).
- Understand the distinction between cause and effect (Section 5.2.1).
- Do not confuse means with ends (Section 1.9).
- A company that says, "We have no problems" is the same as one that says "We have plenty of problems"; neither knows which of its problems are serious (Section 4.3.1).
- Control cannot exist without policies, targets, and objectives (Section 7.12).
- Standardization can only make progress and control can only be implemented when management policy has been decided.
- All leaders and those in positions of responsibility have a policy (Section 1.5.2).
- Correct policies can only be formulated on the basis of correct information.
- Are your policies and plans specific? Are evaluation criteria provided? (Section 7.12.)
- Are your policy-deployment and communication methods good? (Section 7.12.)
- Are the policies of superiors and subordinates adequately connected? Is policy consistent from the top to the bottom of your organization? (Section 7.12.)
- Does policy permeate every corner of your organization? (Section 7.12.)
- Does policy become more specific and concrete the further down the organization it goes? (Section 7.12.)
- How quickly can you act with caution? (Sections 1.5.2 and 4.2.2.)
- Serious problems are few, insignificant ones many ("vital few, trivial many"; Sections 1.4.4, 1.5.2, and 2.6).
- Usually only two or three major causes seriously affect a job or process.
- Our aim is quality; we must use intrinsic technology, statistical techniques, and control techniques in managing it and promoting effective TQC (Section 1.8).

- Good standardization and control are impossible without intrinsic technology.
- Practice preemptive control (Section 1.5.2).
- Clarify who should check what.
- Control charts and graphs should be seen and used by managers at every level.
- Control without checking is the ideal form of control.
- The span of control is 100 people. One person can control 100 others (e.g., an orchestra conductor; Section 1.5.2).
- Control that does not check the results of plans, orders, and actions is incomplete control.
- Always think what action to take. Control without action is simply a hobby.
- When accidents always occur for the same reasons, control is not being exercised.
- Do not confuse regulation and adjustment with removing the causes of abnormalities (Sections 1.5.2 and 5.2).
- Rather than simply eliminating symptoms, place priority on removing immediate causes and root causes and on preventing symptoms from recurring.
- To err is human. It is wrong to get angry at subordinates' mistakes (Section 1.5.2).
- When a worker makes a mistake, it is usually one-quarter to one-fifth her or his fault and three-quarters to four-fifths management's fault (Section 1.5.2).

(8) Analysis and improvement

- Without adequate analysis and reliable technical knowledge, improvements and standardization cannot be carried out, good control cannot be effected, and control charts suitable for control cannot be prepared (Sections 4.1 and 4.6.1).
- Good quality control is impossible without intrinsic technology. The motors powering the search for the causes of defects are research, technology, and skill (i.e., experience and training). However, technology will improve dramatically when quality analysis and process analysis are carried out by the QC approach using statistical methods (Sections 1.8 and 4.7.1).
- Good standardization and control are impossible without process analysis.
- When you believe you have no problems, progress stops and a backward slide begins (Section 1.7.1).

- A problem cannot be solved if the key points and objectives are not understood (Section 4.4).
- When key points and objectives are understood, the problem is already half-solved (Section 4.4).
- Determine the priority problems and attack them en masse.
- Engineers should tackle problems whose solution would save at least half a million dollars a year (as of 1987).
- Giving up is the enemy of improvement and progress.
- Before thinking of causes, first identify the facts. This is the first step in implementing quality control.
- Good process control is impossible without sound process analysis (Section 5.1).
- If control charts cannot be used, it is because real process technology and analysis are lacking (Section 4.1).

(9) Data and statistical methods

- Good quality control is impossible without knowing statistical methods.
- Dispersion exists in every type of work.
- The basis of control is accurate data and information. Abolish false data!
- Data are for using and acting upon. Don't collect data unaccompanied by action.
- From now on, statistical methods are an essential part of every engineer's knowledge.
- Discussion based solely on intrinsic technology and experience is like riding from Tokyo to Kyoto in a palanquin. Using statistical methods in conjunction with these is like making the same trip in the bullet train.
- Good standardization and control are impossible without statistical techniques.
- Ninety-five percent of a company's problems can be solved using simple statistical methods.
- Almost all problems can be solved with Pareto charts and cause-and-effect diagrams.
- Good control and analysis are impossible without good stratification (Sections 1.5.2 and 2.2).
- When false data emerge from the workplace, it is the fault of those in charge.

(10) Control charts and process capability

- Quality control begins and ends with control charts.
- Control charts should not be used for checking people. They should be

used for helping people with their work and making it go better (Section 5.6.2).
- Control leads to predictability and reliability (Section 6.5).
- The state of statistical control itself is the basic problem of reliability (Section 6.5).
- Process capability (quality) research is the foundation of quality control (Section 4.7.6).
- How can quality control be implemented if the capability of the process is not known? (Section 4.7.6.)
- How can you design a product without knowing the process capability? (Sections 4.7.6, 5.2, and 7.4.)
- How can you set materials standards without knowing the process capability?
- Process capabilities improve tenfold when they are properly researched (Section 1.6.4).

(11) QC circle activities

- Quality control only succeeds when foremen and line workers take responsibility for their processes.
- QC circle activities cannot be kept alive unless they are promoted as an integral part of TQC.
- A common misunderstanding is that carrying out QC circle activities equals implementing TQC (Sections 1.1.3 and 1.10).
- A common misunderstanding is that QC campaigns consist of QC circle activities (Sections 1.1.3 and 1.10).
- A common misunderstanding is that QC circle activities are a form of labor management (Section 1.10).
- QC circle activities and QC team activities are separate (Section 4.5.2).

(12) Sales and other activities

- Sales are the entrance and exit of TQC (Sections 1.6.2 and 7.7).
- A company whose sales activities develop no QC awareness will not grow.
- Has your company succeeded in developing several new products from suggestions made by the sales department? (Section 1.6.2.)
- If you think that sales have no connection with TQC, you do not understand TQC or QC.
- Merely selling things cheaply does not need sales activities. Sell through quality (Section 7.7).
- Salespeople must never say things like, "This product is absolutely safe" (Section 6.6).

- If proper cost control is carried out, the effect of quality control will rapidly improve.
- If quality control is done well, cost control will become real cost control.
- A well-controlled company is one in which production plans are not revised.
- When quality control is carried out, manpower control goes well. When manpower control improves, quality control goes well. How can you carry out quality control without knowing the true figures?
- Before introducing new machines and equipment, use the full capabilities of what you already have (Section 1.6.4).
- TQC means using the capabilities of old machinery and equipment, qualitatively and quantitatively (Section 1.6.4).
- Technology means using poor materials to make good products (Section 5.2.1).
- What do good salespeople say?

Chapter 2
THE STATISTICAL APPROACH, WITH SOME SIMPLE STATISTICAL TOOLS

2.1 Statistical Methods Used in Quality Control

I started studying statistical methods in 1948 because I believed that those who had to make judgments on the basis of data had to be masters of statistical philosophy and methods. Since that time, I have obtained excellent results through my efforts to disseminate the use of statistical methods widely, not only for quality control but also for business management. As a result of the wealth of experience I have gained in doing this, I generally find it best to teach statistical methods under the following headings:

1) The statistical approach (see Section 2.2)
2) Statistical theory
3) The use of statistical methods (introductory, intermediate, and advanced)

Since an understanding of the statistical approach is essential, it must be taught to everybody, and everybody must also acquire a feeling for statistics. Statistical theory should not be taught at the introductory stage, and only a few of the most basic aspects should be touched on at the intermediate stage. I go into the theory in slightly more detail at the advanced stage. People working in companies or society at large have no need to become specialist statisticians; it is sufficient if they are able to use the tools known as "statistical methods" skillfully. It is the same with the tools known as "measuring instruments"; even ordinary workers are perfectly able to use them in factories without knowing the theory of metrology. However, those in charge of planning the installation of measuring equipment have to know a little about the theory of metrology, while those researching, designing, and developing measuring instruments themselves must study metrology in depth.

Caution is needed, since people often want to study statistical theory when they start studying statistical methods, but amateurs who become engrossed in the study of statistical theory end up as neither good theoreticians nor good practitioners. Statistical methods are their aim, but they forget to use them as tools or become unable to use them properly, while at the same time tending to let theory and methods dictate their actions rather than the other way around.

(1) Statistical tools

Statistics and statistical methods are continuing to make great advances, but it is not necessary to know all about them to promote quality control and business management. On the contrary, it can in fact be harmful to teach too much about them, and statistical methods courses should be divided into introductory, intermediate, and advanced grades to suit the level of the students, with due regard to the actual conditions in the workplaces where the methods will be used. This book deals only with introductory and some intermediate methods; those wishing to study the remaining methods should consult more specialized works.

Introductory (aimed at all employees from top management through middle management down to ordinary workers) statistical tools include:

1) Pareto diagrams (see Section 2.5).
2) Cause-and-effect diagrams (not strictly a statistical tool).
3) The idea of stratification (mentioned in every chapter).
4) Check sheets (see Section 2.6).
5) Histograms and frequency distributions (see Sections 2.5, 2A.2, and 2A.3).
6) Scatter diagrams (see Sections 2.8 and 4A.8) (the concepts of correlation and regression).
7) Graphs and control charts (see Section 2.7 and Chapter 3).

The common characteristic of the above Seven QC Tools is that they are all visual, in the form of charts, graphs, or diagrams. They were named the Seven QC Tools after the famous seven weapons of the Japanese Kamakura-era warrior-priest Benkei which enabled Benkei to triumph in battle; so too, the Seven QC Tools, if used skillfully, will enable 95% of workplace problems to be solved. In other words, intermediate and advanced statistical tools are needed in about only 5% of cases.

Also, when one is teaching the seven tools, it is too much to try to teach them all at the same time. Tools 1–4 or 5 should be taught first, and the rest may be taught when the students have learnt to use these properly and want to learn more. It is wrong to try to teach too many of the tools from the beginning.

The following **intermediate** (aimed at general engineers and young workplace supervisors) methods should be taught in addition to the introductory methods:

1) Distribution of statistics, statistical estimation, and testing.
2) Sampling estimation, statistical error theory, additivity of variance.
3) Statistical sampling inspection.
4) The use of binomial probability paper.
5) An introduction to design of experiments (simple use of orthogonal arrays including contingency tables; analysis of variance).
6) Simple correlation and regression analysis.
7) Simple reliability techniques.
8) Simple sensory testing methods.

I regard the above methods as essential knowledge for engineers from now on. However, depending on the students, some of the above eight methods can be left out. When a person can use all of the above methods freely, he or she will have become a fully fledged engineer and will be able to solve many problems.

For the **advanced** level (aimed at specialist engineers, and some quality control engineers), the following methods should be taught in addition to the introductory and intermediate methods:

1) Advanced design of experiments.
2) Multivariate analysis.
3) Advanced reliability techniques.
4) Advanced sensory testing methods.
5) Time-series analysis, OR methods.
6) Other methods.

The above methods should be taught to selected people as necessary. They should also be taught in combination with the use of computers.

(2) Where are statistical tools used?

In Japanese-style companywide quality control, statistical tools are used extensively in a wide variety of fields and at all organizational levels. The Japanese are the best in the world at this, and it is one of the main reasons why so many Japanese products now dominate world markets. Various analyses are carried out and excellent results obtained through the wide use of the introductory-level tools in combination with specific technology in all departments and at all organizational levels, as well as through the use of more advanced tools in combination with computers and specific technology. By themselves, statistical tools are useless; they can only produce significant results when used in combination with specific technology, i.e., theory, technology, and experience concerning the work being performed.

In quality control and business management, statistical tools are used in the following areas:

1) Surveys: market surveys, surveys of measuring methods.
2) Policy setting and goal setting.
3) Analysis and improvement: process analysis and quality analysis.
4) Control and management: process control, work management, business management.
5) Quality assurance and inspection: quality assurance (including reliability assurance), statistical sampling inspection, inspection control.

2.2 The Statistical Approach

The science of statistics has recently made enormous strides and is still advancing. It is a rather forbidding field of study, but we do not need to know the theory in order to implement quality control; it is enough if we understand the philosophy, as well as the methods. At the very least, we should understand the philosophy. If we chase after the theory and methods without understanding the basic statistical approach, we will simply be indulging in games with figures and formulas.

In this section, I would like to focus on the statistical approach from the standpoint of quality control and the application of statistical methods to ordinary businesses.

2.2.1 The Statistical Approach

To start, we should understand the following four things about the statistical approach:

1) The results of any work we do always contain variation and follow a certain distribution pattern.

Human work and industrial processes are affected by an almost infinite number of different factors, and sampling, measurement, testing and surveying are also all subject to error. This means that data inevitably contain dispersion, and the results of the work and processes that produce the data follow a certain distribution pattern. This distribution pattern must be considered when making judgments about a process as part of process control, for example, since all processes have their own distribution patterns.

2) Error is a basic concept; the data produced by business enterprises and organizations in society include dirty data, abnormal values and false data.

3) Data are always collected with an intention to take action.

Modern statistics can be called a "science of action" from one viewpoint, since it aims at obtaining accurate data appropriate to the purpose, analyzing it statistically, and taking action (the object of the exercise).

4) Stratification is another basic concept; everything should be thought of in a stratified way, and everything must be stratified (i.e., segregated into meaningful streams or groups) for data collection and analysis. Totals and averages generally contain hidden information about dispersion. Stratifying individual, raw data in various ways can reveal this dispersion and its causes.

In this section, I would like to concentrate on the first and third of the above four ideas.

(1) The target of collecting data is to take action

When data are collected, we must always have some aim in mind; we must always intend to examine it and take some kind of action. In other words, we do not need to collect data that will not be used. The aim of collecting data should usually be one of the following:

a) for analysis or survey
b) for control
  (i) for policy-setting
  (ii) for adjustment
  (iii) for checking
c) for inspection.

For example, when we measure humidity and adjust a valve, check the quality of a product to see whether design and production are proceeding smoothly, or check quality, sales amounts, profit ratios, etc., to see whether a business is being managed skillfully, we must collect data. However, Japanese offices and factories today have a lot of data whose purpose is unclear and that are not accompanied by action. In the past, much of the data collected was simply for reassurance, for historical records, for preparing accounts, or for inspection.

The first thing we should do is look at all the data we are collecting and re-examine the purpose of collecting them. One of the first steps in the statistical approach is to remind ourselves thoroughly that we do not collect data for their own sake but for using them and taking action. Simply reflecting on this point will completely change the way data are collected. Unnecessarily large amounts of data are usually accumulated, much of them simply for the purpose of putting people's minds at ease. We should review our systems of responsibility and authority for collecting data and abolish as much detailed data collection as possible.

The next step in the statistical approach is to collect "action data" in various fields in this way and consider how to create an information-collecting network, provide feedback, and utilize the data for action. These points should be considered in conjunction with reporting systems.

(2) There is a relation between data, sample and population (the aim of action)

We do not collect data in order to obtain knowledge about and perform action on samples; we do so in order to obtain information about and perform action on the population from which the samples are taken.

Some points about sampling finished or half-finished products and performing measurements on those samples are as follows:

(a) We take a sample from a product lot in order to find things out about the lot and to take action with respect to that lot. We obviously do not perform measurements or experiments simply in order to obtain knowledge about a sample.

(b) Inspection omissions and measurement errors can never be entirely eliminated. This means that the data obtained from a sample will never represent the true values for the lot, even if the lot is subjected to 100% inspection. For example, try counting the number of letter 'd's on one page of this book within five minutes. Now try counting them again. You are unlikely to obtain the same figure. Even 100% inspection is always subject to error. We are thus trying to find things out about and take action with respect to lots and processes through a veil of measurement, sampling, experimental, and other errors.

(c) Testing products produced by a process means collecting data in order to find out about the state of the process.

(d) Compiling figures on daily or monthly results means collecting data in order to check whether or not the process of managing a company or factory is proceeding well and to take action if necessary.

(e) Performing an experiment to obtain data is also intended not simply to

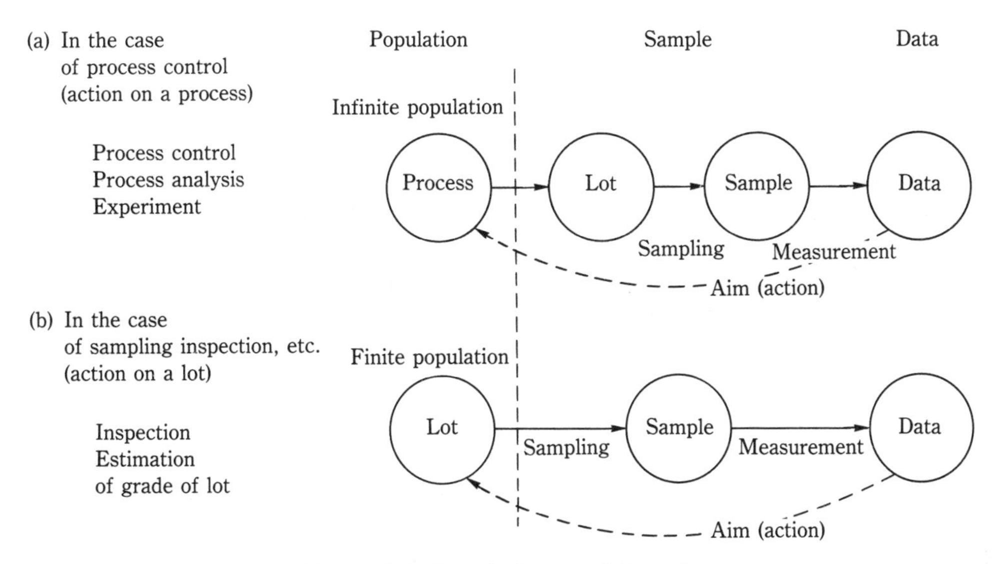

Figure 2.1 Populations and Samples

get the data from that experiment, it is collecting data in order to find out actual values under conditions like the experimental conditions and to decide whether those values can be used or not.

Our aim is thus to obtain information about finished or unfinished product lots, about the state of a job or process, or about actual values under certain experimental conditions, and to take appropriate action. These are the entities we are interested in; in statistics, they are called "populations" or "universes." In short, we take samples and perform measurements in order to find things out about populations and take action with respect to them. The relationship among data, sample, and population is depicted in Figure 2.1.

In process control, we try to control a process in order to produce good products, so we must always think of the process as the population. We take a lot or part-lot (i.e., whatever we are passing from our area of responsibility to the next person responsible) as a sample of the population which we call a "process," and measure its characteristics in order to check whether or not the process (the population, i.e., the way in which we are carrying out our work) is going well. We then try to take rational action with respect to the process. We try to put the population we call a "process" in a controlled state and keep it there before it can produce bad products, so that it produces good products. Also, when an experiment is carried out by deliberately altering the conditions, the data will be scattered as a result of the various errors present. There may be a different population distribution for each of two different sets of experimental conditions, and the experiment is carried out in order to find the difference between the two population distributions and select the best set of conditions. Again, in sampling inspection, samples are taken from a population called a "lot" in order to decide whether that lot should be accepted or rejected.

As the diagram shows, a population may either be a "finite population," regarded as consisting of a finite number of items, or an "infinite population," regarded as consisting of an infinite number of items. In process control or experiments, what is known as a "process," i.e., work carried out under set operating standards, is thought of as a population. Since such a process is theoretically the source of an infinite number of products, it is treated statistically as an infinite population, and the product lots that emerge from it daily are regarded as samples from this population.

(3) All data are dispersed

The data we obtain are always scattered and are never a single, unchanging value. If a group of data all had the same value, it would in most cases be false; in fact, data that are not scattered are useless. For example, if we took samples of a certain product and measured their strength, the results we would obtain would be scattered, e.g., 25, 20, 28, 30, 32kg/cm^2. In the past, we would simply have taken the average of these values, 27.0kg/cm^2, and based our judgments on this. In other words, our thoughts were confined to the world of averages. Nowadays, we also consider dispersion. In other words, we enter the world of dispersion when estimating the quality of a lot and deciding whether to accept or reject it or when judging whether or not there is any abnormality present in a process or job. In statistical terminology, this is called "testing."

The number of factors causing dispersion in any industrial process is theoretically infinite. Since we can only control a tiny fraction of these by technology, there will inevitably be dispersion in the characteristics of the products of our processes. Sampling and measurement are also subject to error, and the results will always be scattered even if the same thing is measured several times. What happens if we simply accept this dispersion for what it is?

(a) Identifying a distribution

When data are scattered, they follow a certain distribution pattern. In other words, the population (e.g., a process) from which the data have come also has a distribution. If we measure the thickness of two hundred rolled-steel sheets, for example, we will obtain the kind of results shown in Table 2.1.

These data do not by themselves tell us anything, but, if we divide them up into classes or "cells" with a width of 0.03mm, e.g., 3.695–3.725mm, 3.725–

Table 2.1 Thickness of Steel Plate (units: mm)

3.88	3.88	3.84	3.82	3.83	3.93	3.86	3.84	3.90	3.97
3.84	3.85	3.90	3.87	3.94	3.89	3.87	3.87	3.86	3.87
3.84	3.84	3.85	3.88	3.89	3.96	3.84	3.79	3.81	3.84
3.88	3.83	3.84	3.85	3.93	3.81	3.87	3.83	3.89	3.87
3.81	3.91	3.90	3.86	3.83	3.90	3.87	3.90	3.86	3.86
3.78	3.92	3.98	3.74	3.88	3.81	3.94	3.91	3.97	3.75
3.88	3.94	3.90	3.88	3.85	3.87	3.90	3.78	3.86	3.87
3.88	3.79	3.80	3.80	3.79	3.82	3.86	3.84	3.92	3.83
3.90	3.90	3.83	3.84	3.95	3.84	3.97	3.89	3.86	3.90
3.84	3.81	3.84	3.98	3.99	3.86	3.85	3.79	3.87	3.78
3.93	3.84	3.88	3.85	3.91	3.89	3.84	3.88	3.89	3.97
3.83	3.90	3.93	3.87	3.90	3.92	3.91	3.70	3.79	3.73
3.97	3.89	3.78	3.83	3.87	3.90	3.84	3.76	3.81	3.82
3.85	3.83	3.81	3.83	3.76	3.77	3.90	3.79	3.83	3.90
3.89	3.86	3.84	3.89	3.83	3.80	3.86	3.80	3.89	3.83
3.90	3.77	3.79	3.83	3.85	3.85	3.89	3.84	3.83	3.95
3.88	3.87	3.81	3.91	3.89	3.84	3.79	3.86	3.78	3.89
3.81	3.77	3.73	3.85	3.80	3.77	3.78	3.83	3.75	3.83
3.94	3.90	3.75	3.77	3.83	3.79	3.86	3.89	3.84	3.99
3.83	3.94	3.84	3.93	3.85	3.79	3.84	3.88	3.83	3.80

Table 2.2 Frequency Distribution Table

Cell number	Cell boundaries	Cell midpoint	Tally	Frequency	Relative frequency (%)	Cumulative frequency
1	3.695-3.725	3.710	/	1	0.5	1
2	3.725-3.755	3.740	卌 /	6	3.0	7
3	3.755-3.785	3.770	卌 卌 卌	13	6.5	20
4	3.785-3.815	3.800	卌 卌 卌 卌 卌	25	12.5	45
5	3.815-3.845	3.830	卌 卌 卌 卌 卌 卌 卌 卌 卌	45	22.5	90
6	3.845-3.875	3.860	卌 卌 卌 卌 卌 卌 卌 ///	37	18.5	127
7	3.875-3.905	3.890	卌 卌 卌 卌 卌 卌 卌 卌 ///	43	21.5	170
8	3.905-3.935	3.920	卌 卌 ///	13	6.5	183
9	3.935-3.965	3.950	卌 ///	8	4.0	191
10	3.965-3.995	3.980	卌 ////	9	4.5	200
				200	100.0	200

3.755mm, etc., and count the number of measurements in each cell, we obtain Table 2.2. This is called a "frequency distribution table." Arranging the data in the form of such a table clarifies the way in which the data are scattered, i.e., the shape of their distribution. It is usually possible to see the general shape of a distribution if a hundred or more values are collected and a frequency distribution table with ten to twenty cells is prepared.

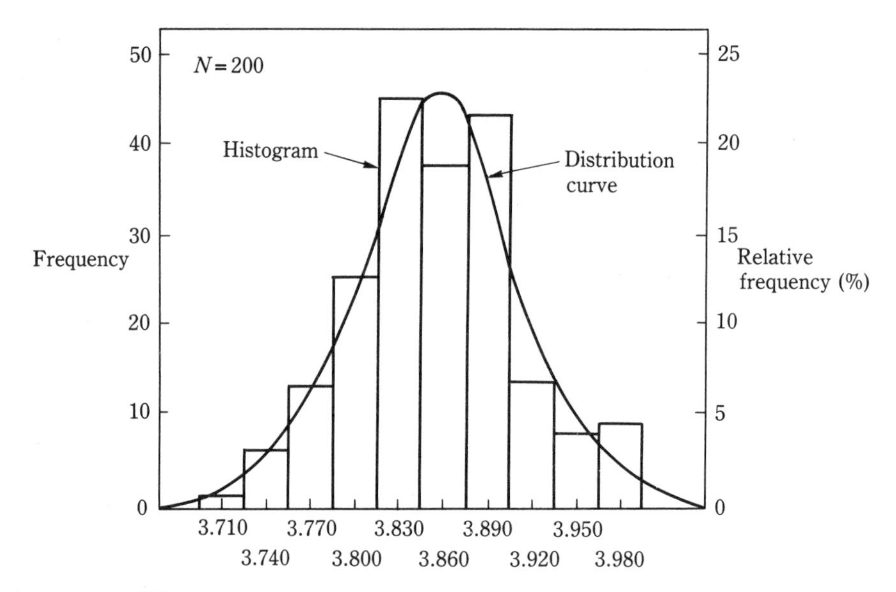

Figure 2.2 Histogram of Thicknesses of Steel Plate

The distribution can be shown even more clearly in the form of a diagram (Figure 2.2*). This kind of data diagram is called a histogram.

This histogram clearly shows that the thickness of the steel plates follows a distribution. That being so, what kind of population is the process that made the plates? We can regard it as having the kind of distribution that would be obtained if an infinite number of these measurements were taken; for example, the kind of distribution shown by the curve in Figure 2.2. The point is that the population follows a certain distribution. When judging whether our work is being carried out well or not, we should always remember that the population—the process—and therefore the products emerging from that process, follow a distribution.

(b) The quantitative expression of distributions

Measures such as mean, range, standard deviation, variance, etc., are used for quantitative expression (see Sections 2.3 and 2A.2).

(4) Random sampling

That a population has a distribution means that we must be careful to take any samples at random. For example, if the surface of a polished product has good and bad areas, which experienced workers can distinguish between, the worker

---

* See Section 2A.1 for details of how to prepare frequency distribution.

will probably want to choose the good areas as samples. Inspectors, on the other hand, tend to prefer the bad areas. Such samples are not random, and they cannot be said to represent faithfully the process from which they are taken. Deliberately choosing good or bad samples is not random sampling.

We take samples in order to find out the state of a process, so it is generally unsatisfactory to let people choose samples; it is best to take samples at random. Also, statistical theory tells us what kind of values and distributions the sample average, range, standard deviation, and other statistics should have when random sampling is carried out (see Section 2A.3, Distribution of Statistics), and we perform random sampling and evaluate the data according to the principles of the distribution of statistics.

Random sampling is easy to say but not so easy to carry out; in practice, however, taking samples of a product at fixed intervals is generally a good approximation to random sampling.

Random sampling is the most important factor in implementing quality control or other types of control using statistical methods. If it is not done properly, the results obtained will be virtually meaningless no matter how much statistical analysis is subsequently performed on the data.

(5) The two types of cause of dispersion in processes

There are two types of cause affecting processes and causing variation in the product, and there are therefore also two types of variation. One type of cause is that which gives rise to variation in the product (the result of the process) even when everyone concerned with the process is operating it in exactly the same way as always, and everyone is working correctly, i.e., according to standard.

These are the causes that are not yet under technical control, but are present in theoretically almost infinite numbers. They are called unavoidable causes or "chance causes," and the variation produced by them is called "controlled variability." As managers, we cannot blame our workers for them on the basis of existing work standards or drawings.

In contrast, the other type of cause is the type that produces some abnormality in the process and results in a particularly large variation, e.g., when something not covered by the work standards happens or the work standards are disobeyed. Such causes can be eliminated through technology if everybody involved makes a cooperative effort; they are called avoidable or "assignable causes," and the variation due to them is called "uncontrolled variability." If work standards and other standards were perfect, the presence of uncontrolled variability would mean that the workers were not doing as they had been told. This type of variability indicates that non-standard materials are entering the process, jigs and tools are worn, measuring instruments have gone out of calibration, etc. When

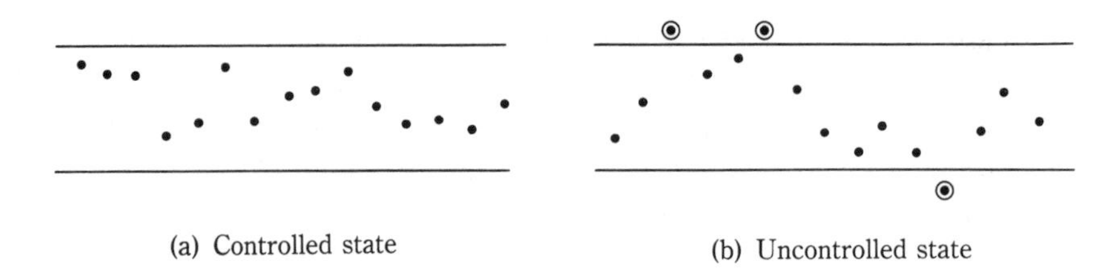

Figure 2.3 Two Types of Dispersion in Products Produced by a Process

uncontrolled variability occurs, managers are responsible for admonishing the workers and taking action to ensure that the work is carried out according to the standards. However, standards are generally imperfect, and these causes usually require action by managers and staff.

As mentioned above, these two types of cause produce two types of variation in the results of processes (i.e., products, etc.). When work is proceeding according to standard and random sampling is carried out and controlled measurements taken, the variation in product quality due to chance causes will generally be found to have a fixed distribution, usually the normal distribution. This situation is known as "controlled variability." The state of a process that produces results in which the only dispersion is controlled variability is called the "controlled state" (e.g., the state shown in Figure 2.3(a)).

In contrast to this, when an assignable cause arises, e.g., when an abnormality occurs in a process, the dispersion in the results of the process (i.e., in the product) becomes abnormally large. Such an abnormally large dispersion is called "uncontrolled variability," and the state of a process that produces results having such a dispersion is called the "uncontrolled state." For example, when a process is in a state like that shown in Figure 2.3(b), with points lying outside the control limits and forming a particular pattern, the process is said to be "out of control."

These ideas are not restricted to quality; they can be applied in exactly the same way to yield, production costs, unit costs, sales volumes, and other quantities, all of which are the results of processes.

In the past, processes were controlled by using gut feelings and experience to distinguish between the two types of variation. Now, however, we use statistical tools to enter the world of dispersion, distinguishing between the two types of variation objectively and economically, and banishing assignable causes from our processes. To make the distinction, we use "control limit lines" on "control charts." The pairs of straight lines in Figure 2.3 are equivalent to these. The study of the use of control charts centers mainly on how to calculate these control limit lines so that they can be used as efficiently, economically, and easily as possible, and on how they should be used for control.

So far, I have been discussing the two types of cause mainly in terms of whether work standards are being observed or not. However, when there are no reasonable work standards, it is impossible to classify the two types of cause simply in this way. In such a case, we could start by classifying them according to whether the work was being carried out as before or whether some change had been introduced. We could also categorize the causes into those that produce a relatively large variation and those that produce only a small variation. In this case, if we eliminated the most important assignable causes, we could then sort out the relatively large variations from among those that had up to then been considered relatively unimportant. In other words, we could sift out the causes in the order of the size of variation they produced and eliminate them one by one, starting with the largest.

(6) Statistical judgment

When data are dispersed and we want to distinguish between two different distributions in order to take some action, we must understand the concepts of probability and type I and type II errors.

Probability may sound difficult, but making judgments and taking action according to probability is in fact exactly what we have always done, i.e., deciding and acting on the basis of common sense. This is easily illustrated with an example.

Let us imagine that you and I are throwing a die and betting. We decide that I win if an even number comes up and you win if an odd number comes up, and away we go. I throw the die five times and announce that I have got five even numbers in a row. What do you think when I tell you that I have won five times running like this? Most people would probably tell me I must be cheating. Let us analyze the thought process behind this. If I were throwing fairly, the probability of obtaining an even number on one throw of the die would be 1/2. The probability of obtaining five even numbers in a row would therefore be

$$\frac{1}{2}\times\frac{1}{2}\times\frac{1}{2}\times\frac{1}{2}\times\frac{1}{2}=\left(\frac{1}{2}\right)^5=\frac{1}{32}\approx 0.03$$

Obtaining five even numbers in five throws of the die is therefore an event that will occur on average only about three times in a hundred. You probably think it strange for an event with such a small probability as this to happen, and therefore conclude that the probability of obtaining an even number on one throw of this particular die is not equal to a half as it should be. You make the intuitive judgment that I am cheating in some way or other.

Ordinarily, most people would use their common sense to infer that I was cheating without actually carrying out the above calculation. The only difference between this kind of mental process and statistics is that the latter tries to make

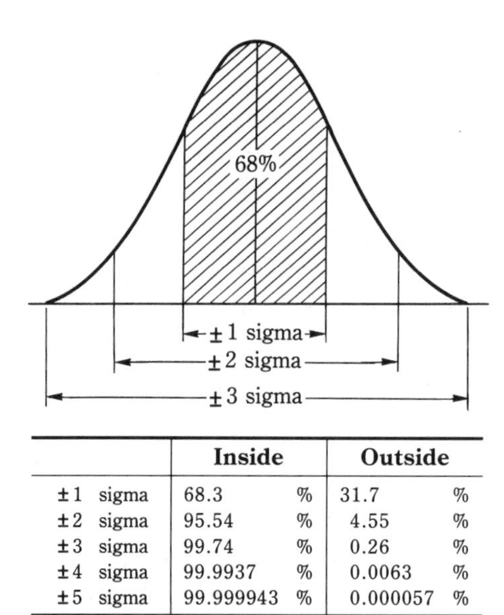

	Inside	Outside
±1 sigma	68.3 %	31.7 %
±2 sigma	95.54 %	4.55 %
±3 sigma	99.74 %	0.26 %
±4 sigma	99.9937 %	0.0063 %
±5 sigma	99.999943 %	0.000057 %

Figure 2.4 The Normal Distribution with Probabilities

judgments on the basis of precisely calculated probabilities; in other respects, it is exactly the same as common-sense judgment based on experience. Statistical tools are used for this kind of probability calculation.

Similarly, if data are randomly sampled from a distribution, we can expect more of the sampled values to be close to the average or peak of the distribution, and fewer to come from near the tails. In other words, the probability of obtaining values lying outside certain limits is small. Thus, if we do in fact obtain such low-probability values, we may decide that they come not from this particular distribution but from some other distribution. When a distribution has controlled variability, low-probability values from the far ends of the tails or outside the tails cannot be regarded as being part of it, and must be regarded as indicating uncontrolled variability. In other words, it is judged on the basis of probability to be data not from the controlled distribution but from a separate process or distribution.

Most of the distributions we normally come into contact with approximate the bell shape shown in Figure 2.4 — the normal distribution. If we use the standard deviation (sigma) to divide up the area under the curve, as shown, we can find the areas of each part as a proportion of the total area under the curve. This proportion is the probability of obtaining the values in each part of the distribution when samples are taken at random from a population having this kind of distribution. As the figure shows, the probability of obtaining a value within one sigma either side of the mean is approximately 68%, while the probability of ob-

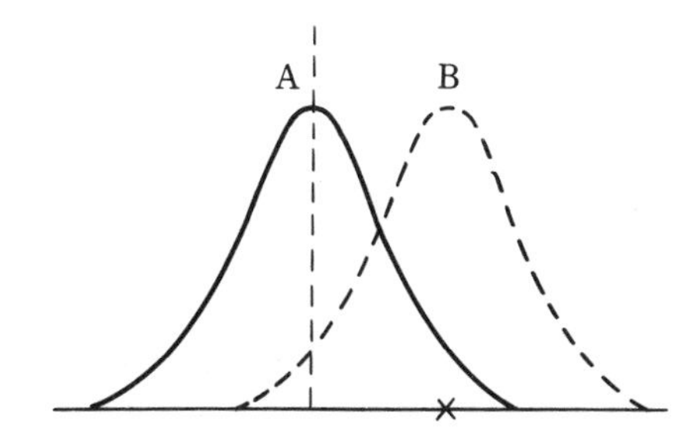

Figure 2.5 Judged to be Two Distributions

taining a value outside these limits is approximately 32%. Similarly, the probability of values falling outside the ±3 sigma limits is extremely small; only 0.3%, or 3 in 1,000. Since such a low-probability event usually occurs only rarely, we usually conclude when data fall outside the 3-sigma limits that the distribution has changed, the process has changed, and there is some abnormality in the process.

Going back to the example of the die, if you accused me of cheating every time I claimed to have thrown five even numbers in a row, would you always be right? On average, there is a 1 in 32 chance of obtaining five even numbers in five throws even when the die is fairly thrown. In other words, if you accuse me of cheating, you have a 1 in 32 chance of being wrong, so you will not always be right. In statistics, this type of error is called a "type I error." On the other hand, how about if I claim to have thrown three even numbers running, and you simply say, "Lucky, aren't you?" without accusing me of cheating. I might in fact have been cheating without your noticing it. In statistics, this type of error is called a "type II error." The fact that whenever we make a judgment, there is always the possibility of our making these two types of error is the basis of statistical testing.

Let us consider what happens when we are thinking of taking action with respect to a process that is showing dispersion. If we obtain some data lying in the tails of a controlled distribution A,—e.g., the value shown by the cross in Figure 2.5,—we might conclude from the low probability of obtaining such a value that some factor giving rise to an abnormal distribution had occurred in the process. However, this might be a type I error, in which we mistakenly conclude that the distribution is an uncontrolled one (B), while it is actually a controlled one (A). However, we cannot use this possibility as an excuse for claiming that the process is in the controlled state (A). If we did this, we might be making a type II error and failing to notice that some abnormality had occurred in the process and that the mean of the distribution had shifted to B. If we accept that there will be some variation in our products even if we are working as usual, we must also accept that those responsible for controlling the process cannot completely eliminate the possibility of making these two types of error.

If our sole concern is not to let a single process abnormality slip by unnoticed, and we decide that something is wrong with the process even when the data from it show only a slight variation, we will be magnifying the type I error. Even though there is no abnormality in the work or process and it is being carried out according to the standards, we will be scurrying hither and thither hunting vainly for causes of abnormalities. For this reason, we could also call this type of error a "hot-headed" type of error. To avoid committing it, we could simply leave things as they were and do absolutely nothing, however scattered the data became. But this would mean failing to take action even if there really were an abnormality. It would magnify the type II error and might be disastrous for the process. We could call this type of error a "wool-gathering" type of error.

Many Japanese, particularly executives, managers, and supervisors, have no sense for dispersion. They easily lose their nerve and run about in confusion, now over the moon and now down in the dumps, whenever there is a small variation in the data. For example, they sound off whenever sales drop slightly and start berating people if yields fall a little. All this behavior does is encourage the workplace to put out falsified data.

In controlling a process, we will have problems if we are either too hot-headed and hasty or too slow to react. The control limit lines on a control chart take account of the possibility of making the two types of error and are drawn so as to distinguish empirically between two different distributions. We use them as a basis for making judgments. The lines are drawn at the 3-sigma points of the within-subgroup variation; in other words, they are drawn so as to make the probability of committing a type I error approximately 0.003, or 0.3%. Thus, as the common-sense ideas about probability mentioned earlier also tell us, a variation that breaks through the limits (in the example of the die, this would be the same as obtaining a run of seven or eight even numbers) shows that there is probably something wrong with the process. In such a case, we can therefore discharge our control responsibilities with (statistical) confidence, searching out and eliminating the causes of the abnormality using engineering and statistical techniques. Control limit lines are thus a type of rule setting a basis for action, and their use enables anyone to make judgments objectively.

If managers make the "hot-headed" type of error and scold their subordinates even when they are doing their jobs properly, the effect on what happens later will be extremely bad, with falsified data emerging from every workplace. This is why the control chart sets the probability of making this type of error at the extremely low value of 0.3%.

In the general statistical testing of hypotheses in non-control situations, the probability of making a type I error is usually set at 5% or 1%. In statistics, this probability is called the "significance level."

(7) Take action with regard to the population

When we have proceeded as described above and made a decision based on sample data, we must then take action with regard to our primary objective, the population. In process control, since we can say with confidence that an abnormality has occurred in the process, this means searching out its causes, eliminating them, and preventing the abnormality from recurring. As mentioned before, statistics could from one point of view be called a "science of action." We only fulfil the purpose of collecting data as described above when we take definite action based on statistical conclusions.

The above is a rough outline of the statistical approach. We use this kind of approach to try to rationalize our organizations by controlling management, quality, and other factors on a statistical basis.

2.2.2 Precautions from the Control Viewpoint

From the control viewpoint, there are a number of important considerations concerning the statistical approach.

(1) The reliability of data

Whether in control, inspection, or analysis, false, deliberately doctored, or otherwise unreliable data from the workplace are worse than useless. In traditional companies, particularly those with a more centralized bureaucracy, more "artisan directors" and "artisan department managers," and a poorer appreciation of statistical variation, more false data emerge from the workplace and the farther such data are from the true facts. This was the same in the era of the managed economy. Under management based on the view of mankind as fundamentally evil, the reliability of data decreases.

We could use all sorts of statistical methods to analyze such false data or spend hours discussing the details of sampling and measurement errors, but doing so would be meaningless. Before applying statistical methods, we need to ensure that true data emerge from the workplace, and this requires the following:

(a) Everybody, particularly upper-level managers, must recognize that all data contain dispersion.
(b) Discussion must be objective, with due consideration given to statistical estimation and testing.

(c) Authority must be delegated.
(d) Upper-level managers should not be too concerned with the fine print. Directors and department managers must act like executives, not artisans or craftsmen.

Particularly when large numbers of items are being handled (e.g., large numbers of bottles or parts), control of production volume and cost, let alone control of quality, is obviously impossible if the numbers cannot be correctly identified.

Put the other way round, issuing unreasonable orders, loss of temper, dictatorial control, excessive control by head office, misleading statements by superiors, lack of understanding of dispersion on the part of superiors, poor work standards and evaluation methods, and insufficient checking all encourage the production of false data. When quality control is skillfully implemented across an entire company, false data disappear and everyone becomes able to speak their mind freely.

(2) Clearly showing the history of data and product lots — Stratification

As the above discussion has made clear, the statistical approach is in effect a process of examining facts, making judgments, tracking down the causes of abnormalities, and taking action. For example, it means looking at the way in which the results of a process (i.e., the quality of the product) are dispersed, searching out the assignable causes from among the many factors affecting the process, and taking action.

Thus, if the history of data (in other words, the history of product lots) is unclear, it will be difficult for us to identify the causes of variation in the process. Clarifying the history of data and product lots is absolutely essential in quality control, and we must therefore do the following:

(a) Clearly define what we mean by a lot.
(b) Stratify product lots.
(c) Record who performed the sampling and measurement.

This can be done relatively easily by contriving some kind of card system and by devising suitable containers and conveyers. If it is not pushed through strongly, even though there may be some strain on the system at first, quality control will not make progress and any data collected will not be very useful. Without stratification, quality control will stick in its tracks.

(3) Estimating an in-control process

To identify what kind of entity a process is when it is under control, we must estimate that process's (i.e., the population's) distribution.

If a process has been in the controlled state for a long time, the variation in the result of the process (i.e., the product) must also be under control. Thus, if the variation in the results of a process has continued in the controlled state for a long time, we can estimate the distribution in the future. This is extremely important in process control, since it means we can assume that if we continue to control a process in the same way in the future, the distribution of the result of the process (i.e., the product of the future in-control process) will also be the same. In other words, if we take control limit lines calculated from an analysis of past data and simply extend them without modification into the future, the process variation will remain between the limits, provided that the process remains under control. This means that we can also estimate the shape of the distribution of the results of the process in the future and use this estimate for assuring future quality and reliability.

In this way, quality can be assured even without inspection if a process is kept in a well-controlled state, and we can confidently predict what sort of product will be produced. Conversely, if the product characteristics lie between the limit lines in the future, this indicates that we are justified in assuming the process to be under control. It also means that if values appear outside the lines, we can assume that some change has occurred in the process. Thus, if data break through the limits in the future, we must immediately investigate the abnormality in the process and take action.

The aim of controlling processes using control charts in this way is really to keep processes under control in the future and thus to be able to provide society with reliable products.

2.3 The Different Types of Data

The measurements that we make are of two types: metrical and countable. For example, measurements such as "three defective bolts out of one hundred," "five blemishes in one sheet of cloth," or "x breakdowns in one month at the factory," are obtained in the form of integers (i.e., "one bolt," "two bolts," "one blemish," "two blemishes," "one breakdown," "two breakdowns"), not in the form of fractions or decimals, such as "1.6 breakdowns." Measurements that jump from one whole number to the next in this way are called "discrete" or "discontinuous," and discrete values like the above are called "attributes." Also, when, for example, 3 sheets of a product out of 200 are found to be defective, this can be expressed as $(3/200) \times 100 = 1.5\%$, i.e., a percent defective of 1.5%. Percentages such as this are not integers like the above examples, but they do jump (from 1 to 1.5 to 2 to 2.5 in this case) without intermediate values (such as 1.1, 1.2, etc.) and are thus not continuous. Percent defectives of this kind are therefore also attributes.

In contrast to this, measurements of, for example, thickness (mm), weight (g), moisture content (%), strength (kg/cm^2), yield (product weight/raw material weight, %), or work time (h) are continuous values. In a series of measurements such as 1.50mm, 1.51mm, etc., the values might appear at first glance to jump from one number to the next, but all we actually mean by 1.50mm is that, because of the limited precision of the measurement, we have rounded off all readings from 1.495mm to 1.505mm to the nearest one-hundredth of a millimeter. The actual object we are measuring can have an infinite range of values, and measurements that can take on a continuous range of values in this way are called "variables."

The statistical nature of attributes and variables is different, and they often require the use of different statistical tools and control charts. Attributes also form discrete distributions, while variables form continuous distributions.

The attributes that we usually meet in quality control are also of two kinds: distributions of number of defectives and fraction defective, obtained by classifying products as either conforming or not conforming to specifications, are statistically different from distributions of number of defects and number of defects per unit, obtained by counting the number of defects in a single product.

Statistically speaking, distributions of number of defectives and fraction defective form binomial distributions. If the product is classified into three or more grades (e.g., first grade, second grade, third grade, etc.), the distribution will be multinomial. In contrast, distributions of number of defects and number of defects per unit follow the Poisson distribution.

Note: In quality control, the terms "acceptance," "rejection," "defective," and "defect" have strict definitions. "Acceptance" and "rejection" are used for the acceptance and rejection of product lots, while "defective" and "non-defective" are used for evaluating individual inspection units. A "defect" is an area or item on one inspection unit that fails to meet the specifications or requirements.

2.4 EXPRESSING DISTRIBUTIONS QUANTITATIVELY*

The fact that sample data and populations are statistically distributed was mentioned in Section 2.1. The approximate shapes of such distributions can be seen if the data are shown in the form of tables or diagrams such as Table 2.2 and Figure 2.2, but it is also often convenient to express them quantitatively.

A distribution is determined by its position (or "central tendency"), spread (or "dispersion"), and shape; in other words, whether its peak is displaced to left or right ("skewness"), or is sharp ("peakedness" or "kurtosis"), or flat ("flatness"). Generally, it is sufficient to express the central tendency and dispersion of a distribution quantitatively, and show its shape (i.e., its skewness and peakedness) by a histogram or similar diagram. Such quantitative values are called "measures."

(1) Measures of central tendency

The arithmetic mean (also referred to as "average") or the median is normally used for measuring central tendency.

Average, $\bar{x}$ (read "x bar"): if, for example, we take the first five values (3.88, 3.88, 3.84, 3.82, 3.83) from the data shown in Table 2.1, their average is given by

$$\bar{x} = (1/5)(3.88 + 3.88 + 3.84 + 3.82 + 3.83) = 3.850$$

The average can be expressed by the following general formula:

$$\bar{x} = (1/n)(x_1 + x_2 + \cdots + x_n) = (1/n)\sum_{i=1}^{n} x_i = (1/n)\Sigma x_i$$

Median, $\tilde{x}$: this is the middle value when the data are arranged in order of magnitude. For example, the median of the above values (3.88, 3.88, 3.84, 3.83, 3.82) is 3.84. When there are an even number of values, the median is the average of the two middle values.

Mode: the value corresponding to the peak of the distribution. Some distributions may have more than one mode.

---

* See Section 2A.2.

(2) Measures of dispersion

Range, R: the difference between the highest value, x_{max}, and the lowest value, x_{min}. In the above example,

$$R = x_{max} - x_{min} = 3.88 - 3.82 = 0.06$$

R is generally used when the number of data points is ten or less.

Sum of squares of deviations, S: a "deviation" is the difference between a particular value and the mean, and the sum of the squares of these differences is called the sum of the squares of the deviations.

$$\begin{aligned} S &= (x_1 - \bar{x})^2 + (x_2 - \bar{x})^2 + \cdots + (x_n - \bar{x})^2 \\ &= \Sigma(x_i - \bar{x})^2 \\ &= \Sigma x_i^2 - (\Sigma x_i)^2/n \end{aligned}$$

The second term of the above expression is called a "correction term" (CT).

Sample variance, or unbiased estimate of population variance, V : the sum of the squares of the deviations divided by $n-1$, where n is the number of data points.

$$V = S/(n-1)$$

Standard deviation, s or $\hat{\sigma}$: the positive square root of the variance.

$$s = \sqrt{V} = \sqrt{S/(n-1)}$$

When the number of data points is large, dividing by n instead of $n-1$ will give approximately the same result.

2.5 INTERPRETING AND USING FREQUENCY DISTRIBUTIONS*

Frequency distributions are the most simple, commonly used, and effective statistical tool. Anyone who cannot even use frequency distributions will certainly not be able to use the more advanced statistical methods.

---

* See Section 2A.1 for methods of preparing frequency distributions and calculation methods.

We are surrounded by masses of data, but simply listing them as in Table 2.1 does not tell us very much. Expressing them in the form of a frequency distribution as in Table 2.2 or Figure 2.2 gives us a variety of insights. Since frequency distributions are used extremely often in quality control, companies should prepare blank frequency distribution forms that also provide space for notes and for calculating and recording averages, standard deviations, etc.

(1) The aim of preparing frequency distributions

The main aim of preparing a frequency distribution is usually one of the following:

(i) To make the state of a distribution easily visible, i.e., to identify the shape of the distribution.
(ii) To identify process capability.
(iii) For process analysis and control.
(iv) For determining the average, standard deviation, and other measures of a distribution.
(v) To test what type of mathematical distribution an empirical distribution can be fitted to statistically.

Since frequency distributions make it easy for anyone to see and intuitively understand the shape of a distribution and the state of a process, they are used particularly often for purpose (i) above. They have a wide range of applications: for example, to report daily, weekly, monthly, or annual results to superiors in a readily understood form, or to analyze the causes of dispersion.

Frequency distributions and histograms are more easily interpreted if values such as standard deviation, target value, specified value, $\bar{\bar{x}} \pm 3\bar{R}/d_2$ calculated from control charts, or $\bar{x} \pm 3s$ obtained from the histogram are included.

(2) Interpreting frequency distributions

When looking at frequency distributions, attention should be paid to the following points:

(i) Is the distribution (i.e., its average) in the appropriate position?
(ii) What is the spread (i.e., the dispersion) of the distribution like?
(iii) What is the relation between values such as standard deviation, target value, and specified value?
(iv) Do any values lie outside the $\bar{\bar{x}} \pm 3\bar{R}/d_2$ or $\bar{x} \pm 3s$ lines?
(v) Are there any gaps like missing teeth or sudden ups and downs like the teeth of a comb in the distribution?

(vi) Are there any outlying points isolated from the main body of the distribution?
(vii) Are the maximum and minimum values of the distribution acceptable?
(viii) Is the distribution lopsided, with one tail extending farther than the other, or is it symmetrical?
(ix) Is the left-hand or right-hand edge of the distribution cliff-like in appearance?
(x) Does the distribution have more than one peak?
(xi) Is the peak of the distribution too pointed or too flat?

If several stratified histograms are prepared from the same data and comparisons of the above points are made, more information can be obtained. Readers should study the interpretation of frequency distributions using Figures 2.6–2.9 below.

(3) How to use frequency distributions

Frequency distributions are used for finding out about the shape of a distribution or the state of a process when interpreted as described above. They have the following applications:

(1) For reporting: if monthly quality reports and other reports are presented not just as an array of data but in the form of histograms with the number of values N and their average and standard deviation appended, they can be easily understood by anybody.
(2) For analysis: if stratified frequency distributions are segregated according to likely assignable causes (e.g., operator, machine, raw material, month/day, etc.) as shown in Figures 2.9 and 2.10, the differences can be immediately identified. Stratification is the secret of preparing good frequency distributions, while interpreting them as described in Figure 2.8 may allow the causes of outlying data to be discovered. It also allows the relationships between specifications, standard values, and target values to be readily identified.
(3) For surveys of process capability and equipment capability: frequency distributions are often used to show the capabilities of processes, machinery, and equipment both qualitatively and quantitatively.
(4) For control: displaying frequency distributions in the workplace (and in some cases stratifying them and updating them daily) conveys the concept of dispersion to workshop foremen and ordinary workers and helps to enhance their control-mindedness. Also, if $\bar{\bar{x}} \pm 3\bar{R}/d_2 = \bar{\bar{x}} \pm E_2\bar{R}$ lines are drawn on frequency distributions (see Section 3A.2), they can be used for control in the same way as the control lines on a control chart.

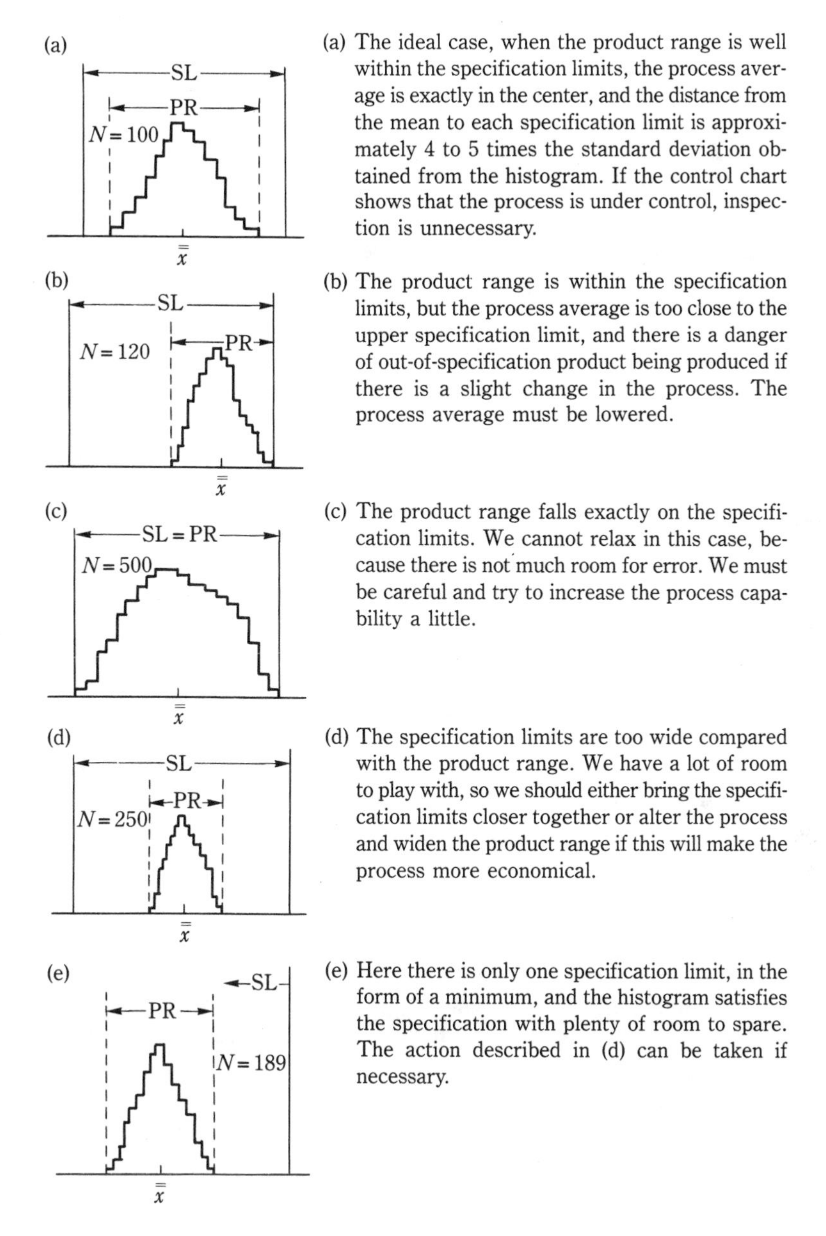

(a) The ideal case, when the product range is well within the specification limits, the process average is exactly in the center, and the distance from the mean to each specification limit is approximately 4 to 5 times the standard deviation obtained from the histogram. If the control chart shows that the process is under control, inspection is unnecessary.

(b) The product range is within the specification limits, but the process average is too close to the upper specification limit, and there is a danger of out-of-specification product being produced if there is a slight change in the process. The process average must be lowered.

(c) The product range falls exactly on the specification limits. We cannot relax in this case, because there is not much room for error. We must be careful and try to increase the process capability a little.

(d) The specification limits are too wide compared with the product range. We have a lot of room to play with, so we should either bring the specification limits closer together or alter the process and widen the product range if this will make the process more economical.

(e) Here there is only one specification limit, in the form of a minimum, and the histogram satisfies the specification with plenty of room to spare. The action described in (d) can be taken if necessary.

SL: Specification Limits PR: Product Range

Figure 2.6 Comparison of Specification and Histogram (in a case where specifications are satisfied)

(a)

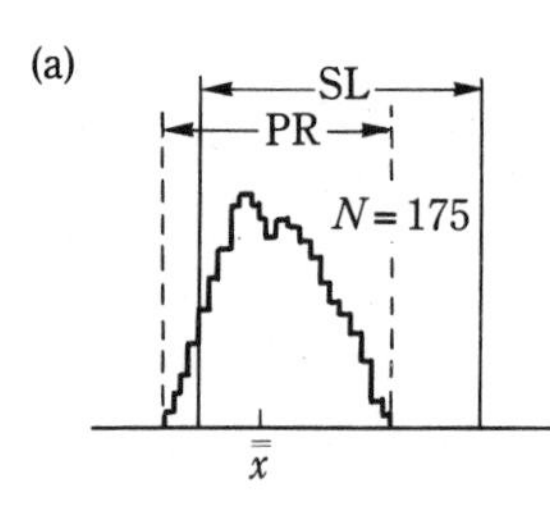

(a) The process average has shifted too far to the left. When it is a simple engineering problem to change the process average, it should be altered to the midpoint of the specification limits.

(b)

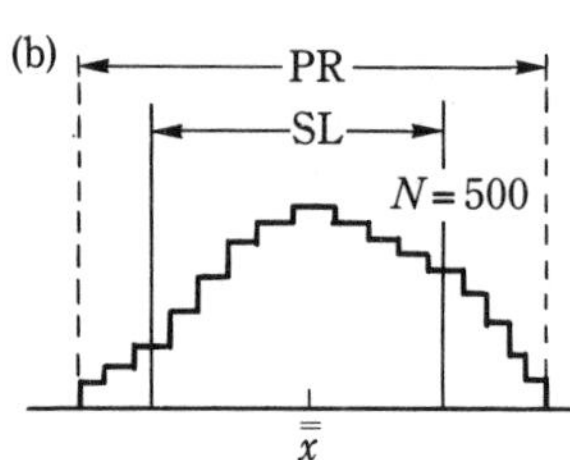

(b) The dispersion in the process is too great. Either the process must be changed or the specification limits must be changed. 100% screening of the product is needed.

(c)

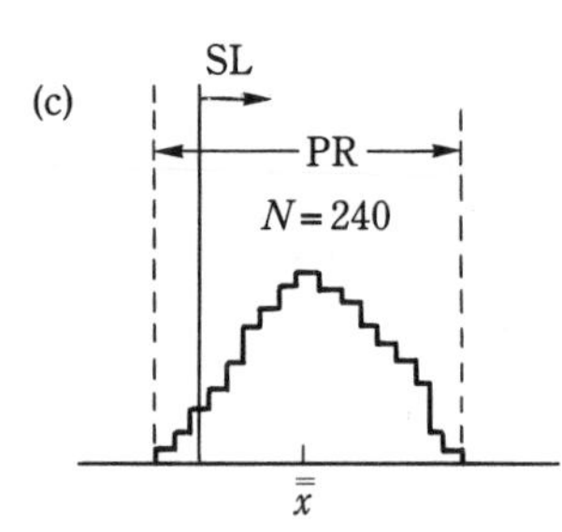

(c) In this case, the specification is in the form of a maximum, e.g., at least x kg/cm^2. The process average $\bar{\bar{x}}$ must be raised, the dispersion reduced, or other changes made.

(d)

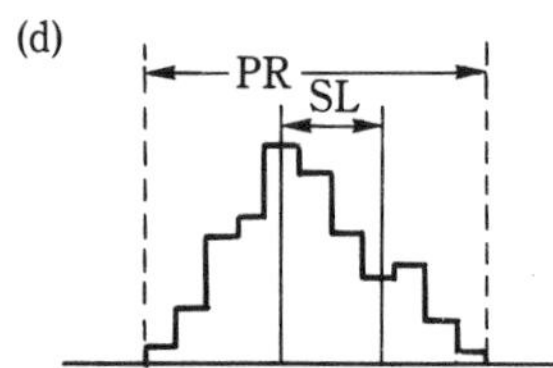

(d) Here, the process capability is totally inadequate to meet the tolerance in the specification. In such a case, if there is no way to change the specification or the process, the product must be stratified in various ways, and 100% screening must be carried out to select the acceptable product.

Figure 2.7 Comparison of Specification and Histogram (in a case where specifications are not satisfied)

(a)

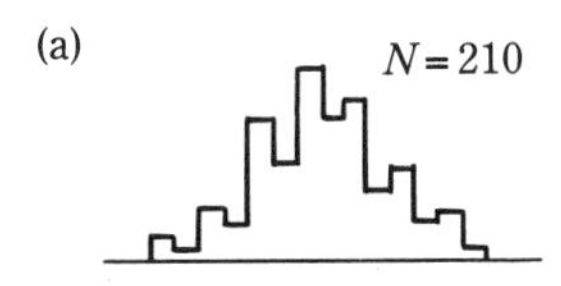

(a) Gap-toothed or comb-like histogram. Check whether there is anything wrong with the measurement or calculation methods. Were the data appropriately grouped when preparing the histogram?

(b)

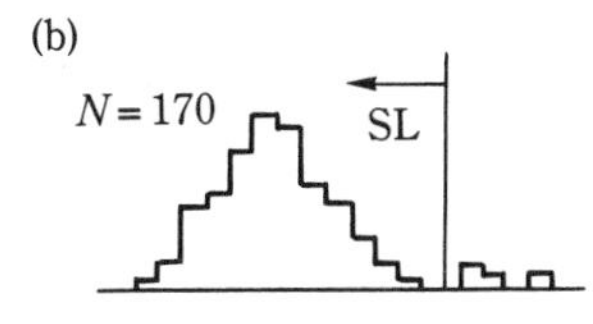

(b) Histogram with "outlying islands." This type of histogram shows some abnormality, and its cause must be sought out and eradicated. If the cause of the outlying data can be eliminated, product well within the specification limits can be made.

(c)

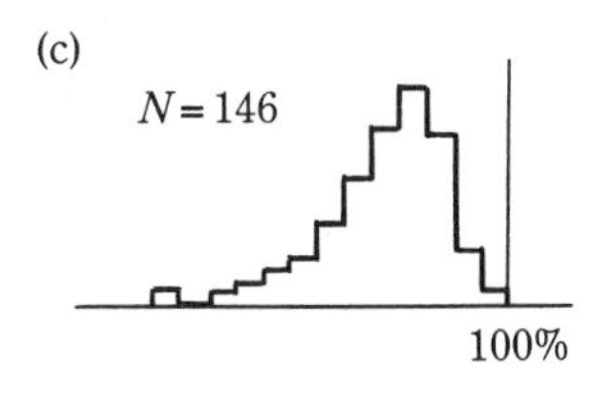

(c) Histogram with extended left-hand tail. This often occurs when there is an upper limit—e.g., a theoretical maximum or a specification limit—beyond which the data cannot go; for example, when a process yield or the purity of a product is close to 100%. The average yield, purity etc., will improve in such a case if action is taken to remove the cause of the left-hand tail.

(d)

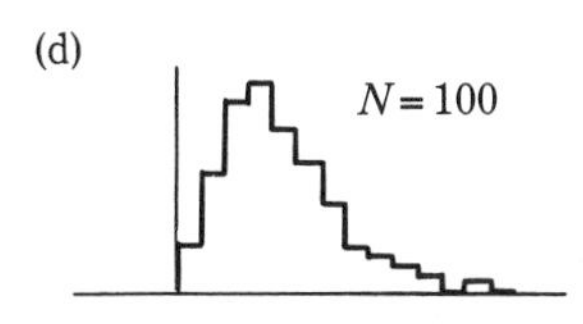

(d) Histogram with extended right-hand tail. This occurs when there is a lower limit beyond which the data cannot go, e.g., with impurity values close to 0%, or defectives or defects close to zero. When the distribution tails off as shown, the technical reasons for this should be investigated.

(e)

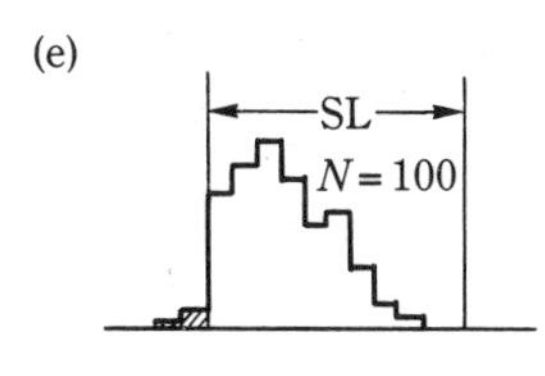

(e) Cliff-like histogram. This shape is often obtained when a process has inadequate capability and its products are subjected to 100% inspection to select only that which meet the specifications. The few values lying outside the lower limit represent products that have slipped through the screening net as a result of measurement or inspection error.

Figure 2.8 Various Shapes of Histogram

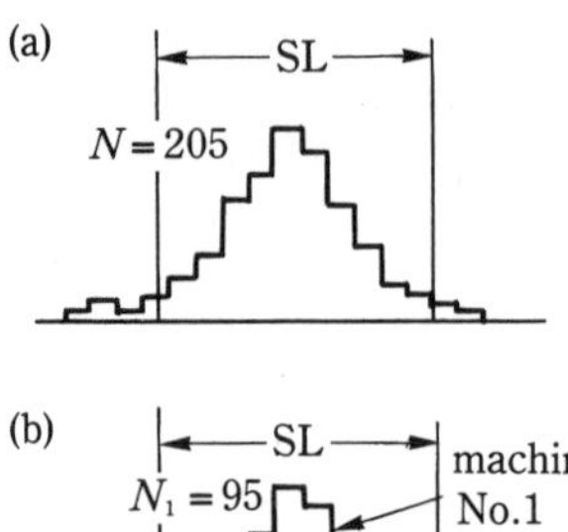

(a) A histogram prepared from data on products produced by two machines. It looks like a nice normal distribution, but a lot of the product is outside the specification limits.

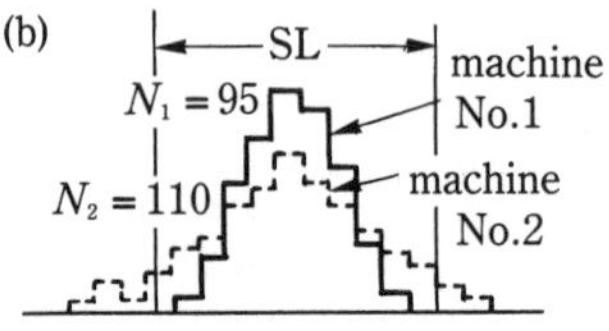

(b) Histograms prepared separately for each machine using the same data as in (a) above. This shows that the product from machine no.1 meets the specifications, but that machine no.2 is insufficiently precise, i.e., its process capability is insufficient.

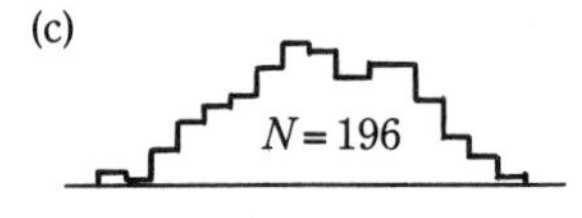

(c) A histogram prepared by pooling the data obtained by two laboratory technicians testing standard samples. The peak gives the impression of being a little too flat.

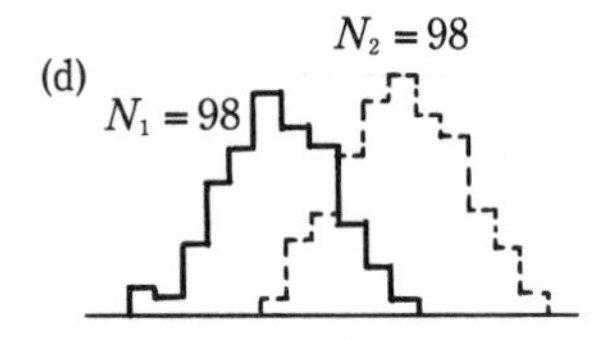

(d) Histograms prepared separately for each laboratory technician using the same data as in (c) above. The dispersion is approximately the same in each case, but there is bias in the averages.

Figure 2.9 Stratified Histograms

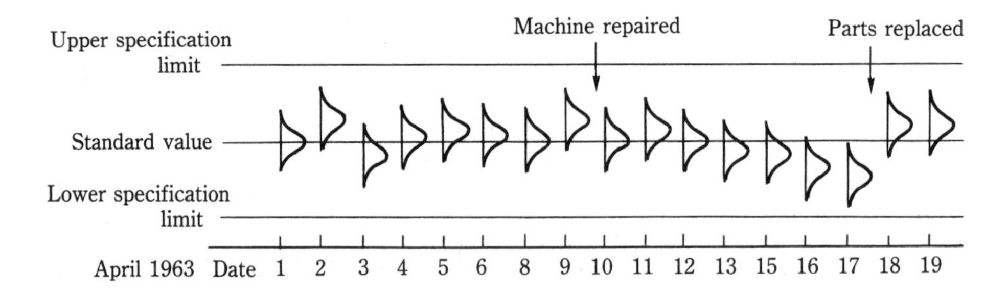

Figure 2.10 Frequency Distributions Arranged in Time Sequence, Clearly Showing Trends in Mean and Dispersion

(4) The shortcomings of frequency distributions

As discussed above, the range of applications of frequency distributions is extremely wide, and they are very effective. However, they do have the following disadvantages:

(a) They do not show changes with time. Since all the data for each lot, month, etc., are lumped together to make a frequency distribution, this method cannot be used to identify the causes of variations within each lot or month. To overcome this limitation, we can either stratify the data by time as much as possible when preparing frequency distributions and check sheets, or we can use other methods (such as the control charts described later) to check the state of control at the same time.

(b) To express limitation (a) above in slightly more statistical terms, using the language of control charts, we can say that frequency distributions do not convey much idea of within-a-group or between-groups variation.

(c) To prepare a frequency distribution and identify its shape requires a lot of data—at least fifty values, and if possible, more than a hundred. This is unavoidable if we are to identify the shape of the distribution and other essential factors.

In spite of these disadvantages, frequency distributions are extremely useful if we employ stratification and other devices, and if we become proficient in their preparation and interpretation and apply them sensibly.

2.6 Pareto Diagrams and Pareto Curves

(1) What is a Pareto diagram?

The Pareto diagram is a type of frequency distribution. It is prepared by collecting data on, for example, numbers of different types of defects, rework, scrap and claims, or on dollar losses and percentage losses together with their various causes, and plotting them in descending order of frequency, as shown in Figure 2.11.

When we arrange data in this way and plot the cumulative totals as shown by the solid line in the figure, we often find that the first two or three types of defect, for example, account for at least 70 to 80% of the total. It is clear that, if we eliminate these particular defects, we will have eliminated the majority of the defects, and the fraction defective will drop dramatically. Although there is generally a huge variety of defects, losses, accidents, and other problems in com-

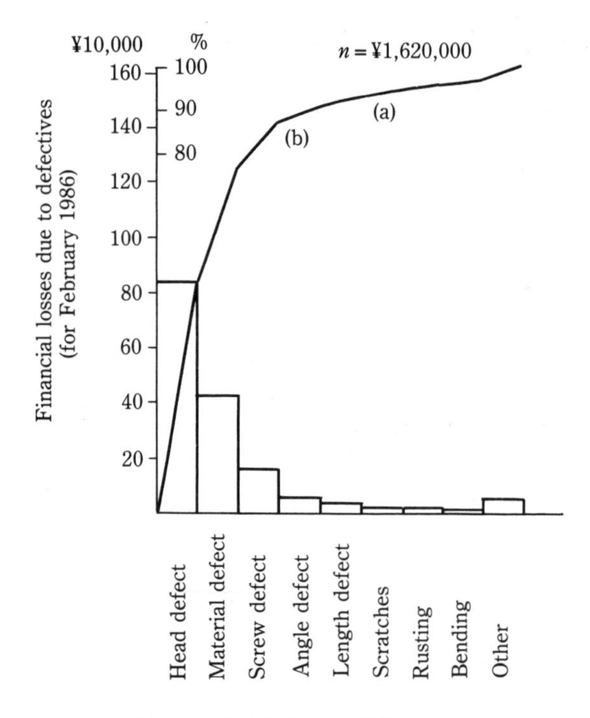

Figure 2.11 Pareto Diagram

panies, with a multitude of different causes, by far the greater part of the undesirable effects is often due to only two or three main problems or causes. This is called the Pareto principle ("vital few, trivial many"), which says that there are many unimportant problems but only a few serious ones. Pareto diagrams enable us objectively to identify the serious problems we are currently facing and take up the really important ones as a matter of policy.

Identifying and eliminating these problems enables us to achieve tremendous benefits. If we do not identify the really important problems in this way, we will waste a lot of effort; for example, if we say that our problem is the fifth item in Figure 2.11 (length defects) and make great efforts to solve it, we will only save around ¥50,000 ($400) a month even if we succeed. This type of effort is "nitpicking effort." Incomparably better results can be achieved if every department cooperates to eliminate the biggest problems. Unfortunately, the biggest problems are often dodged because they involve a large number of departments and are considered too troublesome, while the small problems are tackled with gusto. In quality control, it is important to build up a system of cooperation in which everybody works together to eliminate problems and causes in order, starting with the biggest ones.

(2) Some points to note when preparing Pareto diagrams

(a) Always record the total number of items, monetary amounts, and dates or times when the data were collected.
(b) As far as possible, stratify the data by different causes, types of defect, etc. The method of stratification will depend on the purpose of collecting the data.
(c) If possible, express losses in monetary terms rather than as numbers, amounts, defect ratios, etc. Depending on the problem, the dispersion contributed by each cause may also be expressed in terms of variance (in the form of a contribution ratio).
(d) Think of the purpose of preparing the diagram when deciding the period for which data are to be collected. This period should not be too short nor should it be so long that it includes the results of various corrective actions.
(e) If any action is taken, draw up Pareto diagrams before and after in order to check the results.
(f) As far as possible, stratify Pareto diagrams by time, machine, etc.
(g) Break down the biggest problems in further detail and prepare individual Pareto diagrams for them.

(3) Some points to note when interpreting and using Pareto diagrams

(a) Always start with the problem that will bring the biggest benefits if solved.
(b) Form teams of people from all relevant departments, have each department discuss proposals for solving the problem, and have them cooperate in finding a solution.
(c) Prepare Pareto diagrams for each month and each accounting period.
  (i) If the most frequent defects or losses suddenly decrease, this shows that either the cooperative improvement effort has succeeded, or that the process or other factors have suddenly changed even though nothing was done about them.
  (ii) If the different types of defect or loss decrease in an approximately uniform manner, this generally shows that control has improved.
  (iii) If the most frequent defect or loss changes every month, but the overall defect or loss ratio does not decrease very much (in other words, if the Pareto diagram is unstable), this shows lack of control.

The Pareto diagram is a simple but extremely useful tool. For this reason, it should be used widely, not only in quality control but in every possible situation.

2.7 CHECK SHEETS

(1) What are check sheets?

Recording figures one by one is a troublesome task when collecting data in the workplace, and stratifying and collecting data during inspection adversely affects inspection efficiency. Data on damage stratified by location, for example, are difficult to collect in practice. Check sheets are extremely useful in such cases, particularly when stratifying data. They enable data to be segregated and collected in groups simply by making check marks.

(2) How to prepare a check sheet

(a) Frequency distributions. Even with continuous data, collecting a large amount of data and then using it to prepare a frequency distribution is a duplication of effort. In such cases, individual values are often not really needed, and it is sufficient to know the shape of the distribution and whether it meets the standard. We can therefore record the values beforehand on a blank frequency distribution form and have the workers make tally marks, as shown in Table 2.2. This simplifies the recording of the data and produces a ready-made frequency distribution once the measurements have been completed. Such a form is also a type of check list. When only a small number of values are taken at a time, having the workers take them at specific times is also a convenient way of seeing temporal changes.

(b) Frequency distributions for individual defects. When there are various types of defect, simply knowing the overall number of defects will not give us a clue for action. However, listing the different types of defect or cause on blank inspection sheets and having inspectors place check marks against the appropriate items, thus stratifying the defect data, will be useful for both analysis and the planning of countermeasures. In such cases, if defectives with more than one type of defect are found, analytical inspection should be carried out, with all characteristics of the item being checked for the purpose of analysis. It is also good to arrange for defects to be recorded according to time of production.

(c) Positional check sheets. When the problem is scratches, cracks, and other blemishes, knowing the position of the damage often simplifies tracing the causes and taking corrective action. In this type of situation, sketches or development diagrams of the product should be prepared and divided up into a number of different zones. Inspectors should record their inspection results directly on these diagrams using color coding or other symbols. When this is done, the zones should be, as far as possible, of equal area. When the check marks are totalled, we can easily see whether the defects are concentrated in particular positions, in particular products, or at particular times, or whether they are randomly scattered, and we can then take appropriate action.

(d) Cause-and-effect diagram check sheet. A cause-and-effect diagram that can be easily understood by those on the shop floor can be prepared showing assignable causes, different types of defect, etc. This is passed to the shop floor, where workers are told to put a check mark beside the appropriate arrow whenever they come across a particular cause or situation. This soon tells us which causes we should control.

The above are only a few examples of the many possible types of check sheet. Preparing well-thought-out check sheets according to the conditions in each workplace enables stratified data to be readily obtained and Pareto diagrams to be prepared. Check sheets are an extremely useful tool, which I recommend be used extensively. The skillful use of check sheets in conjunction with tools like frequency distributions, Pareto diagrams, and cause-and-effect diagrams enables 80% to 90% of workplace problems to be solved.

2.8 Process Capability Diagrams

Process capability diagrams show the capability of a process with respect to quality. In the case of machinery and equipment, we use the term "machine capability." The question of process capability is an important problem in quality control, and it is discussed in detail in Section 4.7.7. Here I would like to mention only the method of preparing process capability diagrams and to note some points to be kept in mind when preparing them. The three methods listed below are often used when illustrating process capabilities diagrammatically. We try to identify the actual quality capability of a process through these diagrams.

(a) Frequency distributions
(b) Control charts
(c) Graphs showing specified values (see Figures 2.12 and 2.13).

Frequency distributions clearly show the distribution of the capability and enable its average and standard deviation to be calculated easily, but do not show how it changes with time. Since the data are plotted in production order on con-

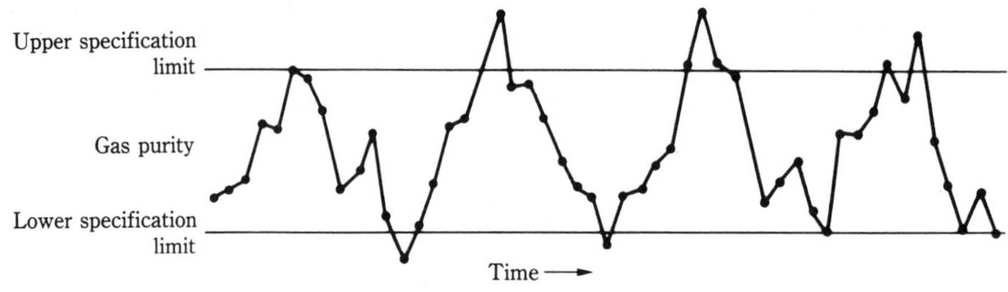

When single measurements are taken at fixed time intervals, there is a large periodicity in the process, showing that the process capability is unsatisfactory. The diagram indicates that the process capability will become fairly good if this periodicity is controlled.

Figure 2.12 Process Capability Chart (1) for Showing Temporal Changes

trol charts and graphs, they clearly show temporal changes, but do not clearly show how the process capability is distributed. However, both these objectives can be achieved if a frequency distribution is plotted at one end of a chart or graph, as shown in Figure 2.13. A process is said to be exercising its full capability if a correctly plotted control chart shows that it is in the controlled state.

When we are using (a), (b), or (c) to find the capability of a process, we must use data obtained after the process has been fully analyzed and is well controlled, and when the process or equipment is exercising its optimum capability. A process

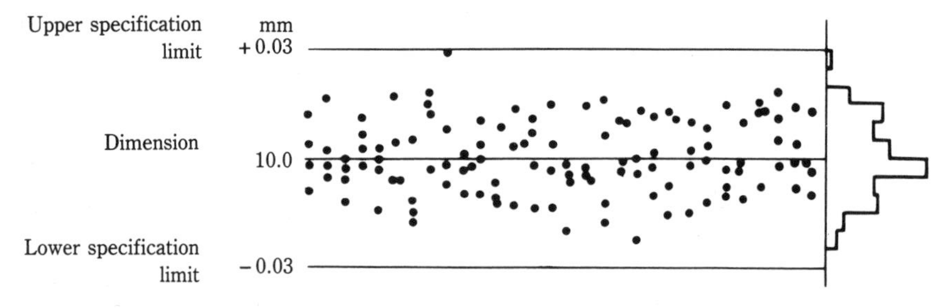

When data are obtained on n random samples ($n = 4$ in this case) taken at fixed time intervals, the diagram shows that the values lie just within the specifications with no room to spare. We must either operate the process with extra care or raise the process capability somewhat.

Figure 2.13 Process Capability Chart (2) for Showing Temporal Changes

capability calculated only from data from a process not in the controlled state cannot be said to be the true process capability.

Process capability is expressed by the process capability index, C_p (see Section 4.7.7).

2.9 SCATTER DIAGRAMS (CORRELATION DIAGRAMS)

Methods such as plotting frequency distributions enable us to identify the approximate shape of the distribution of a set of data of one type, but they do not show a relationship between two different sets of data. To identify the relationship between two different sets of data, we can use the scatter diagram. For instance, scatter diagrams can be used for corresponding sets of measurements such as temperature and yield, dimensions before and after processing, raw material composition and fraction defective, product hardness and tensile strength, etc. When this is done, it is important to collect the data in pairs (this is called "correspondence"); possessing data on raw material composition and fraction defective, for example, will not enable us to draw the scatter diagrams described below and perform an analysis unless we know what fraction defective occurred when each raw material composition was used. As I have already frequently mentioned, lot stratification is indispensable for this.

If we have data paired in this way, we can draw scatter diagrams and correlation tables using the methods described below.

Table 2.3 is a table of values of the hardness of a steel product and the percentage of a certain constituent in the raw material used. The data were obtained by measuring the average hardness of the product corresponding to particular raw material lots. When the values are plotted on a graph, they appear as shown in Figure 2.14, clearly indicating that the average hardness increases as the percentage of the raw material constituent increases. The graph in Figure 2.14 is a scatter diagram. Although the raw material composition clearly affects the hardness of the product, various other factors also affect it, even at the same composition. This is why the same composition does not always yield the same hardness. Nevertheless, simply plotting the data on a diagram like this tells us various things and gives us more information than we can obtain from an array of values like the data in Table 2.3.

Table 2.3 Raw Material Ingredient (%) versus Product Hardness

(N=100)

Percentage of ingredient x	Mean hardness y	Percentage of ingredient x	Mean hardness y	Percentage of ingredient x	Mean hardness y	Percentage of ingredient x	Mean hardness y	Percentage of ingredient x	Mean hardness y	Percentage of ingredient x	Mean hardness y
0.52	26.2	0.45	23.5	0.70	27.2	0.99	29.4	0.35	23.8	0.36	23.1
0.58	25.4	0.73	28.4	0.41	23.3	0.07	19.8	1.10	30.7	0.62	29.2
0.66	24.2	0.28	23.6	0.40	26.4	0.93	27.7	0.18	22.7	0.65	26.3
0.18	22.7	0.45	26.2	0.65	26.4	0.97	30.0	0.18	21.6	0.93	28.5
1.00	30.0	0.38	21.9	0.63	27.1	0.76	27.0	0.40	22.1	0.11	24.0
0.71	26.9	0.67	25.4	0.87	30.5	0.10	22.8	0.36	23.9	0.65	28.1
0.87	27.0	0.37	23.6	0.18	21.4	0.69	28.1	0.58	27.6	0.82	29.0
0.36	25.3	1.03	28.4	0.88	29.5	0.35	24.5	0.32	21.8	0.79	27.3
0.62	25.6	0.29	23.9	0.44	23.3	0.54	25.0	0.20	22.4	0.36	24.4
0.73	27.3	0.70	24.5	0.94	30.1	0.65	26.0	0.80	29.0	0.08	20.8
0.76	28.7	0.58	25.1	1.13	28.6	0.96	27.9	1.11	29.6	0.21	20.2
0.40	24.6	0.59	26.5	0.25	24.7	0.85	29.4	0.18	23.1	0.91	31.5
0.24	22.4	0.20	24.1	0.27	22.5	1.07	30.5	0.42	25.4	0.79	27.1
0.94	31.0	0.18	20.1	0.60	25.8	0.37	20.4	0.71	24.4	0.29	21.8
0.94	29.8	0.21	23.5	0.76	28.4	0.42	25.6	0.52	24.3	0.92	30.0
0.90	30.3	0.45	26.4	0.62	28.3	1.09	29.2	0.95	30.5	1.11	29.8
0.52	25.1	0.93	31.8	0.11	20.1	0.72	27.3				

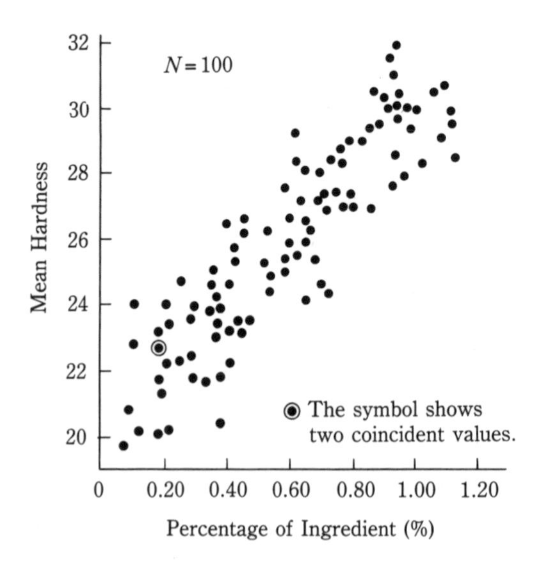

Figure 2.14
Example of Scatter Diagram (Correlation Diagram): Relation between Percentage of Ingredient in Raw Material and Mean Hardness of Product

(1) Some points to note when preparing scatter diagrams

(a) When investigating correlation, the more pairs of data the better. There should be at least fifty pairs, and, if possible, more than one hundred.
(b) The data thought to be the cause should be plotted on the horizontal (x) axis, on a scale increasing in value from left to right.
(c) The data thought to be the effect should be plotted on the vertical (y) axis, on a scale increasing in value from bottom to top.
(d) The x and y scales should be chosen so as to give an approximately equal width of scatter in the x and y data. Data thought to be abnormal should be isolated and considered separately. The data should be stratified as much as possible and data from different sources should be plotted on separate scatter diagrams or on the same diagram in different colors.

(2) Correlation tables

Scatter diagrams clearly show the relation between two sets of data, but correlation tables (two-dimensional frequency distributions) such as Table 2.4 can be used for the same purpose. When a correlation table is drawn up from the data of Table 2.3, it appears as in Table 2.4.

Table 2.4 Example of Correlation Table

Percentage of Ingredient (x)

Hardness (y) Cell number	Cell boundary values y \ Cell number / Cell boundary values x	1 0 – 0.105	2 0.105 – 0.205	3 0.205 – 0.305	4 0.305 – 0.405	5 0.405 – 0.505	6 0.505 – 0.605	7 0.605 – 0.705	8 0.705 – 0.805	9 0.805 – 0.905	10 0.905 – 1.005	11 1.005 – : :	Frequency f_y
12	30.05 -									//	卌	//	9
11	29.05 - 30.05							/		卌	////	///	13
10	28.05 - 29.05							///	///		//	//	10
9	27.05 - 28.05						/	/	卌 /	/	/		10
8	26.05 - 27.05				/	//	//	///	/				9
7	25.05 - 26.05				/	//	///	////					10
6	24.05 - 25.05			//	//	/	/	/	//				9
5	23.05 - 24.05		//	///	////	///							12
4	22.05 - 23.05	/	//	///	/								7
3	21.05 - 22.05		//	/	//								5
2	20.05 - 21.05	/	//	/	/								5
1	19.05 - 20.05	/											1
	Frequency f_x	3	8	10	12	8	7	13	12	8	12	7	100

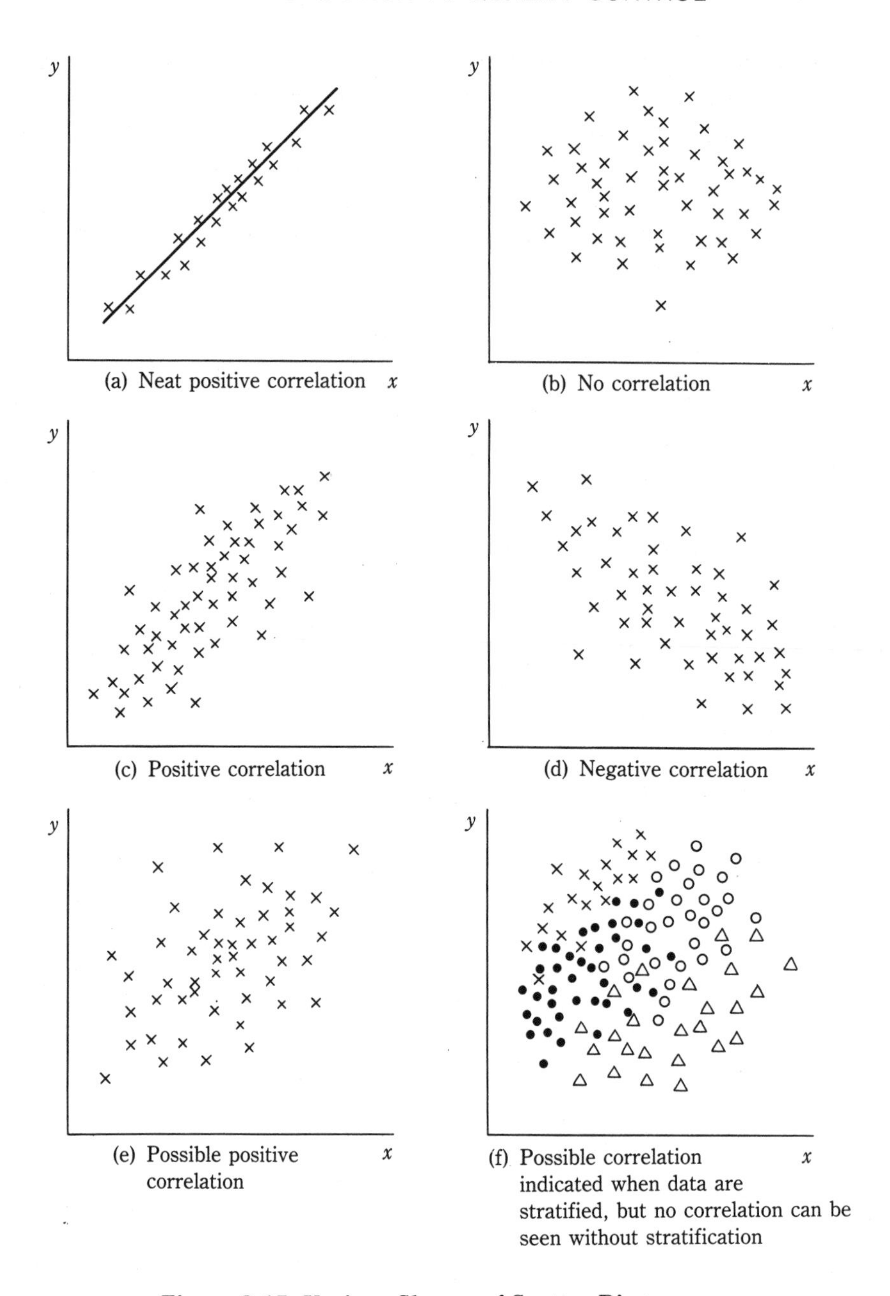

Figure 2.15 Various Shapes of Scatter Diagram

The scatter diagrams shown in Figure 2.15 illustrate various types of correlation. When the data are neatly lined up as in Figure 2.15(a), we immediately see the relationship between them. A relationship like this in which y increases as x increases is called "positive correlation," while a relationship in which y decreases as x increases (and vice versa) is called "negative correlation." Figure 2.15(b) shows no correlation. In Figure 2.15(c), the data are scattered, but a positive correlation is indicated. Figure 2.15(d) shows a negative correlation. When the data are scattered to the extent shown in Figure 2.15(e), there is doubt as to whether or not we can claim a positive correlation. In fact, we should also be cautious about claiming a correlation when the data are scattered to the extent shown in (c) and (d). In such cases, we will make all sorts of mistakes unless we base our judgments on statistical tests (see Section 4A.8).

When looked at as a whole, Figure 2.15(f) appears to show no correlation, but when the different raw materials are plotted with different symbols, each set of data appears to show a positive correlation. Preparing scatter diagrams requires care, since we may overlook a correlation if we do not stratify the data in this way.

2.10 WHAT IS ERROR?

It goes without saying that it is advantageous to base our discussions of various matters on data, but it is also dangerous to exaggerate data. Examples of this danger follow.

(1) If we take a random sample of 20 items from a lot containing 1,000 items and find that none of the sampled items is defective, can we say that there are no defectives in the lot?
(2) Until today, the average fraction defective was 10%, but today it is 12%. Can we say that today's result is particularly bad?
(3) A certain chemical product was analyzed and was found to have a purity of 87.5%. Can we really say that the product is 87.5% pure?
(4) A thermometer gives a reading of 850°C. Can we say that the temperature of the furnace from which the reading was taken is really 850°C?

The answer to all of the above questions is "no," because the process of obtaining data introduces various errors, such as sampling error, measurement error, computation error, rounding-off error, etc. We have to try to identify the true state of affairs through a veil of such errors.

Conventionally, however, either such errors have been totally ignored or the word "error" has been bandied about lightly and its meaning has been extremely vague. From now on, we must think of errors under the following headings:

(a) Sampling errors
(b) Measurement errors
(c) Computational and other errors.

If our sampling methods are poor and we do not know why we are collecting data, or if the sampling error is too large, process variations and other variations will be masked by the errors and we will remain unaware of them. Thus, when implementing various types of control numerically using data, we must first rationalize our sampling methods.* People are often said to be poor at numbers, because they easily believe a set of figures when they are shown them, forgetting about the large sampling errors behind them. Especially when implementing quality control, we must rationalize our sampling methods and prepare the ground for accurate data collection.

Similarly, analysis, measurement, and experimental errors are often relatively large. Such errors are particularly awful with data taken on different days, at different locations, by different people, or on different equipment. In implementing control, these errors must be kept small. To achieve this, we must carry out comprehensive measurement control and analysis control in their broad sense.

Whenever figures are being handled, transcription and calculation mistakes will always occur. This means that, while ensuring that great care is taken in handling data, we must establish systems that allow these mistakes to be readily discovered.

Errors can also be classified as:

(i) Reliability errors,
(ii) Precision errors, and
(iii) Bias and accuracy errors.

This method of classification stresses taking corrective action to minimize the sampling errors, measurement errors, and other types of error described above. First, let us give some definitions:

---

* For details of this, see Kaoru Ishikawa: *Sanpuringu-hō Nyūmon* (An Introduction to Sampling Methods), pub. JUSE Press (in Japanese).

Error is the difference between a measured value and the true value of the targeted population.

Reliability is the degree to which data can be trusted; in other words, whether the same sampling method was used for all the data, and whether the analytical and experimental work was free of assignable causes such as mistakes and omissions. Reliability can also be considered under the separate headings of reliability of precision and reliability of accuracy. Whichever method of classification is used, ensuring reliability is a question of controlling the work of sampling and measurement.

Remarks such as, "There's something odd about these data," "The sampling wasn't done properly," "The analysis is poor," and "The calculation is wrong" all indicate lack of reliability and control. Unreliable data are good for nothing except giving a false sense of security.

Precision involves the degree of data dispersion: if something is measured an infinite number of times by the same method, or an infinite number of samples are taken from the same lot by the same sampling method, there will always be dispersion in the data; the degree of this dispersion (the spread of the distribution) is called precision. Precision can be indicated in a variety of ways, such as by standard deviation, variance, $2 \times$ standard deviation, range (R) control limits, average value of R, etc. Statements heard in the past such as "The error is about $\pm 0.5\%$" are ill defined and therefore extremely vague, and it is not at all clear what they mean.

Bias, or accuracy, is the difference between a true value and the mean of the distribution of the values obtained from an infinite number of measurements by the same method. For example, the statement, "Our values seem on average to be $0.5 kg/cm^2$ higher than theirs" indicates bias.

Note: Since there is some confusion about the use of words like "error," "accuracy," and "precision," misunderstandings will arise when reading the literature unless careful attention is paid to their definitions.

Securing precision and accuracy is chiefly a problem of measurement and sampling techniques and statistical studies. When investigating error, it is best to do so in the order (i) to (iii) above, i.e., starting with reliability, then moving on to precision, and finally to bias.

Classifying and analyzing errors under the above headings tells us what we should do to minimize them. When data are unreliable, it may be a question of controlling the collection of data by formulating good sampling method standards and ensuring that these are implemented; when there is bias, we must search out the cause of the deviation in the mean and take corrective action; and when precision is poor, we must take action to reduce the dispersion.

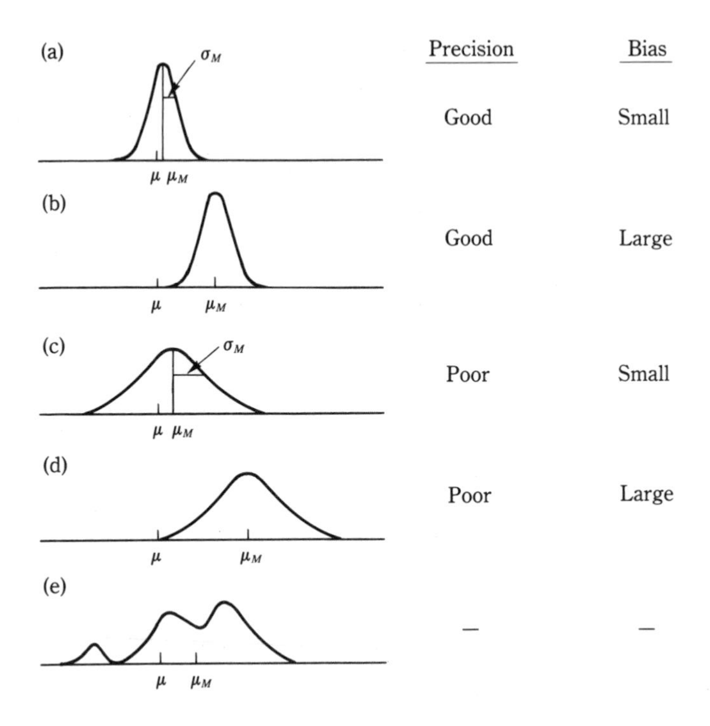

Figure 2.16 Types of Error

The relation between the different types of error is shown diagrammatically for ease of understanding in Figure 2.16.

2A.1 Preparing Frequency Distributions

Frequency distribution tables and histograms, like control charts, are important tools in organizing various kinds of data, particularly in quality control. I will therefore explain briefly how they are prepared.

When drawing up frequency distributions, we must consider the following:

(a) How many cells to have.
(b) How to decide the width of the cells.
(c) How to determine the cell boundaries.

(1) Number of cells

To show the shape of a distribution, the number of cells should be chosen as shown in Table 2A.1. In practice, the precise number will be determined naturally if the

Table 2A.1 Recommended Number of Cells for Frequency Distributions

Number of values	Number of cells
50-100	6-10
100-250	7-12
≧250	10-20

cell width and boundaries are decided with reference to this table. A larger number of cells is better for calculating statistics, but the approximate numbers shown in Table 2A.1 are satisfactory when sampling error is taken into account.

(2) Cell width

The cell width is determined as follows:

Step: 1 Find the maximum and minimum values of the data, but do not include abnormal outlying data when doing this. The maximum and minimum values of the data in Table 2.1 are 3.99 and 3.70.

Step: 2 Divide the difference between the maximum and minimum values by the number of cells. In this example, we have $(3.99-3.70)/10=0.029$.

Step: 3 Set the cell width at a convenient value close to the value obtained in Step 2 and an integral multiple of the smallest unit of measurement. In this example, the smallest unit of measurement is 0.01, so we set the cell width at 0.03.

(3) Cell boundaries

The cell boundaries are determined as follows:

Step: 1 Take half of the smallest unit of measurement as the unit for the cell boundaries.

Step: 2 Set the cell boundaries so that the maximum and minimum data values lie approximately equidistant from the respective outer cell boundaries. However, there is no need to be too strict about this. In this example, if we set the cell boundaries at $3.695-3.725,...3.965-3.995$, we have $3.70-3.695=0.005$ and $3.995-3.99=0.005$.

Note: When we want to make comparisons with standards or other values, it is convenient to set the cell boundaries so that they coincide approximately with these values.

(4) Table preparation

Once we have decided on the number, width, and position of the cells as described above, we arrange the data in the form of a table. As in the two left-hand columns of Table 2.2, the cell numbers and cell boundaries are recorded in ascending order from the top to the bottom of the table. The cell midpoints, which represent the values of the cells, are taken as the averages of the cell boundaries. In this example, we have $(3.695+3.725)/2=3.710$, etc.

If all we are doing is preparing a frequency distribution table, we only need to go as far as the tally and frequency columns (with relative frequency and cumulative frequency if needed). If we are calculating the mean, standard deviation, or other values, we must also include the extra columns shown in Table 2A.2.

(5) Tallying

Tallying consists of scanning the raw data and putting check marks in the tally column. The check marks are made as follows: /, //, ///, ////, ~~////~~. Since it is easy to make mistakes when carrying out this tallying operation, it should always be done twice. Table 2.2 was prepared by the above procedure.

2A.2 Methods of Calculating Statistics*

Table 2A.2 Calculation of $\bar{x}$ and s from Frequency Distribution Table

Cell number	Cell midpoint	Frequency f_i	u_i	f_iu_i	$f_iu_i^2$	Cumulative frequency	Relative cumulative frequency (%)
1	3.710	1	−5	−5	25	1	0.5
2	3.740	6	−4	−24	96	7	3.5
3	3.770	13	−3	−39	117	20	10.0
4	3.800	25	−2	−50	100	45	22.5
5	3.830	45	−1	−45	45	90	45.0
6	3.860 = a	37	0	(−163)	0	127	63.5
7	3.890	43	1	43	43	170	85.0
8	3.920	13	2	26	52	183	91.5
9	3.950	8	3	24	72	191	95.5
10	3.980	9	4	36	144	200	100.0
Total	—	200	—	(129) −34	694	—	—
Mean	Divide by 200			−0.170	3.470		

*The calculations described below can nowadays be performed easily using minicomputers or pocket calculators.

(1) Methods of calculating the mean

The mean may be calculated in the usual way by totaling the data values and dividing by the number of values. Although this is quite easy, it can be further simplified by choosing one of the following formulas:

$$\bar{x} = (1/n)\sum_{i=1}^{n} x_i \quad (2A.1)$$
$$= a + (1/n)\sum (x_i - a) \quad (2A.2)$$
$$= a + (h/n)\sum (x_i - a)/h \quad (2A.3)$$

where a and h are suitable constants.

Example:

Formula 2A.1	Formula 2A.2 ($a = 184$)	Formula 2A.3 $\begin{pmatrix} a = 184 \\ h = 1/10 \end{pmatrix}$
x_i	$x_i - 184$	$(x_i - 184) \times 10$
184.2	0.2	2
183.8	−0.2	−2
185.1	1.1	11
184.7	0.7	7
185.3	1.3	13
$n = 5\overline{)\ 923.1}$	$n = 5\overline{)\ 3.1}$	$n = 5\overline{)\ 31}$
184.62	0.62	6.2
$\bar{x} = 184.62$	$\bar{x} = 184 + 0.62$	$\bar{x} = 184 + 6.2 \times 1/10$
	$= 184.62$	$= 184.62$

When calculating manually, formulas 2A.2 and 2A.3 simplify the calculation greatly and reduce the size of the error if a mistake is made in the computation.

(2) Methods of calculating dispersion

Various statistics are used to express dispersion, including e.g., range (R), sum of squares of deviations (S), variance (s^2), unbiased estimate of population variance (V), sample standard deviation (s), and square root of unbiased estimate of population variance ($\sqrt{V}$). Some methods of calculating are explained briefly here.

(a) Range, R: $R = \text{maximum value} - \text{minimum value}$

$$= x_{max} - x_{min}$$

Example: For the set of values 8.8, 8.2, 8.4, 8.8, 8.3, $R = 8.8 - 8.2 = 0.6$.

(b) Sum of squares of deviations, S (also known simply as "sum of squares"): This calculation is the most time-consuming one, but it can be made extremely simple by choosing a suitable method of calculation.

$$S = \sum_{i=1}^{n} (x_i - \bar{x})^2 \tag{2A.4}$$

$$= \Sigma x_i^2 - \frac{(\Sigma x_i)^2}{n} = \Sigma x_i^2 - T^2/n = \Sigma x_i^2 - \text{CT} \tag{2A.5}$$

where $T = \Sigma x_i$ = total of all the data values. The term CT is given by $\text{CT} \equiv T^2/n$ and is called a "correction term."

$$S = \Sigma(x_i - a)^2 - \frac{\{\Sigma(x_i - a)\}^2}{n} = \Sigma(x_i - a)^2 - T^2/n \tag{2A.6}$$

where $T = \Sigma(x_i - a)$.

$$S = h^2[\Sigma\{(x_i - a)/h\}^2 - \{\Sigma(x_i - a)/h\}^2/n] \tag{2A.7}$$

Using formula 2A.4 in the above example, we have

$S = (8.8 - 8.50)^2 + (8.2 - 8.50)^2 + (8.4 - 8.50)^2 + (8.8 - 8.50)^2 + (8.3 - 8.50)^2 = 0.30^2 + 0.30^2 + 0.10^2 + 0.30^2 + 0.20^2 = 0.32$

When calculating by hand, the calculation becomes more and more trouble some as the number of figures in $\bar{x}$ increases; a minicomputer greatly simplifies this calculation.

In the case of formula 2A.5,

$S = 8.8^2 + 8.2^2 + 8.4^2 + 8.8^2 + 8.3^2 - 42.5^2/5$

$= 361.57 - 1{,}806.25/5 = 361.57 - 361.25 = 0.32$

This calculation is quite a nuisance.

In the case of formula 2A.6, we can considerably simplify the calculation if we take $a = 8$:

$S = 0.8^2 + 0.2^2 + 0.4^2 + 0.8^2 + 0.3^2 - 2.5^2/5$

$= 1.57 - 1.25 = 0.32$

In the case of formula 2A.7, taking $a = 8$ and $h = 1/10$:

$S = 1/10^2\ (8^2 + 2^2 + 4^2 + 8^2 + 3^2 - 25^2/5)$

$= 32/100 = 0.32$

(c) Sample variance V

$V = 1/(n-1)\Sigma(x_i - \bar{x})^2$
$= S/(n-1)$

In the above example, $V = 0.32/4 = 0.08$. When n is large, $n-1$ can be approximated by n.

(d) Sample standard deviation s

$$s = \sqrt{V} = \sqrt{1/(n-1)\Sigma(x_i - \bar{x})^2} = \sqrt{S/(n-1)}$$

In the above example, $\sqrt{V} = \sqrt{0.08} = 0.283$.

(3) Method of calculating mean and standard deviation from frequency distribution tables

The procedure for this calculation method, for manual calculation, is:

Step 1: Prepare a table like Table 2A.2.

Step 2: In the u_i column, denote the value thought to be approximately equal to the mean by 0 and denote the values above this by -1, -2, -3, etc. and those below it by 1, 2, 3, etc.

Step 3: For each cell, multiply the frequency f_i by u_i, and record the result in the f_iu_i column. Leave the $u_i = 0$ row blank. In this example, the value of f_iu_i for cell No.1 is $1 \times (-5) = -5$.

Step 4: Add up all the (negative) figures above the $u = 0$ line, and record the total in the $u_i = 0$ space. Add up all the (positive) values below the $u_i = 0$ line, and record the result as shown in the table. Add the two totals together and record the result in the appropriate space in the "totals" row. In this example, we have $-163 + 129 = -34$.

Step 5: Divide the value obtained in Step 4 by the total number of values (the sum of the figures in the *fi* column), and denote this by E_1:

$E_1 = 1/n\Sigma(x_i - a)/h$
$= 1/n\Sigma f_iu_i$
$= -34/200 = -0.170$

Step 6: Calculate the mean from the following formula (see formula 2A.3):

$\bar{x} = a + hE_1$

Where a is the midpoint of the $u_i = 0$ cell ($a = 3.860$ in this example), h is the cell width ($h = 0.03$ in this example), and E_1 is the value obtained in Step 5 ($E_1 = -0.170$ in this example):

$$\bar{x} = 3.860 + (0.03)(-0.170) = 3.860 - 0.0051 = 3.8549$$

Step 7: For each cell, multiply f_iu_i and u_i together and record the results in the $f_iu_i^2$ column. All these values will either be zero or positive.

Step 8: Add together all the $f_iu_i^2$ values. In this example, the total is 694.

Step 9: Divide the value obtained in Step 8 by the total number of values and designate the result E_2.

$$E_2 = 1/(n-1)\Sigma\{(x_i - a)/h\}^2$$
$$\approx 1/n\Sigma\{(x_i - a)/h\}^2$$
$$= 1/n\Sigma f_iu_i^2$$
$$= 694/200 = 3.470$$

When $n = 200$ as in this example, we can divide by n instead of $n-1$.

Step 10: Calculate the standard deviations from the following formula (see formula 2A.7):

$$s = h\sqrt{E_2 - E_1^2}$$

$$= 0.03\sqrt{3.470 - (0.170)^2} = 0.03\sqrt{3.441}$$

$$= 0.03 \times 1.855 = 0.0556$$

It should be noted that the above calculation is a simplified method that assumes that all the values within each cell are equal to the cell's midpoint, i.e., it is similar to rounding-off. However, the method is acceptable in practice.

A few remarks concerning these calculations are in order. First, calculating each frequency as a percentage of the total as in Table 2.2 and showing a distribution in terms of the resulting "relative frequencies" enables us to see the shape of the distribution easily and is particularly handy for comparing several distributions prepared from different numbers of data.

Second, the running totals of the number of data points over a certain value (boundary value), as in the second column from the right in Table 2A.2, are called "cumulative frequencies." They are convenient for making comparisons with specifications and for calculating distribution curves statistically. The percentage cumulative frequencies relative to the total are shown in the far right-hand column; these are called "relative cumulative frequencies." In some cases, it is more convenient to calculate these starting from a larger value and moving up the table, e.g., when a specification is in the form of a maximum permissible value.

And third, when the calculation is done by computer, u_i should be set to 0 for cell no.1 in step 2, and the remaining cells should be assigned the values 1, 2, 3, etc. down the table.

2A.3 Distribution of Statistics

When samples are taken at random from a population, the sample data will be scattered. The values of the mean, range, fraction defective and other statistics taken from the samples will therefore also be scattered. This distribution of statistics follows certain laws.

These distributions are determined by their mean (expectation E( )) and dispersion (standard deviation D( ) or variance V( )) together with their shape. This is shown in Tables 2A.3 and 2A.4.

Table 2A.3 Distribution of Statistics (for Variables)

Infinite population (population mean μ, population variance σ^2)

Statistic	Symbol	Hypothesis	Mean E ( )	Standard deviation D ( )	Variance V ( )	Shape of distribution
Mean	$\bar{x}$	None	μ	$\sigma/\sqrt{n}$	σ^2/n	Approaches normal distribution as n increases
Variance	V	Normal distribution	σ^2	$\sqrt{\frac{2}{n-1}\sigma^2}$	$\frac{2}{n-1}\sigma^4$	Tail extends toward the high-value side[1]
Standard deviation	s	〃	$c_2{}^*\sigma$	$c_3{}^*\sigma$	$(c_3{}^*\sigma)^2$	〃
Range	R	〃	$d_2\sigma$	$d_3\sigma$	$(d_3\sigma)^2$	〃

$c_2{}^*$, $c_3{}^*$, d_2 and d_3 are coefficients for the normal distribution whose values vary with n. They are obtained from tables (see Table 2A.5 and Table 3.3). Their values do not change greatly even if the population distribution is not exactly normal.

Table 2A.4 Distribution of Statistics (for Countables)

Statistic	Symbol	Population	Mean	Standard deviation D ( )	Distribution	Shape of distribution
Fraction defective	p	P	P	$\sqrt{P(1-P)/n}$	Binomial	
Number of defectives	$r=pn$	P	nP	$\sqrt{nP(1-P)}$	Binomial	Tail extends to the right
Number of defects per unit	$u=c/n$	U	U	$\sqrt{U/n}$	Poisson	Approaches normal distribution as n increases
Number of defects	c	C	C	$\sqrt{C}$	Poisson	

This distribution of statistics is one of the important basic characteristics of statistical tools.

Table 2A.5 Coefficients for Distribution of Standard Deviation

Sample size	Mean C_2*	Standard deviation C_3*	Sample size	Mean C_2*	Standard deviation C_3*
2	0.798	0.603	10	0.973	0.232
3	0.886	0.463	15	0.982	0.187
4	0.921	0.839	20	0.987	0.161
5	0.940	0.341	30	0.991	0.113
6	0.952	0.308	40	0.994	0.113
7	0.959	0.282	50	0.995	0.101
8	0.965	0.262	100	$1-1/4n$	$1/\sqrt{2n}$
9	0.969	0.246			

Chapter 3
THE PREPARATION AND USE OF CONTROL CHARTS

3.1 What are Control Charts?

In a broad sense, control charts include all kinds of charts and graphs used for control purposes. They have been used for a long time, ever since Dr. W. A. Shewhart first coined the term in 1926. Let us define them here as "a statistical tool used for control purposes, consisting of graphs with statistically calculated control limit lines." We will not be too concerned here with a more exact definition, and will call any graph obtained by the methods described below a control chart. We must, however, at least draw a clear distinction between adjustment charts and control charts, since the former (described in Section 3.9.1) are often mistakenly used for the latter. Since control charts can be used for all types of control as well as quality control, the term "quality control chart" is best avoided.

The basic role of control charts in the control cycle was mentioned in Section 1.5. However, they also have various other applications.

3.2 Types of Control Chart

There are many different kinds of control chart showing various statistics and data, with control limits calculated by various statistical methods. Here we will talk about those employing 3-sigma control limits, since they are the most basic, practical, and widely used type. Their skillful use enables almost every form of control to proceed well.

As explained in Section 2.3, variables and attributes are statistically different. There are also differences even among attributes; data on fraction defective and number of defectives are distributed differently from data on number of defects, and require different types of control chart. Control charts can be classified into the three main types described below, according to the nature of the data they depict.

(1) The $\bar{x}-R$ chart, $\tilde{x}-R$ chart, and x chart (see Sections 3.3, 3A.1 and 3A.2)

These types of control chart are used when the process characteristic to be controlled is a continuous variable such as length, weight, strength, purity, time, or production volume. However, they can also be used for other types of data.

The $\bar{x}$ chart is used mainly for observing changes in the mean of a distribution. The $\tilde{x}$ (median) chart is sometimes used in place of the $\bar{x}$ chart. The R chart is used for observing changes in the spread, or variation, of a distribution. The s (standard deviation) chart is sometimes used instead of the R chart in very special cases but is not dealt with in this book.

The $\bar{x}$ and R charts are generally used together, since only their combined use enables us to identify the changing state of a process in the form of a distribution. Of all the different types of control chart, these two give us the most technical information, which makes them extremely useful for technical analysis and process capability studies. Either one alone, however, is not enough to show the change in a distribution, i.e., the change in both the mean and the variation. The $\bar{x}-$R control chart is the most basic and useful form of control chart, particularly in the initial stages of quality control. Beginners should start by using this type of control chart in various situations in order to become comfortable with the technique of process control.

The x control chart is used for plotting individual variables of data without further modification, but it is often used incorrectly, and considerable care must be taken in its use.

(2) The p chart and pn chart (see Sections 3.4 and 3.5)

When controlling a process in which the vital characteristic is an attribute such as the number of defective items in a sample of a certain size (e.g., "three defective steel sheets out of 100"), either the p control chart or the pn control chart is used. These charts are also used for plotting attendance rates, data obtained by snap readings, numbers of unserviceable machines, etc. However, since they deal with data expressed as conformance or non-conformance, their use requires considerable technical knowledge of the work.

The p chart is used when the number of defective items in a sample is expressed as fraction defective (p), while the pn chart is used when it is expressed as number of defectives (pn). If the sample size (i.e., the number of products in a sample) is expressed by n, the pn chart is generally used when n is constant and the p chart when n varies. Statistically, the fraction defective (p) and the number of defectives (pn) follow the binomial distribution. Since these types of control chart are easily understood by anyone, and the data they require is often easily collected, they can be used by operators, workplace foremen, factory managers, etc.

Two points must be noted here concerning p and pn charts. First, even when all the products produced in a day are inspected, they constitute a lot that is still only a sample of the process, and a p or pn chart should be used for controlling the process.

Second, even with data expressed in percentages, such as purities or yields, the $\bar{x}-R$ or x chart should be used, not the p chart, when the percentages are continuous and the data cannot be enumerated.

(3) The c chart and u chart (see Sections 3.6 and 3.7)

These charts for attributes are used when we are concerned with the variation in the number of defects in a single item of product, e.g., the number of cracks, splits, scratches, or stains on the surface of a single steel plate, the number of flecks in $10cm^2$ of paper, the number of pinholes in a painted or plated surface, the number of defects in one car, etc.

In addition to product quality, they are also used for investigating discrete data such as numbers of people injured in a factory, numbers of accidents, numbers of calculation errors, numbers of ledger-copying errors, etc.

The c and u control charts closely resemble the p and pn control charts; they differ, however, in that in the latter, when r is the number of defective items in a sample of n items, r can never be larger than n, while the number of defects (c) in a c or u control chart may be larger than n. Statistically, they are used when the data follow the Poisson distribution.

The c control chart is used when the sample size is fixed, e.g., when a plate of fixed area (5m of cloth, 1 television, etc.) is taken as the sample. It is also used for plotting numbers relating to individual people, e.g., number of calculation errors, number of copying errors, or number of pencils or sheets of paper used. In this respect, it is very similar to the pn chart.

The u control chart is used for showing variation in number of defects per unit when the sample size is not fixed, e.g., when the area of a steel plate or sheet of paper taken as a sample changes with time. It may also be used, for example, for numbers of injuries or stationery consumption in different sections at a factory when the size of the sections varies. In this respect, it is very similar to the p chart.

We can decide which type of control chart should be used by considering the nature of our measurements and taking the above points into account.

3.3 PREPARING AVERAGE AND RANGE ($\bar{x}-R$) CONTROL CHARTS

As discussed later in Section 3.9, control charts have various applications. First, however, I would like to discuss the procedure for preparing control charts from existing data, i.e., the steps in drawing up such *charts for the purpose of analyzing past data.*

I will explain how to draw up the most important type of chart, the $\bar{x}-R$ chart, but the philosophy and approach for doing this are exactly the same for the p, pn, c, and u control charts. Preparing good control charts requires a variety of ingenuity and experience, but first and foremost one must know the basic drill.

(1) Collect data

It is necessary to collect at least 100 items of relatively recent data on the process characteristics (i.e., results) that will yield important technical and statistical knowledge about the process from the control standpoint. The data should be obtained under approximately the same technical conditions as are thought likely to exist in the process in the future. If data are scarce, 50 or even 20 values will do, but it is better to collect 100 or more if possible. Control charts drawn with scant data (e.g., with 50 or 20 items) should always be redrawn when more data are accumulated. When redrawing charts, we must, as far as possible, clarify the history of the data and the lots from which they were taken. The quality of the data collected is as important as the quantity.

(2) Stratify the data

The data should be stratified according to factors such as the time of measurement and the order in which the lots were produced, and, if possible, by process. For example, Table 3.1 shows data on the thicknesses of steel plates. Five plates were measured every hour; the data are arranged in order of measurement in 25 groups from left to right, starting at the top left-hand corner.

(3) Organize the data into subgroups

First, the data are arranged in subgroups of three to five items. In control charts, these subgroups are also known as "samples." The number of data points in each subgroup is called the *"subgroup size"* or *"sample size,"* and is usually denoted by the letter n. In Table 3.2, the data of Table 3.1 have been arranged in order in subgroups of size $n=5$. The total *number of subgroups* obtained when the data are arranged in this way, also called the *"number of samples,"* is denoted by the letter k. In Table 3.2, $k=25$.

The next step is subgrouping (see Section 3.9.2), which together with stratifi-

Table 3.1 Plate Thickness (mm)

(Number of data points (N) = 125)

2.1	1.9	1.9	2.2	2.0	2.3	1.7	1.8	1.9	2.1
2.1	2.1	2.2	2.1	2.2	2.0	1.9	1.9	2.3	2.0
2.1	2.2	2.0	2.0	2.1	2.1	1.7	1.8	1.7	2.2
1.8	1.8	2.0	1.9	2.0	2.2	2.2	1.9	2.0	1.9
2.0	1.8	2.0	1.9	2.0	1.8	1.7	2.0	2.0	1.7
1.8	1.9	1.9	3.4	2.1	1.9	2.2	2.0	2.0	2.0
2.2	1.9	1.6	1.9	1.8	2.0	2.0	2.1	2.1	1.8
1.9	1.8	2.1	2.1	2.0	1.6	1.8	1.9	2.0	2.0
2.1	2.2	2.1	2.0	1.8	1.8	1.8	1.6	2.1	2.2
2.4	2.1	2.1	2.1	2.0	2.1	1.9	1.9	1.9	1.9
2.0	1.9	1.9	2.0	2.2	2.0	2.0	2.3	2.2	1.8
2.2	2.2	2.0	1.8	2.2	1.9	1.9	2.0	2.4	2.0
1.7	2.1	2.1	1.8	1.9					

cation, is a vital operation that can make or break a control chart. In most cases, the subgroups should consist of data for each day, shift, process, lot, etc., so that the variation within the subgroups due to technical factors is relatively small, that is, so that the causes that have the biggest effect on the process appear between the subgroups. In this example, since five plates are measured each hour, we have taken the hourly measurements as subgroups, with $n=5$. This is the basic principle of subgrouping, but data can also be subgrouped in production order or measurement order if it is difficult to find a technical basis for subgrouping. In practice, various subgroupings should be tried based on technical considerations, and the method most convenient for controlling the process should be adopted.

The subgroup size (n) should if possible be the same for each subgroup. For example, if four measurements were taken on one day, five on another, etc., the data should be divided into equal subgroups (e.g., of five items) in time sequence provided no particular difference is thought to exist from day to day. However, if there are technical reasons for believing that the day makes a significant difference, the data should be subgrouped by day, with different subgroup sizes ($n=4$, $n=5$, etc.). Since the preparation and use of control charts generally become complicated when the data are arranged in subgroups of different size, the subgroup size should be kept the same whenever possible. For example, if the past data naturally fall into subgroups of size $n=5$ and $n=4$, one value can be removed at random from each of the $n=5$ subgroups, giving a constant subgroup size of $n=4$. Here, I will discuss only the situation when n is constant.

The subgroup size is sometimes taken as $n=6-10$ in special cases, but it is better to split larger subgroups like these into smaller ones of size 5 or less. A subgroup size of $n=2-5$ is most commonly used.

Table 3.2 Example of Data Sheet for $\bar{x}-R$ Control Chart

For $\bar{x}-R$ control chart	Quality control record
Form No.1	No.0208

Factory manager	Department manager	Section manager	Supervisor	Group leader	Foreman

Name: Plate

Quality characteristic: Thickness

Measurer: Tarō Shōwa

Measurement method: Equipment number: No.3

Measurement unit: 0.1 mm

Factory

Section

Inspection group

From: Year Month Day
To: Year Month Day

Subgroup number	Day—Time	x_1	x_2	x_3	x_4	x_5	$\bar{x}$	R	Inspector's initials
1	1—9	2.1	1.9	1.9	2.2	2.0	2.02	0.3	
2	10	2.3	1.7	1.8	1.9	2.1	1.96	0.6	
3	11	2.1	2.1	2.2	2.1	2.2	2.14	0.1	
4	12	2.0	1.9	1.9	2.3	2.0	2.02	0.4	
5	14	2.1	2.2	2.0	2.0	2.1	2.08	0.2	
6	15	2.1	1.7	1.8	1.7	2.2	1.90	0.5	
7	16	1.8	1.8	2.0	1.9	2.0	1.90	0.2	
8	2—9	2.2	2.2	1.9	2.0	1.9	2.04	0.3	
9	10	2.0	1.8	2.0	1.9	2.0	1.94	0.2	
10	11	1.8	1.7	2.0	2.0	1.7	1.84	0.3	
11	12	1.8	1.9	1.9	2.4	2.1	2.02	0.6	
12	14	1.9	2.2	2.0	2.0	2.0	2.02	0.3	
13	15	2.2	1.9	1.6	1.9	1.8	1.88	0.6	
14	16	2.0	2.0	2.1	2.1	1.8	2.00	0.3	
15	3—9	1.9	1.8	2.1	2.1	2.0	1.98	0.3	
16	10	1.6	1.8	1.9	2.0	2.0	1.86	0.4	
17	11	2.1	2.2	2.1	2.0	1.8	2.04	0.4	
18	12	1.8	1.8	1.6	2.1	2.2	1.90	0.6	
19	14	2.4	2.1	2.1	2.1	2.0	2.14	0.4	
20	15	2.1	1.9	1.9	1.9	1.9	1.94	0.2	
21	16	2.0	1.9	1.9	2.0	2.2	2.00	0.3	
22	4—9	2.0	2.0	2.3	2.2	1.8	2.06	0.5	
23	10	2.2	2.2	2.0	1.8	2.2	2.08	0.4	
24	11	1.9	1.9	2.0	2.4	2.0	2.04	0.5	
25	12	1.7	2.1	2.1	1.8	1.9	1.92	0.4	
Total							49.72	9.3	
Average							$\bar{\bar{x}}=1.9888$	$\bar{R}=0.372$	

$\bar{x}$ control chart: (CL) $\bar{\bar{x}}=1.989$

(UCL) $\bar{\bar{x}}+A_2\bar{R}=2.204$

(LCL) $\bar{\bar{x}}-A_2\bar{R}=1.774$

R control chart: (CL) $\bar{R}=0.372$

(UCL) $D_4\bar{R}=2.115\times0.372=0.79$

(LCL) $D_3\bar{R}=$ (not applicable)

$A_2\bar{R}=0.577\times0.372$
$=0.215$

Control chart No.AC103

(4) Prepare data sheets (data record forms)

It is convenient to decide from the start to record data on sheets of specified format. Since it is not only a nuisance to copy data from daily reports and other sources, but uneconomical and error-prone as well , it is best to design daily report forms as shown in Table 3.2 which organize the data into subgroups and allow various calculations to be carried out easily. Such forms should allow space for as much information as possible relevant to the process and the data to be recorded.

(5) Calculate the subgroup average ($\bar{x}$)

The average ($\bar{x}$) is calculated for each subgroup. For group no. 1, the calculation is performed as shown below.

A note about rounding off is in order here: in this calculation, there is no great problem when $n=4$ or 5, since we can divide exactly by these; but in many cases, $n=3$ or 6, leaving a recurring decimal place. For control charts, it is generally sufficient when calculating the subgroup averages to calculate to two more significant figures than the measurements and round off the last figure. For example, with these data, the calculation is as follows for subgroups of size $n=5$ and $n=3$:

$$(2.1+1.9+1.9+2.2+2.0)/5=10.1/5=2.02$$
$$(2.1+1.9+1.9)/3=5.9/3=1.966\approx 1.97$$

To avoid introducing bias when calculating the mean or other statistics, the following procedure should be followed:

(a) The figure to be rounded off should be rounded down when it is 4 or less, and up when it is 6 or more; thus, 1.834→1.83, 1.976→1.98.

(b) When the figure to be rounded off is 5, it should be rounded up or down depending on the value of the figures that follow it:
 i) Round up when the next figures are greater than 0, i.e.:
  2.0451→2.05, 2.04501→2.05.
 ii) When the next figure is 0 or there are no further figures, round down when the figure in front of the one to be rounded off is even and round up when it is odd, i.e.:
  2.0250→2.02, 2.01500→2.02
  2.025→2.02, 2.015→2.02.

(c) Always round off to the required number of figures in one operation. The wrong result may be obtained by rounding off in successive steps, e.g.:

$$2.5498 \rightarrow 2.550 \rightarrow 2.55 \rightarrow \begin{cases} 2.5 \text{ (correct)} \\ 2.6 \text{ (incorrect)} \end{cases}$$

2.4502→2.5

(6) Calculate the subgroup ranges (R)

The range (R) for each subgroup may be calculated by subtracting the minimum value in the subgroup from the maximum value in the subgroup.

For group no. 1, $R = 2.2 - 1.9 = 0.3$.

Note that R is always either 0 or more and is never a negative value. For example, in a group such as $(-1, -3, -5, -4,)$, $R = (-1) - (-5) = 4$.

(7) Calculate the overall average ($\bar{\bar{x}}$)

The overall average ($\bar{\bar{x}}$) is calculated from the averages of each subgroup ($\bar{x}$). Note that the overall average ($\bar{\bar{x}}$) should generally be calculated to three more significant figures than the measurements and rounded off to two significant figures more than the measurements.

(8) Calculate the average of the subgroup ranges ($\bar{R}$)

The average range ($\bar{R}$) is calculated from the values of R for all the subgroups. It is sufficient to calculate $\bar{R}$ to two significant figures more than the measurements. When recording its value on a control chart, an accuracy of one significant figure more than the measurements is sufficient.

(9) Calculate the control lines

The $\bar{x} - R$ control chart requires *control lines* for both $\bar{x}$ and R. Each type of control chart has the following three control lines:

—The upper control limit, UCL
—The central line, CL
—The lower control limit, LCL

The term *"control limit"* refers to the upper and lower control limits. If the points plotted on a control chart lie between the limits, the chart shows a state of control. If any points lie on or outside the limits, the chart shows that some abnormality has occurred in the process.

Control lines are calculated in the following way (see Table 3.2):

(a) Control lines for $\bar{x}$ control chart
—Central line: $\mathrm{CL} = \bar{\bar{x}}$
—Upper control limit: $\mathrm{UCL} = \bar{\bar{x}} + A_2\bar{R}$
—Lower control limit: $\mathrm{LCL} = \bar{\bar{x}} - A_2\bar{R}$.

A_2 is a coefficient whose value depends on the subgroup size, n.

Table 3.3 Table of Coefficients for $\bar{x}-R$ Control Charts

Subgroup size	$\bar{x}$ control chart		R control chart				Relation between $\hat{\sigma}$ and $\bar{R}$ $\hat{\sigma}=\bar{R}/d_2$		
n	A	A_2	D_1	D_2	D_3	D_4	d_2	$1/d_2$	d_3
2	2.121	1.880	—	3.686	—	3.267	1.128	0.886	0.853
3	1.732	1.023	—	4.358	—	2.575	1.693	0.591	0.888
4	1.500	0.729	—	4.698	—	2.282	2.059	0.486	0.880
5	1.342	0.577	—	4.918	—	2.115	2.326	0.430	0.864
6	1.225	0.483	—	5.078	—	2.004	2.534	0.395	0.848
7	1.134	0.419	0.205	5.203	0.076	1.924	2.704	0.270	0.833
8	1.061	0.373	0.387	5.307	0.136	1.864	2.847	0.351	0.820
9	1.000	0.337	0.546	5.394	0.184	1.816	2.970	0.337	0.808
10	0.949	0.308	0.687	5.469	0.223	1.777	3.078	0.325	0.797

Table 3.3 gives the value of A_2 when $n=5$ as 0.577.

$A_2\bar{R}$ should be calculated to the same number of significant figures as $\bar{\bar{x}}$, i.e., to two more significant figures than the measurements. Note that the control limits on the $\bar{x}$ chart depend on R (the within-subgroup variation).

(b) Control lines for R chart
—Central line: $\text{CL}=\bar{R}$
—Upper control limit: $\text{UCL}=D_4\bar{R}$
—Lower control limit: $\text{LCL}=D_3\bar{R}$

D_4 and D_3 are coefficients whose values depend on the subgroup size. For example, if $n=5$, Table 3.3 shows that $D_4=2.115$, while D_3 is not applicable.

The difference between the R control chart and the $\bar{x}$ control chart is that the UCL and LCL in the former are calculated by direct multiplication of $\bar{R}$ by a constant without addition or subtraction. The lower control limit is not applicable when $n \leqq 6$.

$D_3\bar{R}$ and $D_4\bar{R}$ should be calculated to the same number of significant figures as $\bar{R}$, i.e., to one more significant figure than the measurements.

(10) Prepare control chart forms

Control charts are plotted on graph paper; the easiest to use is paper with, for example, 2–3mm divisions horizontally and 1mm divisions vertically. The paper should be ruled as finely and faintly as possible, since it will be hard to see the control lines and points if the ruling is too thick. It is convenient if the forms are designed so that copies can easily be taken after the forms have been completed.

The $\bar{x}$ and R control charts are drawn one above the other, and a height of 15cm is usually sufficient. The paper should be quite long, since control charts often extend over a considerable time. Ample space should be left at the bottom of the chart for recording additional information. The paper should be of the best quality possible, since the chart may need to be used and stored for a long time.

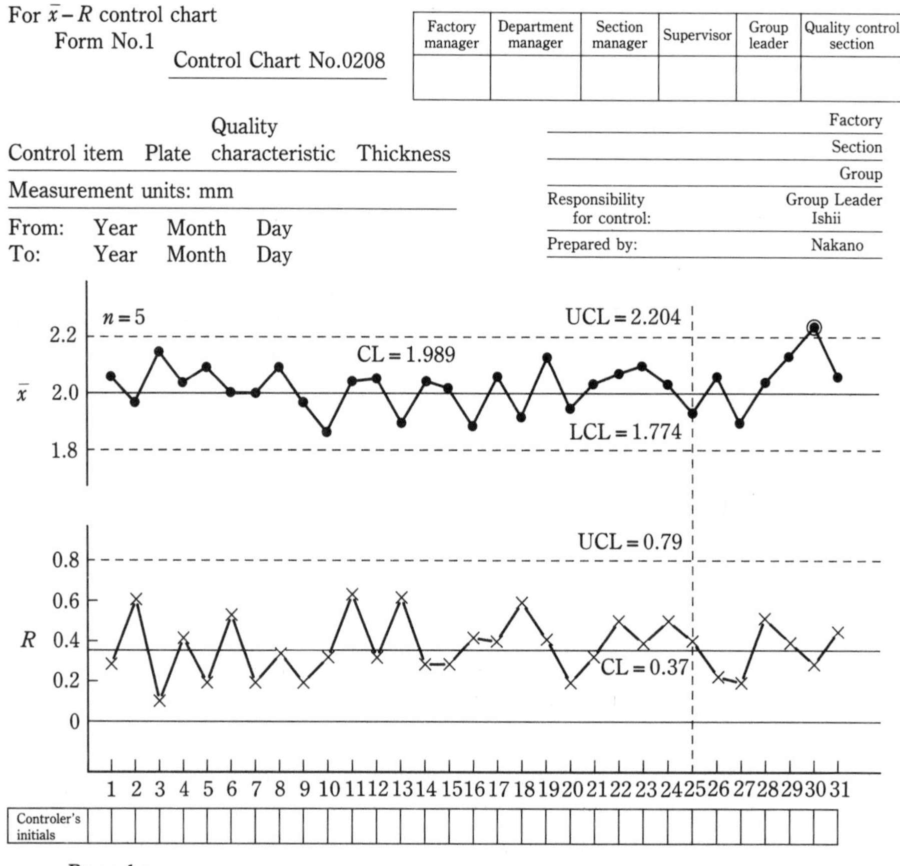

Remarks

Subgroup number	Remarks (reason why points fell outside control limits, action taken, details when causes unknown)	Person in charge
25	Analyze data in subgroups 1 — 25, and begin plotting data daily from subgroup 26 on.	
30	Cause: gage out of order. Action: from now on, check gage every morning.	

Control chart number: AC 103

Figure 3.1 $\bar{x}-R$ Control Chart

When quality control has begun to be carried out in earnest, data sheets and control chart forms should be specially designed and printed.

(11) Draw in the control lines

The $\bar{x}$ control chart is drawn at the top of the form, with the R control chart below it. The subgroup numbers (or date, lot number, etc.) should be recorded on the horizontal axis.

For both the $\bar{x}$ and R control charts, the vertical scale should be chosen so that the width of the control limits (i.e., the distance between the upper and lower control limits) is *approximately 30mm*, and the *units* are then written in. This means that the scales for the $\bar{x}$ control chart and the R control chart may be different. I often come across control charts drawn according to the conventional engineering approach with a gap of 10cm or more between the control limits. With control charts, however, we are usually concerned only with *whether the points lie between or outside the limits*, and it is a mistake to enlarge the vertical scale and thereby focus attention on small movements of the points between the limits. Instead, we should try to make the scale as small and the paper as long as possible in order to show the trend over a long time period. An interval of 2–3mm between subgroups—i.e., between points on the horizontal axis—is sufficient; it is good enough if individual points can be distinguished.

Control charts should be drawn neatly so that they are easy to use and pleasing to the eye, but one should recognize that they will be soiled with grease and grime during serious use. Figure 3.1 shows a control chart prepared from the data of Table 3.2.

When past data have been analyzed and control lines are drawn using such data, the central line should be shown by a solid line — and the limit lines by broken lines ---- on every type of control chart. These control lines should be drawn up to the subgroup number of the last measurement used for analysis.

(12) Plot the points

The average ($\bar{x}$) and the range (R) for each subgroup are plotted in subgroup order on the $\bar{x}$ control chart and the R control chart, respectively, with the R value for each subgroup being plotted directly below the $\bar{x}$ value for that subgroup. The following should be noted when plotting the points:

(a) The points must be plotted clearly. Do not plot tiny points because the scale is small; plot them boldly so that they stand out and enable the pattern to be seen at a glance.

(b) It is better to use different symbols for the points on the $\bar{x}$ and R charts; for example, $\bar{x}$ could be plotted using dots (•) and R using crosses (×).

(c) If the data are subgrouped by shift, machine, team, etc., stratifying the data and using different colors or symbols to distinguish between the different strata makes things easier to see.

(d) Points lying on or outside the control limits (i.e., abnormal points) should be clearly marked by using special symbols such as $\odot$ or $\otimes$ or by plotting them in red.

(e) Points lying close to the central line should be plotted with symbols such as $\underset{\smile}{\bullet}$ or $\overset{\frown}{\bullet}$ to indicate whether they are above or below the line.

(f) When the points have been plotted, they should be joined together with a fine unbroken line in subgroup order. When there are a number of points for each day or each week, the chart will be clearer if only the points for each period are linked, with a gap between one period and the next.

To sum up, the points should be plotted so that they are easy to see and should be stratified if necessary.

(13) Record other necessary information

Write $\bar{x}$ at the left-hand end of the $\bar{x}$ control chart and R at the left-hand end of the R control chart. Above the control chart, write in all the necessary relevant information, e.g., the product, the quality characteristic, the measurement units, the name of the person responsible for controlling the process, the name of the person filling out the chart, the period over which the data were taken, the control chart's reference number, etc. At the top left of the $\bar{x}$ control chart, record the subgroup size, e.g., $n = 5$. Label the control lines UCL, CL, and LCL as shown in Figure 3.1, and write in their values.

(14) Summary

The above explanation of how to draw $\bar{x} - R$ control charts shows that although the study of statistics itself is not so easy, control charts can be prepared using the simple arithmetic operations of addition, subtraction, multiplication, and division. In Japanese factories that are advanced in quality control, workplace supervisors and others responsible for control naturally use control charts, as do ordinary workers, both male and female.

Some differences between control charts and ordinary graphs are as follows:

(a) With control charts, the data are divided into subgroups.

(b) Control charts show changes in both $\bar{x}$ and R.
(c) Control charts show statistically meaningful control limits.

Charting the data in this way is extremely effective and allows the situation in a factory to be identified far better than from the usual daily reports, which are simply a mass of figures. The use of control lines also makes it easy to take action on a process.

It should be noted that the above is the normal procedure for drawing control charts when analyzing past data. However, in company quality control seminars, data from *bowl-drawing experiments* or actual workplace data may be used to provide easily understood explanations of the concepts of variation due to sampling dispersion and control limits. In such situations, the participants should be made to draw control charts according to the following procedure for ease of understanding:

1. Prepare data sheets.
2. Carry out bowl-drawing experiments and subgroup data.
3. Prepare blank control chart forms.
4. Calculate the $\bar{x}$ values.
5. Plot $\bar{x}$ (when this is done, have the participants use a scale that gives a distance between the control limits of approximately 30mm).
6. Calculate the R values.
7. Plot R (also have the participants choose a scale giving a distance between the control limits of 30mm in this case, and have them plot R directly under $\bar{x}$ for each subgroup).
8. Calculate $\bar{\bar{x}}$ and record it.
9. Calculate $\bar{R}$ and record it.
10. Calculate the $\bar{x}$ control lines and draw them in.
11. Calculate the R control lines and draw them in.
12. Record other relevant information.

3.4 Preparing Control Charts for Fraction Defective (p)

The p control charts are used for controlling processes from which the data are collected as values of fraction defective or percent defective, e.g., when 100 sheets or 100 items (or, in general, n items) of finished or half-finished product are tested to see whether they are conforming or non-conforming; if there are 5 defective items out of 100 (or, in general, r or pn items), the fraction defective is given by $p=5/100=0.05$ and the percent defective is 5%. The fraction non-defective (q) may also be used. The preparation of this type of control chart is explained in the following sections:

(1) Collect data

One must collect as much data as possible on the fraction defective. The number of items inspected, n, and the number of defective items, pn, must be known for each fraction defective.*

It is good to have as much data as possible, since this is also convenient for purposes such as process analysis; it is desirable to have data for at least 20 lots, i.e., at least 20 values of the fraction defective (i.e., the number of subgroups). Data may be collected on as many types of defective as desired, but should be stratified as far as possible according to the nature of the defectives and their causes.

(2) Organize the data into subgroups

The data must be divided into rational subgroups as explained in Section 3.3. In general, it is best to form rational lots and subgroup by lot. For example, data should be taken on small lots formed for process control purposes rather than from shipment lots. The data are easier to handle if the lot size is constant. Also, if n is too small, the statistical power of test of the control chart becomes poor. When n is too large, the data should be stratified and subgrouped in various ways.

(3) Calculate the fraction defective for each subgroup, p_i (see Table 3.4)

This is calculated with the following simple formula:

$$P_i = \frac{\text{number of defectives}}{\text{number of samples (subgroup size)}} = \frac{r_i}{n_i}$$

(4) Calculate the average fraction defective, $\bar{p}$

The average fraction defective ($\bar{p}$) is the total number of defectives divided by the total number of items inspected (i.e., the total number of samples). In general, it is not equal to the average of the subgroup fraction defectives ($\bar{p}_i$).

* This is because (as will be seen later from the formulas for control limits) 5 defectives out of 100 items and 10 defectives out of 200 items have different distributions statistically, even though the fraction defective (0.05) is the same for both.

Table 3.4 Example of Data Sheet for Control Charts for Fraction Defective and Number of Defectives

For pn and p control charts

_______________ Factory _______________
Year
Month Day _______________

Product _______________
Product number _______________
Process _______________
Person in charge of process _______________
Inspection method _______________
Person in charge of inspection _______________
Type of defective _______________
Remarks _______________

Subgroup number	Number of items inspected n	Number of defectives pn	Fraction defective p	UCL	LCL
1	50	3	0.06		
2	〃	8	0.16		
3	〃	3	0.06		
4	〃	5	0.10		
5	〃	4	0.08		
6	〃	10	0.20		
7	〃	10	0.20		
8	〃	9	0.18		
9	〃	4	0.08		
10	〃	6	0.12		
11	〃	9	0.18		
12	〃	8	0.16		
13	〃	12	0.24		
14	〃	6	0.12		
15	〃	8	0.16		
16	〃	8	0.16		
17	〃	10	0.20		
18	〃	13	0.26		
19	〃	9	0.18		
20	〃	5	0.10		
21	〃	7	0.14		
22	〃	9	0.18		
23	〃	5	0.10		
24	〃	3	0.06		
25	〃	13	0.26		
Total	1250	187	—	—	—
Average	$\bar{n}=50$	—	$\bar{p}=0.150$	0.302	—

CL $\bar{p}$ $= 0.150$

UCL $\bar{p}+3\sqrt{\bar{p}(1-\bar{p})/n}$ $= 0.150+0.152=0.302$

LCL $\bar{p}-3\sqrt{\bar{p}(1-\bar{p})/n}$ $= 0.150-0.152=$ (not applicable)

However, it is equal to the arithmetic mean of p_i for each subgroup when the subgroups are all the same size.

$$\bar{p}=\frac{\text{total number of defectives}}{\text{total number of items inspected}}=\frac{\Sigma r_i}{\Sigma n_i}=\frac{\Sigma p_i n_i}{N}\quad(\text{where } N=\Sigma n_i)$$

In this example, $\bar{p}=\dfrac{187}{1{,}250}=0.150$.

(5) Calculate the control limits

The 3-sigma control limits for the p chart are calculated from the following formulas:

—Upper control limit: $\mathrm{UCL}=\bar{p}+3\sqrt{\dfrac{\bar{p}(1-\bar{p})}{n_i}}$

—Lower control limit: $\mathrm{LCL}=\bar{p}-3\sqrt{\dfrac{\bar{p}(1-\bar{p})}{n_i}}$

LCL is not applicable when LCL<0. In this example,

$$\bar{p}\pm3\sqrt{\frac{\bar{p}(1-\bar{p})}{n_i}}=0.150\pm0.152$$

—UCL = 0.302
—LCL = (not applicable)

It is clear from these formulas that if n_i varies, the distance between the control limits will change and the control lines will be wavy rather than straight. Thus, when the number of items inspected in each lot (n_i) changes, we will not obtain straight control lines, and we must calculate and mark in individual control limits for each subgroup. When controlling processes, it is therefore simpler to ensure that n_i remains constant as far as possible.

A few points to note are first, that the central line, $\bar{p}$ does not change even if n_i changes.

Second, the distance between the control limits decreases as n_i increases for the same value of $\bar{p}$, and increases if $\bar{p}$ increases (when $\bar{p}\leqq0.5$).

Third, in practice, when the variation in the size of the subgroups (n_i) is so great that n_i becomes more than double or less than half the average number of items inspected in each lot ($\bar{n}$), where $\bar{n}\approx(n_1+n_2+\ldots+n_k)/k$ (e.g., for $\bar{n}=100$, when the maximum $n_i=200$ and the minimum $n_i=50$), the control lines should initially be drawn in for $\bar{n}=100$. Points should then be checked to account for a change in n_i in the following cases only:

(a) When $n_i>\bar{n}$ and a point falls just inside one of the control lines, the control limits should be calculated precisely for that value of n_i. If a point lies even slightly outside one of the control lines, it is always outside the control limits (this is because the separation between the limits decreases as n_i increases).

(b) When $n_i < \bar{n}$, the control limits should be calculated precisely for that value of n_i when a point falls just outside one of the control lines. Precise calculation is not needed when a point lies even slightly inside a control line, because it will always be between the control limits in this case.

Fourth, in the p chart, the width of the limits is determined by $\bar{p}$ itself. This is different from the $\bar{x}$ chart, in which the width depends on $\bar{R}$.

Fifth, when the percent defective is used, the limits are calculated as follows:

$$100\bar{p} \pm 3\sqrt{\frac{100\bar{p}(100-100\bar{p})}{n_i}}\ \%$$

Sixth, since it is a nuisance to calculate control limits for every value of n_i, various charts and tables have been devised to simplify this task (e.g., 'JUSE Statistical Table (A)', published by JUSE).

And seventh, when $\bar{p} \leqq 0.1$—i.e., when the percent defective is 10% or less—$1-\bar{p}$ is regarded as being approximately equal to 1, and the control limits are calculated approximately from the following formula:

$$\bar{p} \pm 3\sqrt{\frac{\bar{p}}{n_i}}, \quad 100\bar{p} \pm 3\sqrt{\frac{100\bar{p}}{n_i}}$$

(6) Draw the control chart

The central line and the two control lines are drawn on the chart, their values recorded, and the values of P_i are plotted. The distance between the control limits should be approximately 30mm as in the $\bar{x}-R$ chart. Since the limits vary with the subgroup size (n_i), the value of n_i should be recorded below the subgroup number when the subgroup size varies.

3.5 Preparing Control Charts for Number of Defectives (pn)

Since this type of control chart closely resembles the control chart for fraction defective (p), I will mention only some particularly noteworthy points here (see Figure 3.2).

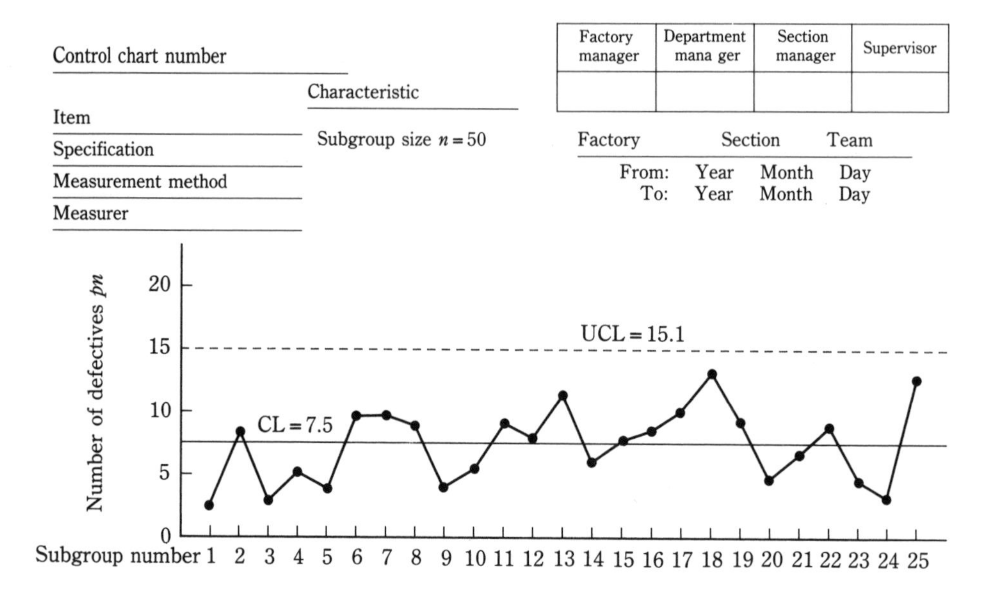

Figure 3.2 ***pn*** **Control Chart**

The control lines for the pn chart are calculated from the following formulas:

$$\text{Central line} = \text{average number of defectives} = \frac{\text{total number of defectives}}{\text{number of subgroups}}$$

$$= \frac{\sqrt{\Sigma r_i}}{k} = \frac{\sqrt{\Sigma p_i n_i}}{k} = \bar{p}n$$

—Upper control limit: $\text{UCL} = \bar{p}n + 3\sqrt{\bar{p}n(1-\bar{p})}$

—Lower control limit: $\text{LCL} = \bar{p}n - 3\sqrt{\bar{p}n(1-\bar{p})}$

As can be seen from the above formulas, the central line in the pn control chart, $\bar{p}n$, varies with n. Thus, when n varies, both the central line and the limits vary, and the positions of the points change greatly. Since this would make the chart very difficult to use, the pn control chart is only used when the subgroup size, n, is constant. If n is constant, the number of defectives (pn) can be plotted as it is, making the chart suitable for use on the shop floor.

3.6 Preparing Control Charts for Number of Defects per Unit (u)

This type of control chart is used when controlling by means of data such as the number of blemishes in a piece of cloth, pinholes in a painted surface, faults (in wire, paper, and other long continuous product, or in machines, electrical equipment, televisions, furniture, and other assembled products), accidents, mechanical breakdowns, specks of dust (in chemicals, solvents, etc.), typographical errors, daily visitors, etc. The c control chart is used when the sample size is constant, and the u chart is used (by converting c to number of defects per unit, u) when the sample size varies. The steps in preparing the chart are as follows:

(1) Collect data

A product is sampled and the number of defects, c, is recorded at the same time as the area, length, weight, volume, etc., when the product is a quantity of steel

plate, yarn, chemical, solvent, etc. When the product is an assembly, the number of defects per assembly is counted. With accidents, breakdowns, etc., the data are recorded for a fixed period, fixed number of people, fixed number of machines, etc. There may be more than one type of defect, but data on two or more types of defect should not be lumped together when there is a correlation between them. As far as possible, defects should be stratified according to their nature and cause when preparing control charts.

(2) Organize the data into subgroups

The data should be organized into rational subgroups, with data taken from the same lot or system being treated as a subgroup. The number of units (n_i) within each subgroup, e.g., the number of meters, square meters, grams, liters, machines, people, etc., need not be constant, but should be clearly indicated.

(3) Calculate the number of defect per unit (u_i) for each subgroup

The formula for calculating u_i is:

$$u_i = \frac{\text{total number of defects } (c_i) \text{ for all units within a subgroup}}{\text{number of units within the subgroup } (n_i)}$$

For example, with a subgroup of 5m^2 and a unit of 1m^2, $n_i = 5$.

(4) Calculate $\bar{u}$

The formula for calculating $\bar{u}$ is:

$$\bar{u} = \frac{\text{total } c_i \text{ for all groups}}{\text{total } n_i \text{ for all groups}} = \frac{\Sigma c_i}{\Sigma n_i}$$

This is the central line.

(5) Calculate the control limits

These are given by $\bar{u} \pm 3\sqrt{\frac{\bar{u}}{n_i}}$

The LCL is not applicable when it is less than 0. The control limits fluctuate from group to group when n_i varies, as in the p chart. The remaining steps are the same as for the p chart.

3.7 Preparing Control Charts for Number of Defects (c)

Since the number of defects, c, is plotted straight onto a c chart, this type of chart is convenient when n is constant. The difference between it and the u control chart is that c_i is plotted directly without calculating u_i and the control lines are calculated as follows:

—Central line: $\bar{c} = \frac{\text{total number of defects for all subgroups}}{\text{number of subgroups}} = \frac{\Sigma c_i}{k}$

—Upper control limit: $\text{UCL} = \bar{c} + 3\sqrt{\bar{c}}$
—Lower control limit: $\text{LCL} = \bar{c} - 3\sqrt{\bar{c}}$
(not applicable when $\bar{c} < 9$)

An example is shown in Table 3.5 and Figure 3.3.

Table 3.5 Example of Data Sheet for Control Chart for Number of Defects

For c and u control charts

______ Factory ______

Year Month Day ______

Product ______ Product number ______

Process ______ Person in charge of process ______

Inspection method ______ Person in charge of inspection ______

Type of defect ______ Remarks ______

Subgroup number	Subgroup size	Number of defects (c)	Number of defects per unit (u_i)	Remarks
1		18		Central line $\bar{c} = 16.8$
2		13		
3		13		$UCL = \bar{c} + 3\sqrt{\bar{c}}$
4		15		
5		21		$= 16.8 + 3 \times 4.1$
6		17		$= 29.1$
7		28		$LCL = \bar{c} - 3\sqrt{\bar{c}}$
8		10		
9		23		$= 16.8 - 3 \times 4.1$
10		16		$= 4.5$
11		15		
12		22		
13		18		
14		12		
15		24		
16		11		
17		19		
18		16		
19		13		
20		14		
21		12		
22		25		
23		16		
24		13		
25		15		
Total		419		
Average		$\bar{c} = 16.8$		

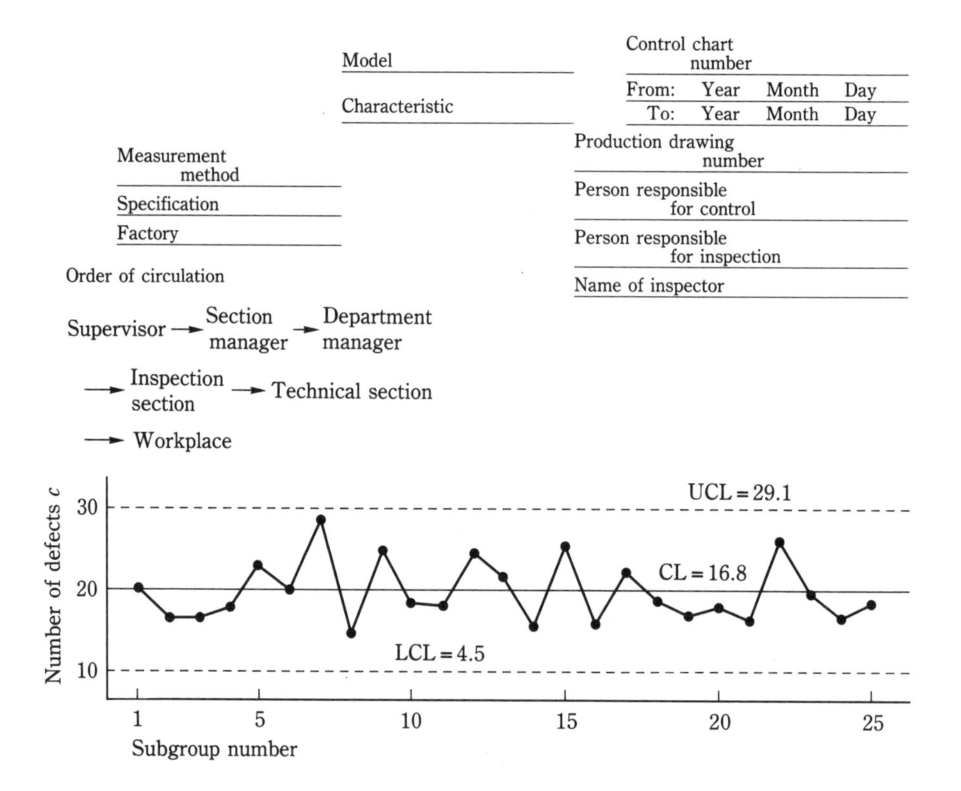

Figure 3.3 *c* Control Chart

3.8 Interpreting Control Charts

Simply drawing up control charts is not very useful; they are not good for anything unless we examine them closely, read off information about the state of quality, process, and work, and search out and eliminate the causes of abnormalities. To do this, we must learn how to read control charts and practice obtaining information from the movements of the points. We must become able to tell at a glance what has happened to a process, what kinds of change have occurred in the distribution, and what sorts of assignable cause have arisen. The principles of reading control charts are described briefly below.

1) The points should be looked at not as individual points but *as a distribution.* In other words, we must think about what has happened to the distribution of the process (i.e., the population) behind the points.

2) It is wrong to pay too much attention to the movement of the points between the control limits. The results will be scattered randomly between the limits even when there are no causes of abnormality and work is proceeding as normal.
3) If the points lie inside the limits, the process is regarded in principle as being in the controlled state. It should be noted, however, that strictly speaking, the controlled state on the $\bar{x}$ control chart is when the points are dispersed randomly between the control limits, forming a normal distribution with the central line as its center (see subsection 6 below).
4) If any points lie outside the limits, some abnormality has definitely occurred in the process and the process is *out of control*. The process is also regarded as being out of control if any points lie actually on a control line. Such a situation is called the "*uncontrolled*" or "*out-of-control*" state.
5) When the points on a control chart being used for analysis satisfy the following conditions, the process is regarded for the time being as in a state of control. The control lines are regarded as representing the process and are extrapolated into the future, enabling the chart to be used for controlling the process. The points should be randomly dispersed and should satisfy the following conditions:
   (a) 25 consecutive points lie within the control limits.
   (b) Out of 35 consecutive points, no more than 1 lies outside the control limits.
   (c) Out of 100 consecutive points, no more than 2 lie outside the control limits.

   In the last two of these cases, the causes of the abnormality must be sought.
6) A number of consecutive points lying on one or other side of the central line is called a "*run*." It is abnormal for a large number of consecutive points to lie above or below the central line. An abnormality is generally considered to be present when a *run of seven or more points* occurs. However, when one control line is absent (e.g., the lower control line in the R control chart when n is 6 or less), no abnormality is judged to be present even if a run of seven or more points occurs on that side of the central line (i.e., a run below $\bar{R}$ in this case).
7) In control charts used for analysis, there is a possibility that an abnormality has occurred in the process when a large number of points appear on the same side of the central line, as described below:
   (a) 10 or 11 points out of 11 consecutive points.
   (b) 12 or more points out of 14 consecutive points.

(c) 14 or more points out of 17 consecutive points.
(d) 16 or more points out of 20 consecutive points.

8) When the points show a rising or falling trend, an abnormality may be present.
9) When over half the points lie outside the control limits, or when most of the points are clustered about the central line in a band half as wide as the control limits, this indicates a control chart for which the subgrouping or stratification of the data was inadequate. When this happens, the chart should be redrawn using a different way of subgrouping or stratification.
10) With the $\bar{x}-R$ control chart, start by examining the R chart.

3.9 Using Control Charts

3.9.1 Applications

From various viewpoints, the control chart can be said to be the central statistical tool for control. Put plainly, it is no exaggeration to say that "quality control begins and ends with the control chart."

The main areas of application of control charts are:

(1) for control
(2) for analysis
(3) as graphs
(4) for adjustment
(5) for inspection.

While control charts may be used for all of the above purposes, their *essential role* is still process control, followed by process analysis. Analysis could also be viewed as a preparatory stage in which control charts useful for process control are prepared. Process analysis is discussed in Chapter 4.

The third of these applications, *using control charts as graphs*, means plotting data in the form of a control chart but not using them as such. Although control limits are drawn in, such charts are only gazed at, even when some points fall outside the limits; the causes of abnormalities are not sought and no action is taken. Such charts are simply drawn mechanically under instructions from superiors, and many of the so-called control charts prepared in factories where process analysis and standardization are inadequate are of this type. They are control charts in form but not in substance, and should really be called graphs. However, plotting the data in the form of a control chart does show the ways in which a process changes with time and can also have quite a good motivational effect. I am therefore not saying that this type of chart should be abolished. If presenting data in the form of graphs produces good results, they should be widely used; but I would not like people to make the mistake of believing that preparing this kind of control chart means they are carrying out quality control or other types of control. Also, boredom inevitably sets in after this type of chart has been used for some time, and people begin to argue that control charts are useless. For this reason, efforts should be made as soon as possible to start analyzing and standardizing processes, reviewing the characteristics to be plotted on control charts, standardizing authority, responsibility, and methods related to searching out assignable causes, and taking action for control purposes, and actually using the charts for control.

The fourth application, *using control charts for adjustment*, means, for example, changing the temperature, the bite of a cutting tool, the raw material composition, or other process condition when a control chart shows that a process is out of control, without necessarily looking for the cause of the abnormality or taking action to eliminate it. This is not the proper use of the control chart, and a totally different approach must be adopted in considering whether or not 3-sigma limits are suitable as adjustment limits (not control limits). In fact, more often than not 3-sigma limits are inappropriate as adjustment limits. To distinguish this type of chart from the control chart, it should be called an *"adjustment chart."* Adjustment limits should be investigated and set in the same way as automatic control is implemented, by considering factors such as the random variation in the process, movement of process average, sampling interval, range of possible adjustments and their effects, feedback time, etc.

The fifth application, *using control charts for inspection*, means using them in various ways from the inspection standpoint: for example, when a chart shows an abnormality in a lot, and the lot is treated differently or subjected to 100% screening, or the downstream inspection methods are changed. Of course, this application can be made quite useful with a bit of ingenuity; for example, when greatly relaxing inspection and switching to check inspection, the results can be plotted on a control chart and the inspection can be tightened up if the chart shows the out-of-control state. However, decisions about the disposition of a lot—e.g., whether to subject it to 100% screening—should not be based on control limits but on the decision criteria for sampling inspection with screening. Using control charts from the inspection standpoint in this way is particularly prevalent in inspection-oriented factories practicing old-style quality control, of which there are many in the heavy electrical and machinery industries, and I urge them to make a point of reviewing this practice. The use of control charts for inspection in this way cannot normally be recommended, except when it has been carefully investigated. However, control charts are extremely useful for controlling inspection operations or inspection processes, and I would like to see them widely used in this way.

3.9.2 Using Control Charts for Analysis

Control charts for analysis can be considered under the following two headings:

(1) Those used for analysis for discovering and eliminating the causes of variation; and,
(2) Those used for analysis for estimating the process capability in preparation for controlling those processes in the future.

The first type of use is discussed in subsections 1–3 below; the second is covered in subsection 4.

The first consists mainly of charts prepared with the aim of discovering and eliminating the causes of variation by devising various ways of subgrouping, stratifying, and modifying data and testing whether a process is out of control.

(1) Subgrouping

Trying out various ways of subgrouping is an extremely important method of discovering the causes of variation. Subgrouping is closely related to sampling; it allows many causes to be discovered, and the skill or lack of skill with which it is done governs the usefulness of control charts used for controlling processes. Some points to be considered when subgrouping are as follows:

(a) When considering control charts, cause-and-effect diagrams or other methods should be used to draw clear technical distinctions between the types of factor that affect the within-subgroup variation and those that affect the between-subgroup variation. For example, when $\bar{x}$ goes out of control, it is usually due to a cause of between-subgroup variation, but when R goes out of control, it is usually due to a cause of within-subgroup variation.
(b) Data on products made under similar conditions should be collected in the same subgroup, so that the data within each subgroup are as uniform as possible and have as small a variation as possible. Put the other way round, this means that data should be subgrouped in such a way as to make the variation between subgroups as large as possible. This is particularly important in process analysis.

(c) Various sampling methods should also be tried in order to satisfy the above requirement.
(d) We should clarify the purpose of the control chart—i.e., the kind of variation we want to discover or control—and group the data in such a way as to exclude that particular kind of variation, as far as possible, from within the subgroups.
(e) The possible causes of variation should be considered from the technical viewpoint, various methods of subgrouping should be tried, and the state of control and the value of $\bar{R}$ and other statistics should be compared.

The best methods of sampling and subgrouping for process control are worked out from the above considerations. Subgroups formed skillfully in this way are called "*rational subgroups*."

(2) Stratification

When a factory has a number of machines, different machines often have their own characteristics and idiosyncrasies. In such cases, it is best to prepare a separate control chart for each machine. Likewise, it is best to segregate the data and prepare separate control charts for raw materials of different types or origins, different auxiliary materials, seasons, months, weather, working conditions, personnel, shifts, work volumes, and other factors that are thought to influence the process in individual ways and to cause variation. Separate control charts should also be drawn up for different types and conditions of defectives, defects, breakdowns, etc. Dividing the data up into different strata in this way is called "*stratification*."

Preparing various stratified control charts in this way for various causes (mainly attribute-type causes) that are considered for engineering reasons to exert particularly significant effects is extremely useful for analysis. The success of control charts for analysis and control could be said to depend on stratification. In most successful examples of analysis and control, the process flow is well stratified from raw material to final product, a variety of data are collected and analyzed, and stratified control charts are used with skill.

In this way, stratified control charts are drawn up, and comparisons are made of the state of control shown by the charts and of the process averages ($\bar{\bar{x}}$, $\bar{R}$, $\bar{p}$, $\bar{c}$, etc.) before and after stratification and between different strata. When doing this, the following points should be noted:

(a) When stratifying, it is best to keep the size of the subgroups as equal as possible.

(b) On the R chart, the value of $\bar{R}/d_2$ decreases if skillful stratification is carried out. If $\bar{R}/d_2$ decreases, it shows that the method of stratification has been effective, and there is often some difference between the different strata. Also, if the subgrouping is rational, stratification often brings out differences in the values of $\bar{R}/d_2$ for the different strata. As a very approximate rule of thumb, if the average values of R for two different strata A and B are $\bar{R}_A$ and $\bar{R}_B$, and if either $\bar{R}_A$ or $\bar{R}_B$ differs by 20% or more from the overall average of R ($\bar{R}$), we can say that the two different strata definitely have different distributions. Please see Section 3A.4 for a detailed method.

(c) In general, if stratification improves the state of control on the control charts, that method of stratification is usually significant and there is a difference between the different strata.

(d) If there is a difference between the averages of different strata, there will be differences between the $\bar{\bar{x}}$ values after stratification. The existence of a difference between the $\bar{\bar{x}}$ values can often be judged intuitively, but in doubtful cases, the method described in Section 3A.4 can be used to perform a statistical test.

(e) If there is judged to be a definite difference between $\bar{R}$ or $\bar{\bar{x}}$ for different strata, the cause must be traced, action must be taken to eliminate the difference, and standards must be revised. After action has been taken, new stratified control charts must always be drawn and examined in order to check the effects.

(f) When there is no way of eliminating the causes of differences between different strata, or if the causes lie outside the range of responsibility for control of the process, the data should be modified to eliminate these differences only. Fresh control charts should then be prepared using the modified data, and the investigation can be continued. However, from the viewpoint of the company as a whole, the responsibility for eliminating these causes lies somewhere within the organization.

(3) Some general points concerning analysis using control charts

The following are some general points to note when performing analysis using control charts:

(a) In process analysis, particularly with the $\bar{x}-R$ chart, one must pay close attention to the subgrouping and to the state of control on the control charts and the behavior of R after stratification. It is advisable to start by trying to make $\bar{R}$ as small as possible and to get R into the controlled state.

(b) Analysis employing stratified control charts is used mainly for investigating the presence or absence of attribute-type causes, checking their seriousness, and deciding on the action needed to deal with them.

(c) In process analysis, the most effective procedure is often to devise various methods of rational subgrouping and stratification and actually try them out. The causes thought for technical reasons to exert significant effects should therefore be analyzed one at a time, starting with the one considered to exert the biggest effect. Then, if there is a difference between different strata, action is taken to eliminate it or the data are corrected to eliminate this difference only, and the next cause is analyzed.

(d) In the world of variation, R (the range) is the basis of process variation. In many cases, if it is possible to adjust the value of $\bar{R}$ freely, it naturally becomes possible to set $\bar{\bar{x}}$ to the desired value. Thus, in many areas of process control, as well as, of course, in process analysis, our aim is to "*stamp out R.*"

The following will help in controlling R:

i) Change the method of subgrouping.

ii) Stratify.

iii) Reduce the variation in sampling (of bulk materials) and measurement.

iv) Implement thorough process analysis and control.

v) When R is not reduced in spite of the above actions, on-site experiments should be carried out using design of experiment methods, and standards should be revised, equipment rebuilt, and other basic technical improvements performed. When this is done, it is best to compare the error variance in the analysis of variance with the square of $\bar{R}/d_2$.

The causes of a large value of R are generally extremely close at hand, in routine operations taking a short time to complete, and operators should seek them diligently in their immediate surroundings.

(e) If R decreases when a different method of subgrouping is tried or stratification is carried out, this generally shows that the subgrouping or stratification has been effective. When this happens, the cause of the between-subgroup variation must be investigated.

(f) The above considerations are more or less the same for the p chart, c chart, and other types of control chart, but the following additional points should be noted:

(i) Attention should be paid to out-of-control points on the good side as well as, of course, to out-of-control points on the bad side. A process will go out of control on the good side when the process really improves, inspection standards become lax, and/or samples are not taken at random and good samples are being preferentially selected, etc. Whatever the reason, we must also track down the causes of lack of control on the good side and use the information obtained to take appropriate action.

(ii) When the method of subgrouping is poor and the subgroups are too large, many points sometimes lie outside the control limits. In this case, a variety of information can be obtained by segregating the data more, stratifying it in various ways and dividing it up into smaller subgroups, or by drawing up stratified control charts.

(4) Procedure for analysis in preparation for process control

This section explains the procedure for analyzing a process and preparing to move on to process control (discussed in the next section; Section 3.9.3). The discussion focuses on the important $\bar{x}-R$ chart, but substantially the same considerations apply to other types of control charts.

(a) Decide on the characteristics to be plotted on the control charts.

As discussed before, we must decide which of the results in our range of responsibility for control we should use as a means for checking the process. In controlling the quality of a product that has a large number of quality characteristics, for example, we must decide which of the characteristics is important and should be checked. When there are many important quality characteristics, all of them may be chosen. We must also consider the important quality characteristics required by the customer (i.e., by the next process). We should use the results of the analysis to select several characteristics for plotting on control charts in order to control the process in the future.

When doing this, many people take the conventional engineering approach and draw control charts for causes. This is a mistake, and most such charts will be no more than graphs. Even so, they may be effective as such. Control charts for causes may sometimes be drawn for analysis as well, but this is not their main purpose.

(b) Decide on the control charts to be used.

Once the control characteristics have been decided, we should consider their nature and decide which control charts ($\bar{x}-R$, pn, c, u, etc.) to use.

(c) Collect data.

In factories it is often sufficient simply to collect past data, but the history of such data must be clear. If the history of past data is completely unknown, a stratified sampling scheme such as sampling by lot must be chosen, depending on our aim in controlling the process, and fresh data must be collected. However, even analyzing data with an unknown history may be useful in some way or other. If possible, at least 100 values should be collected.

(d) Analyze past data using control charts.

This should be done as described in subsections 1–3 above.

(e) Draw up control charts in preparation for controlling the process.

How to draw up a control chart must be decided on the basis of the information obtained in step 4, giving due consideration to the purpose of controlling the process; the chart is then drawn up. If the chart shows an approximate state of control (see subsection 5 of Section 3.8), it can be used for calculating the control limits for a chart to be used for controlling the process in the future. If it does not show a state of control, various methods may be tried to obtain a chart that is as close to the controlled state as possible and is also easy to use. Work standards designed to get the process into this state should be prepared and clearly communicated to subordinates. However, even control charts that do not show the controlled state can be used by extending the control limits into the future, plotting data, and finding and eliminating the causes when any of the points fall outside the control limits.

When standards are newly prepared or revised, about 20 subgroups of data resulting from those standards should be taken and plotted on a control chart. The chart should then be examined and control limits for process control calculated from it.

When this is done, we should have at least 100 values or 20 groups of data. The more data we have, the more precise our estimate of the process (i.e., of the control lines). However, we should calculate trial limits even if we have only a small amount of data, and then recalculate them once more data have been accumulated.

(f) Compare with specifications and targets (see Section 2.4).

If the product specifications and targets have been set on a rational basis (although this is at present a rare occurence), we should use histograms or control charts to check whether the state of control (i.e., the process capability) obtained in step 5 satisfies these standards and targets. When the standards and targets have not been set on a rational basis, they must be decided after discussion with customers, the next process, and top management.

3.9.3 Using Control Charts for Control

After analysis, we move on to control. The procedure is as follows:

(1) Prepare control charts for control
When we have finished analyzing the data, the control lines calculated from the analysis are plotted on control charts using dotted and dashed lines (·—·—·—), in preparation for controlling the process in the future.

(2) Collect data daily and plot on control charts
To ensure that the work is carried out by the methods decided, one must take samples and measurements, calculate $\bar{x}$, R, or other statistics for each subgroup, and plot them on the control charts. The control lines should be drawn in before the data are plotted. Decisions should be taken in advance as to who is to do the sampling and measurement, who is to report to whom and in what form, who is to plot the points, and who is to use the charts.

(3) Decide whether the process is in control (see Section 3.8)
If the points plotted lie between the control limits, the process is in control. If any lie outside the limits, some assignable cause has occurred in the process, and the characteristic that is a result of the process is therefore showing a large variation.

If points on an R control chart fall outside the limits, a type of change increasing the spread of the product distribution has occurred in the process. If points on an $\bar{x}$ control chart fall outside the limits, this shows mainly that a type of change altering the process average has occurred in the process. However, points on the $\bar{x}$ chart will also sometimes fall outside the limits when the variation increases.

If points on a p chart fall outside the upper control limit, an assignable cause giving rise to large numbers of defectives has occurred in the process. If points fall below the lower control limit, either an assignable cause reducing the number of defectives has occurred or inspection has become lax. It should be remembered that, in general, when points fall outside the control limits, something is usually wrong with the sampling, measurement, or inspection.

Normally, in process control we judge that an assignable cause is present only when points fall outside the control limits. However, depending on the situation, the interpretation of runs may also be used. Standards of judgment for deciding when an assignable cause is present should be set for each type of control chart, and who is to look at the control charts and how they should be circulated when necessary should be decided in advance.

(4) Trace causes

When a process is judged to be out of control, the person responsible for controlling the process must immediately track down the cause. Tracking down causes requires a variety of technical knowledge and statistical methods, and the information yielded by control charts is helpful for this. Painstaking analysis should be carried out, and the procedure for tracking down the causes of abnormalities should be laid down in the form of standards.

(5) Take action

Simply tracing a cause and bringing it out into the open is not control. If the cause of an abnormality is clear, the following action must be taken:

(a) Immediately eliminate the cause and bring the process back into a stable state.

(b) At the same time, take radical action to prevent abnormalities due to the same kind of cause from recurring in the future. If this is neglected, an abnormal variation will again arise in the process through the same cause. If, for example, the cause is carelessness on the part of workers, education should be carried out or foolproof jigs and tools devised. If the work standards are inadequate, they should be revised. Thorough, painstaking action must be taken. Seeing that *reliable action is taken to prevent trouble recurring* is the crux of implementing control (see Section 1.5 and Figure 1.14).

(c) For each control chart, decide beforehand matters such as the procedure for taking action (i.e., for eliminating the causes of abnormalities), what action should be taken under whose judgment (i.e., authority), how far such action should proceed, the method of reporting to superiors, and the forms (abnormality report forms) to be used. The action to be taken by group leaders, foremen, supervisors, section managers, etc., is decided on the basis of these reports.

(6) Check the results of action

Even when action to eliminate what is thought to be an assignable cause has been taken and recurrence-prevention measures such as revising standards have been put into effect, one should check again to see whether or not the action was correct and to examine its effects.

Generally, simply taking action and leaving it at that cannot be considered control. One of the principles of control is always to check the results of action taken.

(7) Recalculate the control lines

A process is controlled by drawing control lines and plotting data on a control chart, but we should bear in mind that it is necessary to recalculate the control lines occasionally to ensure that they match the current state of the process. The control lines should be recalculated in the following cases:

(a) When the process has obviously changed for technical reasons.
(b) When a certain period has elapsed since control was started, even if there has been no change in the process (e.g., every month, after every 100 measurements, etc.).
(c) When it is judged from the control chart that the process has obviously changed.

It is wrong to use control charts simply as graphs, not recalculating the control lines for three months or half a year even though the process has changed considerably, and simply gazing at the chart without taking action even when points fall outside the limits or form long runs. The interval at which this recalculation should be carried out, the method of recalculation, and the decision as to when to do it should be specified in standards for the use of control charts. If this is not done reliably, the charts will lose their usefulness.

When control lines are recalculated, points lying outside the limits should be dealt with as follows:

(i) Individual data or subgroups that produce points lying outside the limits but for which the causes are known and against which action can be taken should be omitted when recalculating the control lines.
(ii) Points representing data for which the causes are unknown or against which action cannot be taken should be included in the recalculation.

(8) Formulate control standards

Standards for control are discussed in detail in Chapter 5; here, as mentioned above, we must reiterate that they must be laid down for each type of control chart and must show who is responsible for control and how it should be carried out. Without such standards, control charts cannot be used well and control cannot be effectively implemented. At all events, control charts should be seen and used by those in positions of authority.

The control chart is an effective tool for leaders in every department, not only for quality control but for all other types of control.

3A.1 THE MEDIAN ($\tilde{x}$) AND RANGE CONTROL CHART

(1) Drawing a median control chart

The median of a set of data, expressed by the symbol $\tilde{x}$, is the middle value when the data are arranged in order of magnitude from maximum to minimum. When there are an even number of values, the median is the average of the two middle values. The $\tilde{x}-R$ control chart is used in almost the same way as the usual $\bar{x}-R$ control chart, and the control limits for the $\tilde{x}$ chart are normally calculated from formula 3A.1 or 3A.2 below:

$$\bar{\bar{x}} \pm m_3A_2\bar{R} \quad \text{(3A.1)}$$

$$\bar{\tilde{x}} \pm m_3A_2R \quad \text{(3A.2)}$$

Where $\bar{\tilde{x}}$ is the average of the medians; m_3A_2 is a coefficient for calculating the median control limits from $\bar{R}$. Its value depends on n and is given in Table 3A.1.

Example: For $n=5$, $\tilde{x}=120.020$ and $\bar{R}=2.292$,
$m_3A_2=0.691$
$m_3A_2\bar{R}=0.691 \times 2.292=1.584$

Thus,
UCL = 120.020 + 1.584 = 121.60

LCL = 120.020 − 1.584 = 118.44

(2) When using the median of R ($\tilde{R}$)

Hitherto, we have been discussing methods of estimating variation that utilize $\bar{R}$, but the values of R can also be arranged in order of magnitude and the variation estimated and control limits calculated using the median of these values ($\tilde{R}$), as follows:

For $\tilde{x}$: $\tilde{\tilde{x}} \pm m_3A_3\tilde{R}$ (3A.3)
For $\bar{x}$: $\tilde{\bar{x}} \pm A_3\tilde{R}$ (3A.4)
For R: UCL $= D_6\tilde{R}$ (3A.5)
LCL $= D_5\tilde{R}$ (3A.6)

Where $\tilde{\tilde{x}}$ is the median of $\tilde{x}$; $\tilde{\bar{x}}$ is the median of $\bar{x}$; and m_3A_3, A_3, D_6, and D_5 are coefficients for calculating the control limits using $\tilde{R}$. Their values depend on n and are given in Table 3A.1.

Table 3A.1 Table of Coefficients for $\tilde{x}-R$ Control Charts

Subgroup size n	$\tilde{x}$		When using $\tilde{R}$					
			$\bar{x}$	x	$\tilde{x}$		R	
	m_3	m_3A_2	A_3	E_3	m_3A_3	dm	D_5	D_6
2	1.000	1.880	2.224	3.14	2.224	.954	—	3.864
3	1.160	1.187	1.091	1.89	1.265	1.588	—	2.744
4	1.092	.796	.758	1.52	.828	1.978	—	2.375
5	1.198	.691	.594	1.33	.712	2.257	—	2.179
6	1.135	.549	.495	1.21	.562	2.472	—	2.055
7	1.214	.509	.429	1.13	.520	2.645	.078	1.967
8	1.160	.432	.380	1.07	.441	2.791	.139	1.902
9	1.223	.412	.343	1.03	.419	2.916	.187	1.850
10	1.177	.363	.314	.99	.369	3.024	.227	1.808

It should be noted that $\tilde{\tilde{x}}$, $\bar{\tilde{x}}$, and $\tilde{R}$ are less precise estimators of the population than $\bar{\bar{x}}$, $\bar{\bar{x}}$, and $\bar{R}$, respectively, in the controlled state, but the effect of this is small in the out-of-control state, and the precision of the estimate often improves in this case.

When estimating the population standard deviation from $\tilde{R}$,

$$\hat{\sigma} = \tilde{R}/d_m \tag{3A.7}$$

Where d_m is a coefficient used for σ estimating from $\tilde{R}$. Its value depends on n and is given in Table 3A.1.

(3) Using the $\tilde{x}-R$ control chart

(a) The $\tilde{x}-R$ chart is interpreted and used in exactly the same way as the $\bar{x}-R$ chart.

(b) Since no calculation is required to find $\tilde{x}$, $\tilde{x}-R$ control charts are useful when having foremen and ordinary workers draw them on the shop floor. When this is done, it is better to make n an odd number.

(c) It is also a good idea to have people plot the raw data directly on a control chart, as in Figure 3A.1, and have them find the medians from the chart as shown.

(d) Since individual values are plotted when this is done, the control limits for x, discussed in Section 3A.2, can be used at the same time (see Figs. 3A.1 and 3A.2).

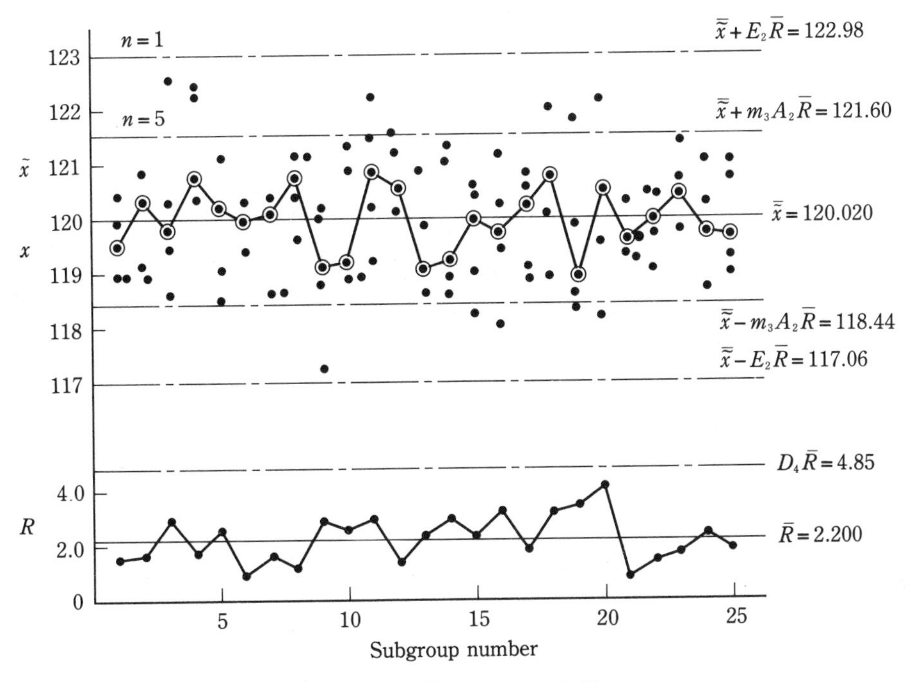

Figure 3A.1 $\tilde{x} - R$ Control Chart

3A.2 Control Charts for Individual Data Points

3A.2.1 Preparing the x Control Chart

A control chart on which individual measurements (x) are plotted is called a "control chart for individual measurements" or an "x control chart." It is generally used in conjunction with a moving-range (R_S) control chart or an $\tilde{x} - R$ control chart.

The problem with preparing this type of control chart is calculating the control limits. Otherwise, it is exactly the same as the normal type of control chart.

(1) Data subgrouping method (see Fig. 3A.2)

In this method of calculating the control limits, rational subgrouping is carried out in the same way as for the normal $\bar{x} - R$ control chart. $\bar{x}$, R, $\bar{\bar{x}}$, and $\bar{R}$ are then found, and the control limits are calculated using the formula given below. This method is suitable in most cases. When the subgroup size is constant, the x control limits are given by:

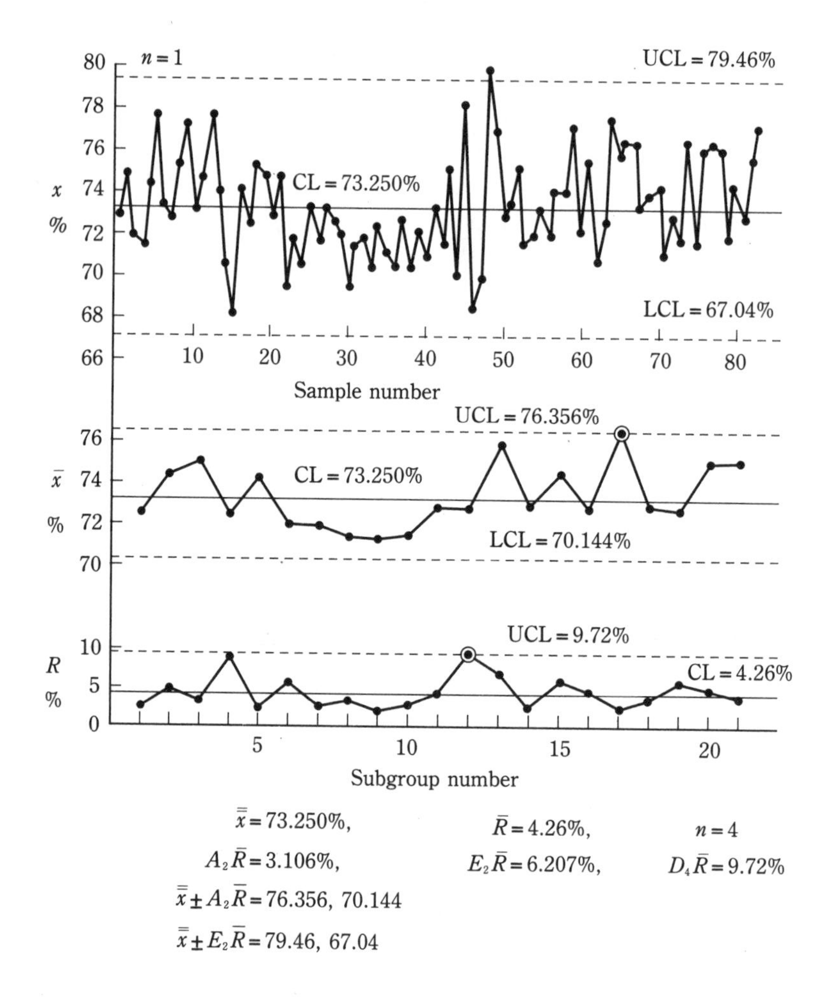

Figure 3A.2 $\bar{x}-R-x$ Control Chart

$$\bar{\bar{x}} \pm 3\bar{R}/d_2 = \bar{\bar{x}} \pm E_2\bar{R} \tag{3A.8}$$

Where the value of E_2 is determined by the subgroup size n and is given in Table 3A.2.

(2) Method using moving range (R_S) (see Fig. 3A.3)

If, for example, the measurements are 18.3, 19.1, 18.5, 18.8, 19.3,... the moving range for $n=2$ is given by $R_S = 19.1 - 18.3 = 0.8$; $19.1 - 18.5 = 0.6$; $18.8 - 18.5 = 0.3$; ....

With the $x - R_S$ chart, the control limits on the x chart are normally calcu-

Table 3A.2 Table of Values of E_2

Subgroup n size	E_2
2	2.660
3	1.772
4	1.457
5	1.290
6	1.184
7	1.109
8	1.054
9	1.010
10	0.975

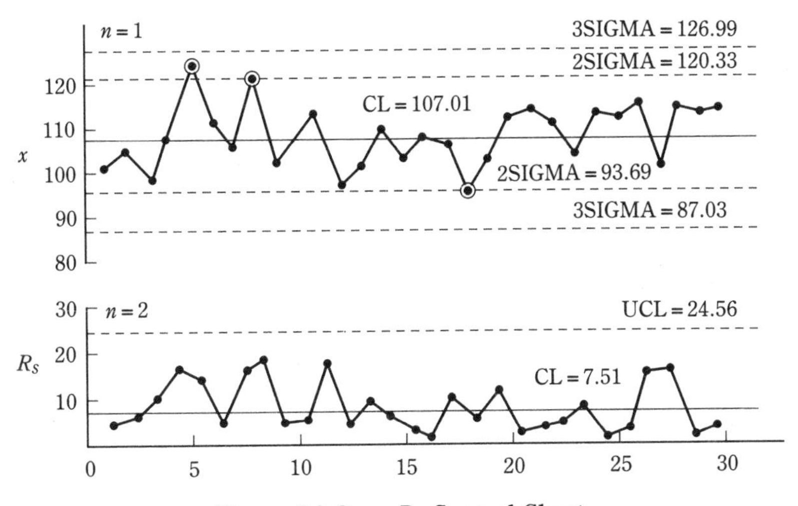

Figure 3A.3 $x-R_S$ Control Chart

lated by the following formula, using the moving range for subgroups of size $n=2$ as above:

$$\bar{x} \pm 3\frac{\bar{R}_S}{d_2} = \bar{x} \pm A_2\sqrt{n}\,\bar{R}_S = \bar{x} \pm E_2\bar{R}_S = \bar{x} \pm 2.66\bar{R}_S \tag{3A.9}$$

Where d_2 and A_2 are normally the values for $n=2$ and are given as 1.128 and 1.880 respectively by Table 3.3.

The control limits on the R_S control chart are found in the same way as those for the R control chart with $n=2$, i.e.:

$\text{UCL} = D_4\bar{R}_S = 3.267\bar{R}_S$
LCL = (not applicable)

The values of R_S are plotted directly below the midpoints between neighboring x points.

The method of calculating the limits from the moving range is used in the following situations:

(a) When rational subgrouping is impossible.
(b) When data can only be obtained at extremely long intervals, e.g., once per week or month.
(c) When the process shows a large fluctuation.

(3) Method using standard deviation obtained from histogram

Because this method is not considered best for our purposes, it is not discussed here.

(4) Method using sampling and measurement error as reference

This is a special method sometimes used for controlling bulk materials, particularly when sampling and measurement error are a problem. For example, when a composite sample is formed by collecting together n increments of the material in an organic batch-type synthesis reaction process or in the processing of materials such as coke, coal, fertilizer, etc., and the sample is analyzed once and one measurement (x) obtained, the control limits for the measurement are calculated using the following formula:

$$\bar{x} \pm 3\hat{\sigma}_S \tag{3A.10}$$

Where $\hat{\sigma}_S$ is the sampling precision of the composite sample. For example, if the variation between increments is $\hat{\sigma}_i$, and n increments are sampled at random, and if the precision of sample reduction and analysis are σ_R and σ_M respectively, a single analysis will give:

$$\hat{\sigma}_s = \sqrt{\sigma_i^2/n + \sigma_R^2 + \sigma_M^2} \tag{3A.11}$$

Checking σ_s requires preliminary experiments and checking experiments over a considerable amount of time. This method is used when the sampling error is comparatively large and the process is relatively well controlled.

It is best to note that when the sampling and measurement error are large and it is difficult to reduce them, or when it is not technically or economically feasible to carry out large numbers of measurements, the so-called "check experiment method" may be useful in some cases. This method consists of taking two samples at random for every n increment, measuring them separately and using each pair of measurements as a subgroup to draw an $\bar{x}-R$ control chart with $n = 2$. However, in a case where the $\bar{x}$ chart indicates the controlled state, the sampling or measurement method is generally too imprecise.

3A.2.2 Using the x Control Chart

(1) Advantages

(a) Since each data point is plotted as soon as it appears, this type of chart enables the state of a process to be evaluated rapidly and quick action to be taken.
(b) The chart shows graphically how a process varies with time. This has a good *motivational effect*, even if such a graph is not very useful as a control chart.
(c) When there is a large fluctuation or periodicity in a process, or when the process average undergoes a sudden large change, this type of chart clearly shows how the change occurs. Its power of test as a control chart may also be better in some cases.

(2) Disadvantages

(a) There is large scope for making the type II error. In other words, the chart's *power of test is poor*. This is because the ability of the $\bar{x}$ chart to detect abnormalities generally deteriorates as the size of the subgroups (n) decreases.
(b) The most important feature of the control chart, *rational subgrouping*, is unclear, and the within-subgroup and between-subgroup variation is obscured.
(c) Since the average is not taken, the pattern of the points is distorted when the population distribution is not a normal distribution, and the probability of making an error is altered.

(3) Method of use

(a) When calculating the control limits (for estimating the variation in the process), the rational subgrouping method should be tried first. If rational subgrouping is impossible, the data should be subgrouped in time order in some meaningful way. If this is difficult, or if the process shows a large periodic fluctuation, the moving-range method should be used. Depending on the process, the sampling and measurement error method may be used to calculate the variation.

(b) Whenever possible, the x control chart should be used in conjunction with the $\bar{x}-R$ control chart. When this is done, the chart must be of sufficient width to allow each group of n points to be plotted in order on the x chart for each $\bar{x}$ and R point. When the $\bar{x}-R$ chart cannot be used in conjunction, the R_s chart should be used instead.

(c) The 3-sigma limits are of course used for $\bar{x}$ and R, and, in principle, they should also be used for x. Since the power of test of the x control chart is poor, 2-sigma limits are sometimes used when type II errors are a particular problem. However, it is wrong to use 2-sigma limits simply on the grounds that the control limits are too far apart, without sufficiently considering methods of estimating the variation.

(d) If any points fall outside the limits on the $\bar{x}$, R, or x control chart, action must be taken to eliminate the cause of the abnormality.

(e) In certain situations, the x control chart may be used in the form of a graph to show periodicity, trends, runs, etc., but care should be taken not to commit type I errors. Also, the shape of the original distribution should be checked by means of histograms, since the pattern of points on the x chart will become distorted if the original distribution is distorted.

(f) x control charts may be used extensively as graphs when plotting data and in this form have a significant motivational effect, but both top management and workplace personnel must be fully schooled in *the difference between graphs and control charts.*

(g) The following misunderstandings often arise when x control charts are used:
   (i) No distinction is made between adjustment and eliminating the causes of abnormalities.
   (ii) People mistakenly believe they should take prompt action against abnormal measurements, instead of giving priority to ensuring that abnormalities do not recur in the future.

3A.3 The Statistical Interpretation of Control Charts

If a process changes, the distribution of any characteristic that results from that process also changes. Such a change in distribution manifests itself as a change in the process average and variation (the within-subgroup variation). Let us consider how this change actually appears in the form of the movement of the points on a control chart.

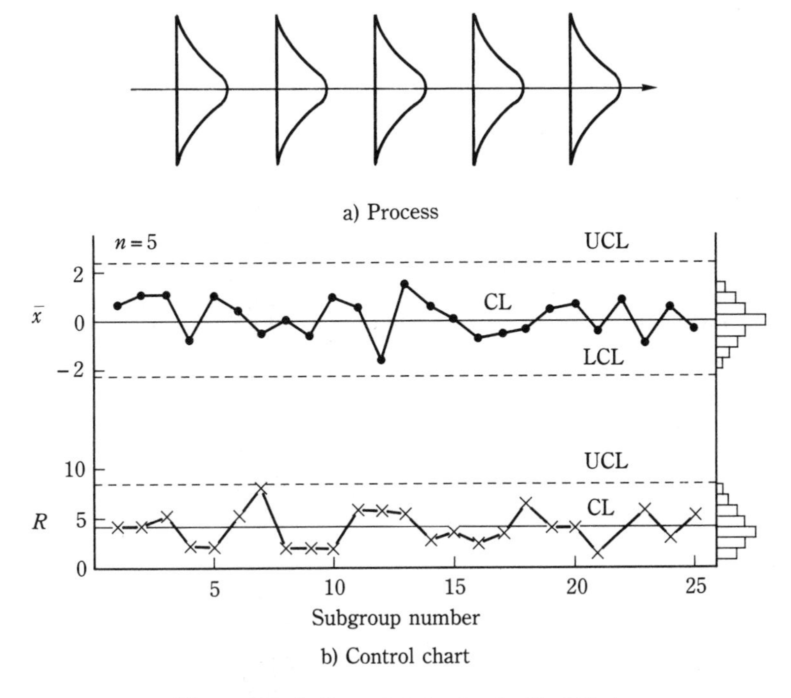

Figure 3A.4 Completely Controlled State

(1) The completely controlled state

When neither the process average nor the process variation (the within-subgroup variation) changes (see Fig. 3A.4),

(a) The points are scattered at random between the control limits. Note that this does not mean that they line up neatly along the central line.
(b) No points fall outside the control limits.
(c) On the $\bar{x}$ chart, most points lie near the central line, but some also lie near the control limits.
(d) On the R chart, more points lie below the central line, showing that the distribution is skewed.

(2) When the process average suddenly changes greatly (see Fig. 3A.5)

Under these circumstances, the following charts are appropriate:

(a) R chart: the same as in Figure 3A.4.
(b) $\bar{x}$ chart: some points fall outside the control limits.

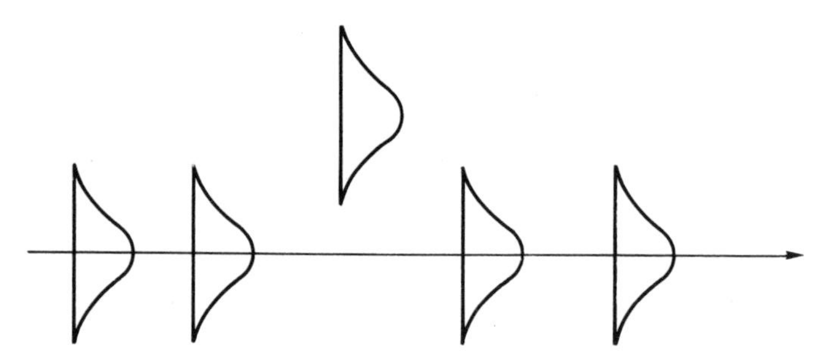

Figure 3A.5 Sudden Large Fluctuation in Process Average

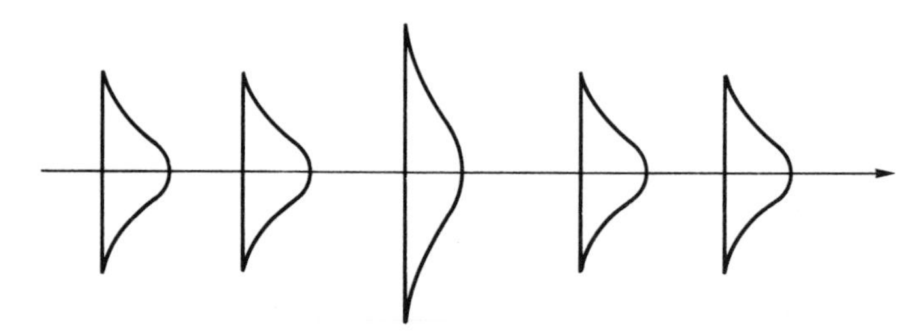

Figure 3A.6 Sudden Large Fluctuation in (Within Subgroup) Dispersion

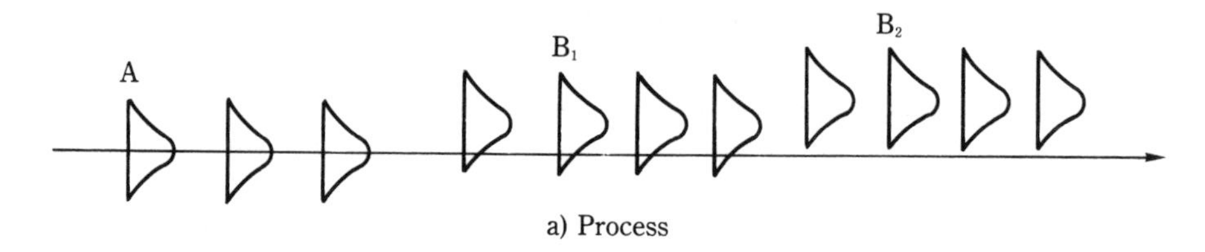

a) Process

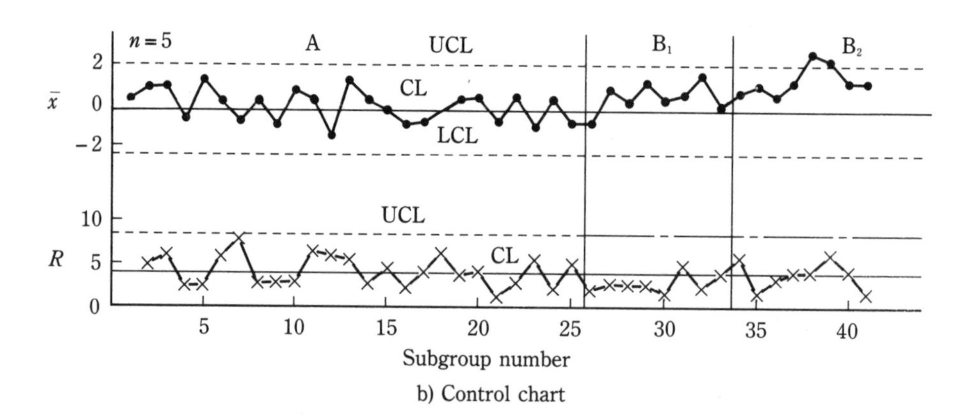

b) Control chart

Figure 3A.7 Stepwise Increase in Process Average

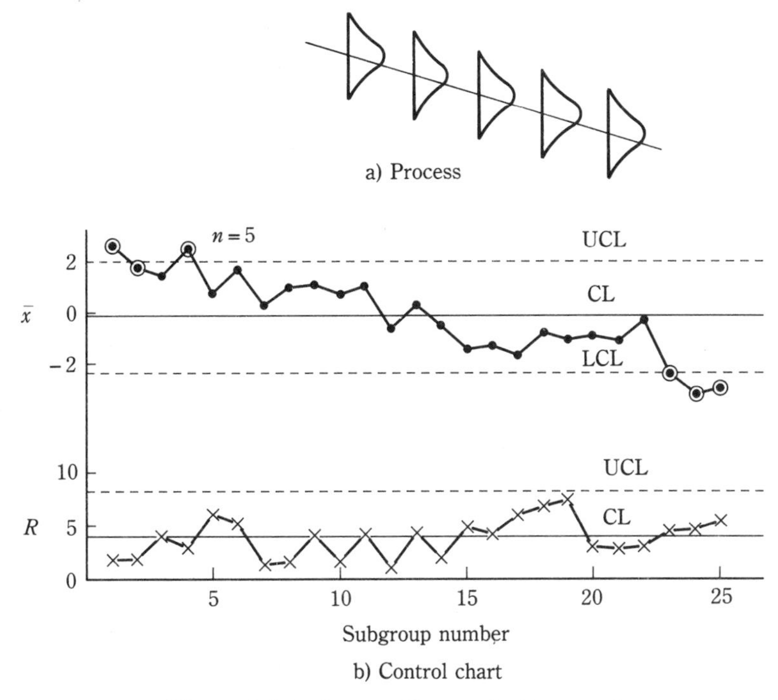

Figure 3A.8 Steady Change in Process Average

(3) When the variation (within subgroups) suddenly changes greatly (see Fig. 3A.6)

Under these circumstances, the following applies:

(a) *R* chart: some points fall outside the limits.
(b) $\bar{x}$ chart: the vertical movement of the points becomes more pronounced and some points fall outside the limits.

(4) When the process average increases stepwise (see Fig. 3A.7)

Here you will note:

(a) *R* chart: no change.
(b) $\bar{x}$ chart: in zone B_1, the overall number of $\bar{x}$ points above the central line increases and runs are observed, but no points fall outside the control limits. In zone B_2, some points also fall outside the control limits.
(c) When this kind of situation occurs, the difference between zones A, B_1, and B_2 will be seen if the data for these zones are stratified and plotted on separate control charts, as explained in subsection (7) below.

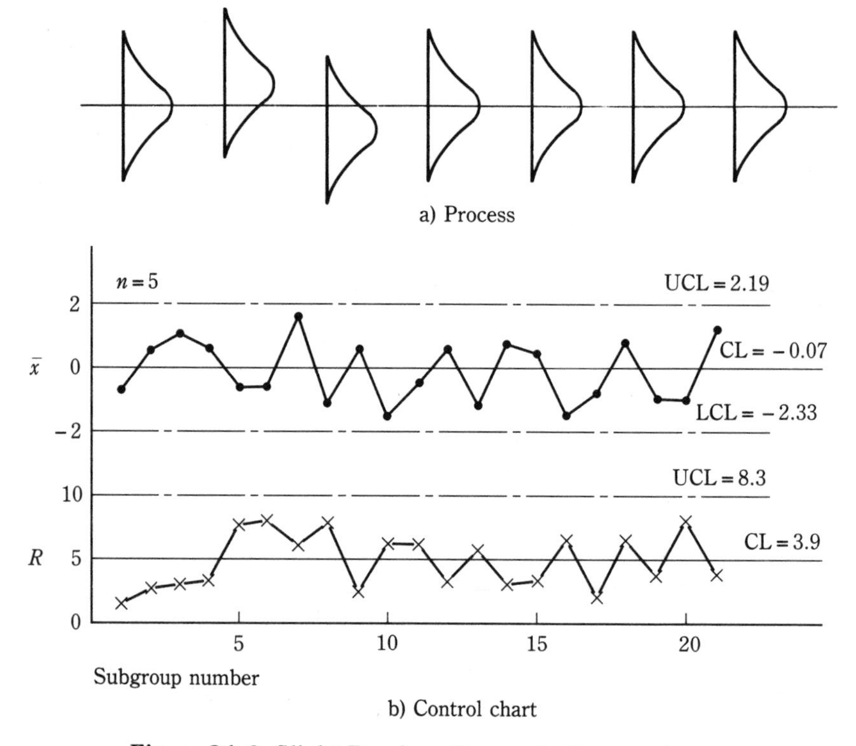

Figure 3A.9 Slight Random Change in Process Average

(5) When the process average changes according to a definite trend (see Fig. 3A.8)

Here you will note:

(a) R chart: no change.
(b) $\bar{x}$ chart: while continuing to scatter vertically, the points gradually sink. Some points fall outside the control limits, and runs appear.

(6) When the process average changes randomly

In this situation, there are two overall possibilities:

(a) The process average shows a slight random change (see Fig. 3A.9) and the following will be noted:
  i) R chart: no change.
  ii) $\bar{x}$ chart: the vertical movement of the points is still random, but it becomes more pronounced, and the number of points lying close to

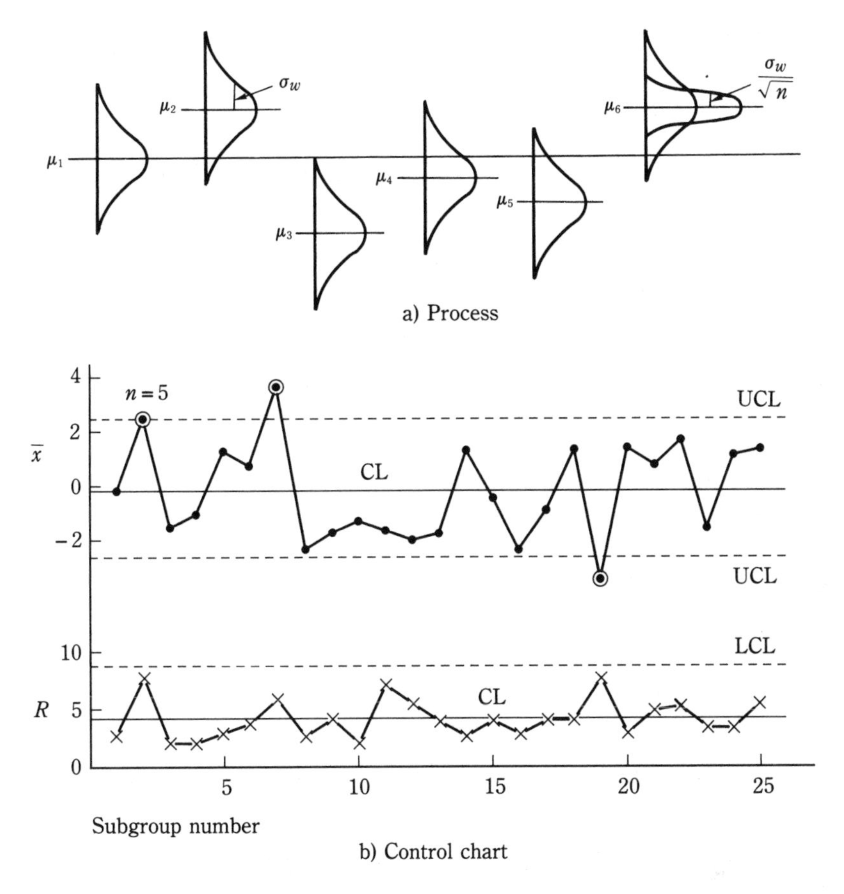

Figure 3A.10 Large Random Change in Process Average

the control limits increases. In this case, no points actually fall outside the limits.

(b) The process average undergoes a large random change (see Fig. 3A.10) and the following may be observed:

i) R chart: no change.

ii) $\bar{x}$ chart: the vertical movement of the points becomes more pronounced and the number of points falling outside the control limits increases.

Let us denote the within-subgroup variation by σ_w (the within-subgroup variance is $\sigma_w{}^2$), and the between-subgroup variation by σ_b (the between-subgroup variance is $\sigma_b{}^2$). The between-subgroup variation is an indicator of the variation in the process average, μ_i. Since the R control chart shows the within-subgroup variation and there is no change in σ_w, the chart shows the controlled state, enabling σ_w to be estimated from $\bar{R}/d_2$. The purpose of the R chart is to control the within-subgroup variation. However, the variation in $\bar{x}$, $\sigma\bar{x}^2$ depends on a combination of the variation due to the within-subgroup variation, $\sigma_w{}^2/n$, and the variation $\sigma_b{}^2$ due to the change in the process average μ_i. The following formula shows this relation:

$$\sigma\bar{x}^2 = \sigma_b{}^2 + \frac{\sigma_w{}^2}{n} \tag{3A.12}$$

In this case, if the process average is absolutely constant and the process is in the completely controlled state, $\sigma_b = 0$, and:

$$\sigma\bar{x}^2 = \frac{\sigma_w{}^2}{n} \quad \text{(completely controlled state)} \tag{3A.13}$$

Since the control limits on the $\bar{x}$ chart are drawn based on $A_2\bar{R}$ (i.e., the within-subgroup variation), the main aim of this type of chart is to detect changes in the process average and the between-subgroup variation σ_b using σ_w as a reference. This is the same for the p, pn, c, and u control charts and is one of the most important characteristics of the control chart.

Also, if we make a histogram from individual measurements (x), and call the standard deviation calculated from this S_H, then $S_H{}^2$ is given approximately by the following formula:

$$S_H{}^2 \approx \sigma_b{}^2 + \sigma_w{}^2 \tag{3A.14}$$

Meanwhile, if we multiply formula 3A.12 by n, we obtain:

$$n\sigma\bar{x}^2 = n\sigma_b{}^2 + \sigma_w{}^2 \tag{3A.15}$$

$$\text{Also,} \quad \hat{\sigma}_w{}^2 = (\bar{R}/d_2)^2 \tag{3A.16}$$

Thus, if we make a histogram of $\bar{x}$ and denote the standard deviation obtained from this by $S\bar{x}$, we have:

$$C_f = \frac{\sqrt{n}\,S\bar{x}}{(\bar{R}/d_2)} \quad \left(C_{f'} = \frac{s_H}{(\bar{R}/d_2)},\ C_{f''} = \frac{\sqrt{n}\,S\bar{x}}{s_H}\right) \tag{3A.17}$$

This usually has a value greater than 1, provided that $\sigma_b \neq 0$. Since C_f is a value that shows the approximate state of control of a process, it is called the "control coefficient." We can normally say that a process is definitely not under control if C_f is 1.3–1.4 or more. Also, if C_f is 0.8–0.7 or less, it means that data from processes with considerably different averages have been combined into a single group, as in the example described in subsection 9 below; in other words, it indicates inadequate subgrouping, with data from different sources mixed in the same subgroups.

(7) When the (within-subgroup) variation changes

The following possibilities exist under this condition:

(a) When the variation increases (see Fig. 3A.11), these changes may be noted:
   (i) R chart: the points show an overall rising trend, and some points fall outside the control limits.
   (ii) $\bar{x}$ chart: the points are still randomly distributed, but their vertical movement becomes more pronounced. However, the points are still distributed more or less equally about the central line. Some points fall outside the control limits.

(b) When the variation decreases (see Fig. 3A.12), these changes are observable:
   (i) R chart: the points show an overall decline, and the number of points below the central line increases.
   (ii) $\bar{x}$ chart: the points are still distributed randomly, but their vertical movement becomes less pronounced. The points are distributed approximately equally above and below the central line, and the number of points close to the central line increases.

   When the within-subgroup variation is thought to have changed in this way, it should be investigated by stratifying the data, preparing two separate R charts, and calculating two values of $\bar{R}$. See Section 3A.4 for how to do this.

(8) Stratified control charts

If the data of Figure 3A.7 are stratified into zones A and B_2, and separate control charts are drawn up and compared (see Fig. 3A.13), we observe the following:

(a) R chart: no difference
(b) $\bar{x}$ chart: both A and B_2 are in the controlled state, but they appear to have different averages. For a statistical test of whether or not there is a difference in the process averages of the two strata when data are

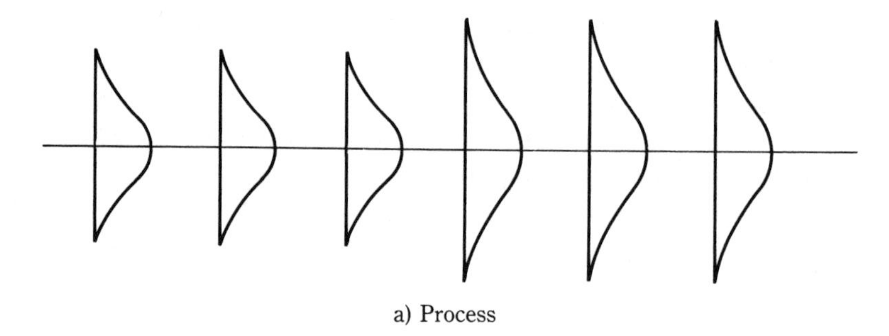

a) Process

$n=5$ UCL CL LCL

$\bar{x}$ 2 0 −2

R 10 5 0

UCL CL

5 10 15 20 25 30 35

b) Control chart

Figure 3A.11 Increase in Process Dispersion

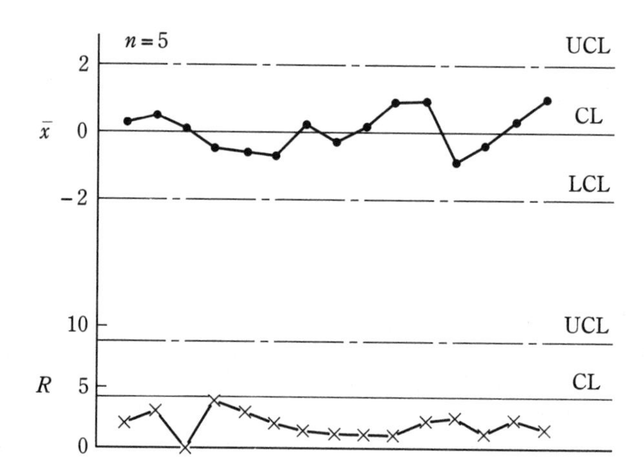

Figure 3A.12 Decrease in Process Dispersion

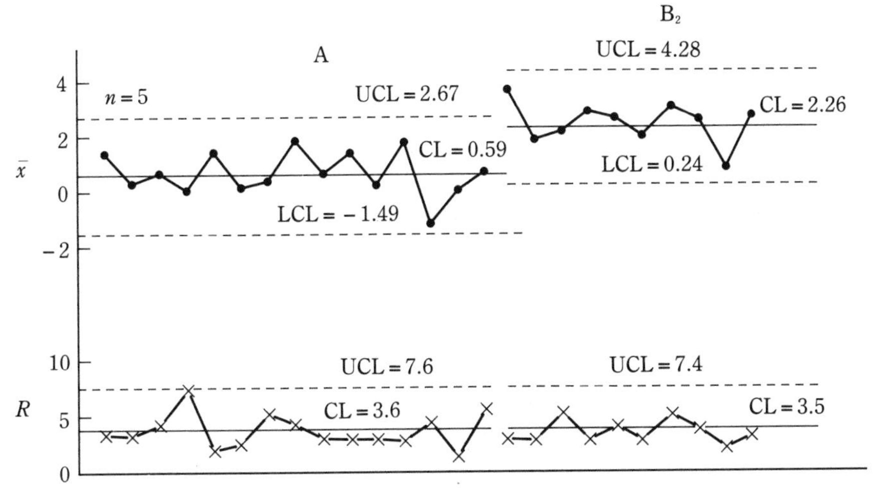

Figure 3A.13 Stratified Control Charts for Zones A and B_2

stratified in this way and separate $\bar{x}$ charts are drawn, refer to Section 3A.4.

(9) When two groups of data with extremely different process averages are grouped together (see Fig. 3A.14)

Under these circumstances, the following may be observed:

(a) R chart: the points cluster around the central line.

(b) $\bar{x}$ chart: the points cluster around the central line.

When this occurs, it often means that data from different sources have been included in the same subgroups, and that the subgroups contain data from processes with widely differing averages. This should be investigated by further stratifying the data within the subgroups in various ways.

3A.4 Methods of Testing Differences in Averages from Control Charts

When several stratified control charts are drawn, it is sometimes possible to judge intuitively whether or not there is a difference between the process averages or within-subgroup variation (σ_w) of the different strata at the time of drawing. However, it is best to use the data to test statistically whether or not we can say that a difference exists.

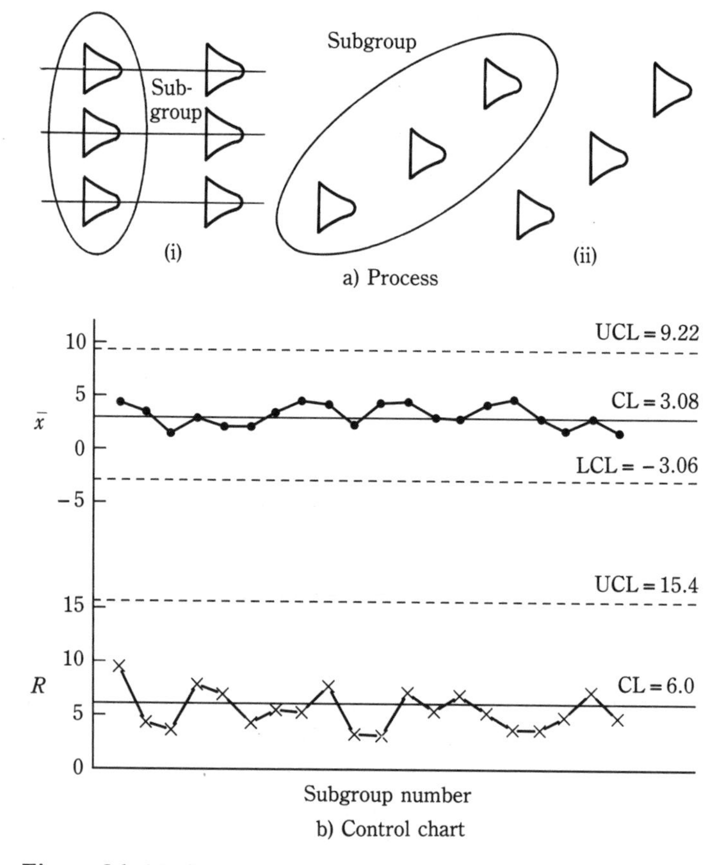

Figure 3A.14 Combination of Data from Populations with Three Very Different Distributions

(1) Testing for a difference in variation (σ_w)

There are three steps in this testing procedure:

Step 1: Prepare separate stratified control charts for data sets A and B (different subgroup sizes may be used), and calculate $\bar{R}_A$ and $\bar{R}_B$.

Step 2: Calculate F_0 from the following formula:

$$F_0 = \frac{(\bar{R}_A / c_A)^2}{(\bar{R}_B / c_B)^2} \tag{3A.18}$$

Where the bigger of $\bar{R}_A/c_A$ and $\bar{R}_B/c_B$ is used as the numerator. c_A and c_B are coefficients that depend on the subgroup sizes n_A and n_B and the numbers of subgroups k_A and k_B; they are given in Table 3A.3.

Table 3A.3 Supplementary Table for Testing Using Range

n \ k	1	2	3	4	5	10	15	20	25	30	$k>5$
2	1.0	1.9	2.8	3.7	4.6	9.0	13.4	17.8	22.2	26.5	$0.876k+0.25$
	1.41	**1.28**	**1.23**	**1.21**	**1.19**	**1.16**	**1.15**	**1.14**	**1.14**	**1.14**	**$1.128+0.32/k$**
3	2.0	3.8	5.7	7.5	9.3	18.4	27.5	36.6	45.6	54.7	$1.815k+0.25$
	1.91	**1.81**	**1.77**	**1.75**	**1.74**	**1.72**	**1.71**	**1.70**	**1.70**	**1.70**	**$1.693+0.23k$**
4	2.9	5.7	8.4	11.2	13.9	27.6	41.3	55.0	68.7	82.4	$2.738k+0.25$
	2.24	**2.15**	**2.12**	**2.11**	**2.10**	**2.08**	**2.07**	**2.06**	**2.06**	**2.06**	**$2.059+0.19/k$**
5	3.8	7.5	11.1	14.7	18.4	36.5	54.6	72.7	90.8	108.9	$3.623k+0.25$
	2.48	**2.40**	**2.38**	**2.37**	**2.36**	**2.34**	**2.33**	**2.33**	**2.33**	**2.33**	**$2.326+0.16/k$**
6	4.7	9.2	13.6	18.1	22.6	44.9	67.2	89.6	111.9	134.2	$4.466k+0.25$
	2.67	**2.60**	**2.58**	**2.57**	**2.56**	**2.55**	**2.54**	**2.54**	**2.54**	**2.54**	**$2.534+0.14/k$**
7	5.5	10.8	16.0	21.3	26.6	52.9	79.3	105.6	131.9	158.3	$5.267k+0.25$
	2.83	**2.77**	**2.75**	**2.74**	**2.73**	**2.72**	**2.71**	**2.71**	**2.71**	**2.71**	**$2.704+0.13/k$**
8	6.3	12.3	18.3	24.4	30.4	60.6	90.7	120.9	151.0	181.2	$6.031k+0.25$
	2.96	**2.91**	**2.89**	**2.88**	**2.87**	**2.86**	**2.85**	**2.85**	**2.85**	**2.85**	**$2.847+0.12/k$**
9	7.0	13.8	20.5	27.3	34.0	67.8	101.6	135.3	169.2	203.0	$6.759k+0.25$
	3.08	**3.02**	**3.01**	**3.00**	**2.99**	**2.98**	**2.98**	**2.98**	**2.97**	**2.97**	**$2.970+0.11/k$**
10	7.7	15.1	22.6	30.1	37.5	74.8	112.0	149.3	186.6	223.8	$7.453k+0.25$
	3.18	**3.13**	**3.11**	**3.10**	**3.10**	**3.09**	**3.08**	**3.08**	**3.08**	**3.08**	**$3.078+0.10/k$**

(ϕ in ordinary type, c in bold type)

Step 3: Compare the value of F_0 with the value of F $(\phi_A, \phi_B; 0.01)$, the value of F for degrees of freedom ϕ_A and ϕ_B and upper 1 percentile point from F distribution tables. If $F_0 \geqslant F$ $(\phi_A, \phi_B; 0.01)$, we can say that there is a difference between the within-subgroup variation σ_w of A and B at a significance level of 2%. If $F_0 < F$ $(\phi_A, \phi_B; 0.01)$, we cannot claim that there is a difference.

In the above,ϕ_A and ϕ_B are degrees of freedom that depend on n_A and k_A and on n_B and k_B. Their values are given in Table 3A.3. The values of F are given in many statistical tables.

(2) Testing for a difference in averages μ_A and μ_B

To carry out this test, one must first prepare an $\bar{x}$-R control chart stratified into zones A and B (in this case, the subgroup sizes must be equal), and test using the following formula:

$$|\bar{\bar{x}}_A - \bar{\bar{x}}_B| \geqslant A_2 \bar{R} \sqrt{\frac{1}{k_A} + \frac{1}{k_B}} \tag{3A.19}$$

When the above inequality is satisfied, we can say that there is definitely a difference between the process averages μ_A and μ_B of processes A and B. k_A and k_B are the numbers of subgroups in A and B.

$$\bar{R} = \frac{k_A\bar{R}_A + k_B\bar{R}_B}{k_A + k_B}, \text{ If } k_A = k_B, \quad \bar{R} = \frac{\bar{R}_A + \bar{R}_B}{2}$$

If the above formula is used, the following conditions must be approximately satisfied:

(a) The stratified control charts must show the controlled state.
(b) $n_A = n_B$.
(c) k_A and k_B must be sufficiently large; they must each be at least 10.
(d) There must be no difference between $\bar{R}_A$ and $\bar{R}_B$ as described in subsection 1 above.
(e) The original distribution must be approximately normal.

If these conditions are not satisfied, the precision of the test will deteriorate, and a different, more troublesome statistical test must be used.

It should be noted that when the difference in process averages is thought significant when carrying out process analysis, one must modify the individual measurements x_B or the averages $\bar{x}_B$ from the poorer set of data by adding $\bar{\bar{x}}_A - \bar{\bar{x}}_B$ so as to give the improved average, which would be obtained by taking action (e.g., $\bar{\bar{x}}_A$), and continue the analysis. A chart obtained in this way may be called a "modified control chart."

Chapter 4
PROCESS ANALYSIS AND IMPROVEMENT

4.1 Process Improvement and Control

As discussed in Section 1.5, if priority is given to recurrence prevention, control in its broad sense is not mere preservation of the status quo but a type of improvement. However, control consists more of fully exercising present capabilities and maintaining them at their optimum level, while improvement consists more of actually raising these capabilities. Control and improvement therefore appear to be different jobs, and this is what leads some people to say things like "Control is line work and improvement is staff work." In reality, however, no such clear distinction can be drawn. Many immediate-vicinity improvements, for example, are implemented through QC circle activities on the line (see Figure 1.18).

Capabilities are often not fully realized because the work at hand is not sufficiently under control and is not being performed reliably. When control is implemented properly, quality, processes, and other factors gradually improve, and process capabilities become fulfilled in terms of manpower, quality, quantity, delivery time, and cost. Furthermore, if work is not controlled (for instance, if there is great variability in routine operations), it will be impossible even to figure out where to start improving, and the resulting benefits will be unclear even if improvements are achieved.

Even when good improvement proposals are made, they often cannot be executed satisfactorily. Everyone rushes around shouting that they are eliminating defectives and increasing production, but in the end nothing is improved. This is because they are confusing control with improvement. If we want to make improvements, we must first have total control. Only when control is sufficiently well implemented do significant improvements become possible.

Also, though they may be scrutinized thoroughly and executed with control, even good improvement plans will come to nothing if efforts are not continued until a state of control has been achieved. An improvement can only be considered complete when it has become possible to maintain the improved work or process in the controlled state for a long period. Traditional improvements often wind up as mere emergency countermeasures, or are not carried through to the point where control has been established and are therefore left only half-finished.

If improvements are carried out at the whim of individuals whenever they happen to have what they think is a good idea, standards will break down. In the past, individuals on the shop floor would often work in non-standard ways on their own initiative, convinced they were doing things better. Far from making an improvement, however, they often made a change for the worse. Improvements should be effected systematically through formal procedures, involving first consultation with the relevant departments, then thorough technical and statistical analysis, and finally formulation of provisional standards or modification of existing ones before implementation.

As this discussion indicates, control and improvement are like the two sides of a coin or the two wheels of a cart; neither can be considered in isolation from the other.

An outline of the improvement process was given in Section 1.7.2. The present chapter describes how to instigate improvements and what points to look out for; the main focus is on processes. Most of the discussion is also applicable, almost as it stands, to quality improvement and new-product development.

4.2 Improvement Types and Procedures

4.2.1 Types of Improvement

There are many types of improvement, but I would like to consider them here under the following headings:

- Passive Improvements: these include reducing rework and adjustment, improving yields, cutting costs, and eliminating assignable causes.
- Active Improvements: these include producing breakthroughs, improving quality, improving process capabilities, highlighting and improving positive qualities with consumer appeal (sales points), and boosting sales.

- Immediate-Vicinity Improvements: these are improvements based on suggestions, improvements by QC circles, and improvements at the workplace.
- Priority-Based Improvements: these are systematic, prioritized improvements to eliminate major problems through cooperation on the company, plant, or section level, improvements by QC teams, and improvements by staff.

Goal-Oriented Improvements: included here are improving quality, halving defectives and simplifying work, increasing process capability and production capability, cutting costs, and shortening delivery times.
Method-Oriented Improvements: these include process improvement, improvements in operating methods and equipment, organizational rationalization, and revision of standards and regulations.

Each of the different types of improvement listed above is explained briefly below:

Passive improvements are a type of improvement that enables a process to exercise its full capability. It consists of doing what should have been done as a matter of course, e.g., reducing defectives and rework, making it possible to get by with simple adjustments, and listening to and satisfying previously neglected requirements of consumers and the next process.

Active improvements mean developing new products and areas of business and improving positive, attractive qualities, i.e., altering qualities, designs, and plans in various ways to improve ease of use and make consumers feel better and thus happy to buy. It is the type of improvement in which product reliability, durability, maintainability, and serviceability are improved by changing the raw materials and other factors, and quality and process capability are raised through the active development of new equipment (e.g., jigs and tools), products, processes, and systems.

There are numerous apparently irrational or inconvenient things in our immediate surroundings which we notice but do nothing about. *Immediate-vicinity improvements* are the type of improvements that result from small but useful ideas, individual creativity and ingenuity, and simple suggestions and opinions. Such improvements must come from shop-floor workers and operators or ordinary office workers. "Many a little makes a mickle," and a steady stream of these small improvements can produce a tremendous effect in time. They may not all be useful, but, like diamonds in the dust, they will contain some excellent ideas. This type of improvement activity primarily concerns improvements in the workplace and is carried out by groups such as QC circles.

Caution is needed, however, since this type of improvement tends to reduce defectives in one process but increase them in the next, or improve quality but lower production volume. Therefore, before suggestions of this type are implemented, they must be discussed with people from the next process and other relevant departments, and formally investigated from the overall standpoint.

In short, an atmosphere must be created in which everyone is constantly searching for improvements in their immediate surroundings and in which a steady stream of suggestions is proffered; suggestion schemes must be used to encourage this. However, people must bear in mind that what is best for their own domain is not necessarily the best from the overall viewpoint; on the contrary, it often actually creates problems in other areas.

Priority-based improvements are the type of improvement in which the most serious problems in a company, branch office, factory, or other organization are investigated systematically and tackled in order of priority as dictated by organizational policy. Problems of this nature must be solved through a team effort by everyone concerned, in all departments including design, materials, research, technical, manufacturing, and sales.

This type of improvement is also the most important type from the standpoint of promoting quality control. If everyone gets a taste of it—i.e., the taste of attacking problems by an all-out, concerted effort—quality control will advance rapidly. With this type of improvement, the company lays down its policy on priorities and devolves it to the various departments (technical, production technology, manufacturing, etc.), which then share responsibility for effecting the improvements. However, it is often a good idea to set up QC teams or project teams to implement the improvements under the heading of "priority control."

Goal-oriented improvements and method-oriented improvements are distinguished somewhat in Table 1.3, which lists some differences between goals and methods. The decision as to what needs to be improved, where the problems lie, and what the goals are will determine the methods through which improvements are to be effected. Goals must be set and needs established before improvements in methods are considered. In other words, goals must always come first. If methods are put first, though many legions of regulations and standards are prepared, or (in office automation) computers are brought into service, the result will be QC or office automation in name only, and nothing much will have been achieved.

People have traditionally been obsessed with improving methods and means. They still tend, for example, to put forward irrelevant improvement plans, reshuffle their organizations, prepare masses of regulations, and formulate and revise operating standards without achieving the slightest benefit in spite of all their hard work. This is why it is so important to clear up the confusion of the ends with the means and clarify our goals.

Setting goals might sound easy, but actually establishing specific goals (which are a type of substitute characteristic) for a workplace requires various technical and statistical analyses. And this is only half the job. The other half is to provide specific methods and tools for achieving the goals, i.e., to improve the processes by which the goals are achieved. If this is not done, we will end up with the old-style "management by objectives" or "exhortational management," which cannot really produce results and make them stick. Specific methods, tools, and process improvements for achieving the goals are essential, and these process improvements require sound analysis. Intelligence, experience, technical knowledge, and statistical methods play a large part in this.

Nevertheless, it is a common assumption that motivation and goals will naturally gel once form and method have been established. Another approach is thus to establish forms and standardization as a starting point during the introductory period of QC, particularly in organizations with an old-fashioned atmosphere. If this is done, however, it is important to move on after a short while to goal-based improvement.

This chapter focuses on the improvement of quality of conformance (see Figure 1.7); the improvement of quality of design and planning is discussed in Chapter 6. The considerations in this chapter apply equally to improving the quality of "hard" manufactured goods and "soft" products such as services.

4.2.2 Obstacles to Improvement

Progress and development only occur when various types of improvement, starting with quality improvement, are implemented. In times of rapid technological innovation and economic change, such as today, maintaining the status quo and failing to carry out innovative improvement actually means slipping back. The history of the rise and fall of companies proves that an organization will drop out of the race if it continues to adopt the outmoded approach of either acting recklessly ("crossing a stone bridge without checking it") or overcautiously ("checking a stone bridge, then not crossing it"). In this day and age, a more apt approach would be, "How quickly can we act with caution?," or "How quickly can we check the stone bridge and then cross it?"

Why, then, are active improvements and advances not implemented? The main enemy of improvement is people, and some of the barriers they erect are listed below:

(1) Negative attitudes on the part of those in authority, starting with company presidents and going on down through operations department directors, factory managers, and sales directors to section managers. This is actually the crux of the problem, but we can go on in further detail.
(2) Believing that everything is hunky-dory and no problems exist.
(3) Believing that things are going better in one's own preserve than anywhere else.
(4) Believing that "the way it's always been done" is easiest and best; trusting only one's own experience and nobody else's.
(5) Being satisfied with the status quo.

(6) Thinking only of oneself and one's own area of responsibility; being unable to listen to the views of others.
(7) The absence of stimuli from outside the department or company.
(8) Resignation, jealousy, envy.
(9) Bad judgment on the part of superiors and directors; fear of losing face.
(10) Sectionalism.
(11) Cutting others out in pursuit of one's own ambitions.
(12) Inadequate technical and statistical knowledge, intellect, resourcefulness, originality, judgment, and practical ability.
(13) Doing nothing through fear of failure, since mistakes often occur when things are changed.
(14) The practice of superiors always criticizing their subordinates' mistakes and never praising them for their successes.
(15) The attitudes of those engaged in office work, and in workplaces and labor unions that lack understanding, since these can be the most old fashioned of all.

These are just some of the many obstacles standing in the way of the desire to carry out improvement, and most of them are erected by people. To break through these barriers requires self-confidence, courage, a spirit of cooperation, an ardent pioneering spirit, and the motivation to make breakthroughs, together with the right tactics, strategy, and techniques, and unceasing effort.

The greatest obstacles to new products, new methods and other improvements are within your own company! Without overcoming this "fifth column," progress is impossible.

4.2.3 The Basic Conditions for Improvement

To promote improvement, we need the opposite of the obstacles discussed in Section 4.2.2. The basic conditions for improvement are chiefly a question of people's attitudes and include the following:

1) Managers must take the lead and demonstrate their desire for improvement. They must communicate basic policy (company policy, etc.) and specific goals and imbue the entire company with an atmosphere charged with a zealous pioneering spirit and an eagerness to achieve improvements and breakthroughs.
2) The right people must be placed in the right jobs, and authority should be extensively delegated.
3) Those in positions of authority must take the lead in the drive for im-

provement, constantly searching for better things and better methods, and superiors must take responsibility for their charges' mistakes. The "protruding nail will be pounded down" type of atmosphere must also be eliminated.

4) Systems for actively recording and carefully investigating complaints and problems from both within and outside the company should be set up, and an atmosphere that encourages this must be created.
5) People should be receptive to stimuli from outside sources such as free competition, a recession, liberalization of trade or capital, the appointment of outside directors, audits and advice by consultants, consumer complaints, or showing things to staff or people from other departments and getting their views.
6) A suggestion system should be started, creativity and ingenuity encouraged, standards revised, and brainstorming sessions held.
7) Personnel should periodically be reshuffled and organizations improved.
8) Systems of reward and punishment, especially awards systems, should be clarified.
9) People should be given the chance to experience the spirit of cooperation and teamwork.
10) Education must be carried out, especially thorough education in QC thinking and methods.

In short, all the employees of your company, or if that is impossible, at least everyone in your own workplace, should be filled with tireless fighting spirit and creative dissatisfaction with the status quo. They must be driven by a pioneering spirit that keeps them moving forward, putting constant pressure on their superiors to break out of the mold. New-product development, process control, and improvement all depend on people. Unless people change the way they think and feel, there will be no continual improvement and progress. However, although attitudes are important, motivation campaigns alone are not enough; you cannot expect to win a battle without putting up a fight. So, as discussed below, proper technology and statistical methods must be used for process analysis so that improvements can be implemented scientifically based on an accurate grasp of the facts.

4.2.4 Process Analysis and Improvement Procedures

To implement improvements, we must analyze processes, which are collections of assignable causes. Let us start by listing the goals of process analysis (this includes analyzing work methods as well as actual processes) so as to avoid confusing the ends with the means, as discussed in Section 4.2.1. From a broad standpoint, the goals of process analysis, which we must focus on when analyzing processes, are as follows:

(1) To formulate business plans.
(2) To implement quality design.
(3) To assure quality and reliability.
(4) To improve processes.
(5) To control processes (stabilization and optimization).

Let us reiterate the procedure for process analysis:

(1) Perform investigations to reveal problems.
(2) Decide which problems to tackle and set targets; identify the current situation.
(3) Decide on the organizational structure and assign responsibility for implementing improvements.
(4) Identify the current situation in detail.
(5) Investigate improvement methods, i.e., plan breakthrough tactics (cause-and-effect diagrams, process capability studies, etc.).
(6) Prepare draft plans and provisional standards.
(7) Perform preliminary trials, check results, revise standards, and implement control.
(8) Check results.
(9) Instigate recurrence-prevention countermeasures, standardize, make permanent fix to prevent slipping back.
(10) Establish control.
(11) Review progress and consider remaining problems.
(12) Make future plans.

In short, an improvement can only be considered complete when we have gone from discovering a problem to achieving the desired situation, and this situation has continued in a state of control for an appreciable time — normally a year. We may not relax until control has been maintained for at least a year.

Depending on the situation, the third step (setting up an organization for improvement) may come before or after the problem has been determined (see Section 4.5). The individual steps are explained in the following section.

4.3 PROBLEM-FINDING INVESTIGATIONS

4.3.1 General Hints

Please refer to Sections 1.5.2 and 1.7.3 for how to discover problems and decide which ones to tackle.

(1) Discovering problems is the duty of managers

People in many businesses and workplaces make no attempt to think about or investigate possible problems; instead they run around in circles doing this and that in an attempt to cope with the unforeseen breakdowns and accidents that crop up daily. Typically, they spend their time scrambling to accommodate demands for temporary production increases or rushing about trying to deal with production decreases, modifications to plans, and trivial complaints.

Many people in leadership positions have forgotten their real job, which is to delegate authority, stay fully in control, and create for themselves as much free time as possible in which they can reflect calmly and deeply and decide what the biggest problems are in their own domains (and, from a broader point of view, for their company as a whole) and what should be done in the future. The higher their position, the more they must think about the future. It is my view that department managers should be thinking about what should be done at least three to five years ahead.

Policy is frequently changed when managers get bees in their bonnets and become deluded into mistaking small problems for big ones because they have only a smattering of superficial information. Drastic solutions of the really serious problems and innovative improvements are impossible with this approach. Those in staff and subordinate positions naturally share responsibility for collecting and organizing relevant information, feeding it to their superiors, and persuading them of the right courses of action; and all employees are responsible for searching out areas that require improvement and reporting these to their superiors. At the end of the day, however, it is vital for those in positions of authority from company president down to group leader to be constantly problem-conscious.

Whether you say your company has no problems or many problems, neither statement makes a distinction between important problems and unimportant ones.

(2) Collection of data and information for exposing problems

It is the responsibility of a manager's staff or subordinates to gather the data and information needed for exposing problems. As things stand, there is either not enough of this type of data, or even if such data are available, they are suppressed or deliberately doctored along the way, while some are passed on too late or only after a major crisis has occurred. No matter how many excellent complaints-handling regulations are laid down or how many complaints are recorded by sales departments, they are only the tip of the iceberg and are often received too late. Consumers' latent complaints are of course not collected, and information on those complaints that are recorded becomes distorted or lost as it passes from retail store to wholesaler to sales representative. Because of strong sectionalism, the suggestions of people who really want to give their opinions do not reach the ears of those who need to hear them. Companies or departments often fail to detect serious problems and to set priority policies because the data and information they need are inadequate or nonexistent.

There is no way quality control or any other type of control can be implemented without having the relevant numbers and amounts (i.e., the facts) at one's fingertips—e.g., if the numbers of defectives or reworked products at a factory are not known, if stock inventories for bottles, raw materials, etc., do not balance, or if the number of products made is unclear. Such a lack of information makes it very difficult to determine where problems lie. To achieve the goal of identifying problems, we must obtain accurate information, even if only on samples at first.

(a) Except in special cases, the information we need consists of data collected over a certain period of time. We must not, as people frequently do, tear our hair over each day's production figures, fraction defective, number of complaints, etc. Problems are turned up from data spanning periods of, for example, a week, a month, an accounting term, or a year; lower-level managers should normally use data from a shorter period, while upper-level managers should use data from a longer period. If this is not done, they will end up being led a merry dance by sudden, unforeseen problems or problems of control, and will find themselves running all over the place dealing with trivialities. The many profitable problems will then tend to be neglected.

(b) The data used to uncover problems should consist mainly of data on characteristics and results (quality, volume, cost, profit, etc.), not causes.

(c) Such data must of course be stratified in a way that facilitates analysis. Even a little each day is sufficient, but it is vital to collect data stratified with the aim of revealing problems. The data are then analyzed using tools such as Pareto diagrams, frequency distributions, check sheets, graphs, and control charts.

(3) Grasping the status quo

To reveal problems, we must obtain a clear picture of the current situation. We must not be misled by untrustworthy data or information; it is vital to take a good look at the actual "scene of the crime" and get a sound grasp of the status quo. People often went wrong in the past through overcompensating because they rushed to deal with causes without really understanding the status (i.e., the effects).

(4) Use of pooled knowledge

It is also a good idea to amass proposals for improvements through the use of a suggestion system and procedures for offering opinions. While the desirability of such a system is obvious, and it should be promoted vigorously, it is also useful to go further back toward the source and have large numbers of people state the problems and difficulties they are facing. To achieve this, I recommend that the suggestion system include statements of problems as well as improvement proposals. The data on these problems should then be organized for analysis in the form of Pareto diagrams. It would probably also be helpful to hold brainstorming sessions to identify the biggest problems.

(5) Problems should be identified in monetary terms

Whenever possible, problems should be expressed in terms of their common denominator, money. Rough estimates are acceptable, but problems should really be identified through cost accounting and cost control, and providing this kind of information is the most important service of a company's accounting and cost control sections. Even when the total number of defectives is small, critical defects often constitute a big economic loss. Cost sheets must be arranged so as to make it easy to see whether unit costs are too high and what factors have the greatest effect on yield, fraction defective, rework, adjustment, machinery operation rates, etc.

(6) Location of problems

Problems exist in places where people have become resigned to the situation or are convinced that nothing is wrong. They exist wherever there are chronic losses (see Section 4.3.3).

4.3.2 How to Stratify

Neither improvement nor control is possible without stratification. I have emphasized over and over that stratification is needed for control, for detecting problems, and for studying improvement measures, and I would now like to discuss the general principles of how stratification is done.

(1) When data are collected, they should be stratified according to the different conditions, causes, locations, or lots that appear likely to give rise

to defectives, losses, and other problems. For example, we could stratify by type of defective, type of defect, raw material, day, shift, time, group, person, machine, process, operating method, weather, measuring instrument, jig, tool, or any of many other factors. Data on overall numbers of defects and rework alone are not terribly useful. We should collect data using check sheets or similar methods and stratify as much as possible.

(2) Parts, products, containers, etc., should be identified with numbers, cards, chits, colors, or symbols to keep them separate, and materials and products should be passed through the process in separate lots or boxes.

(3) Everyone should be careful to keep lots separate by devising suitable methods of moving materials around the factory and designing good product storage layouts and systems.

(4) Defectives, rework, scrap, etc., should be sorted into separate boxes according to the nature or cause of the defective or fault. A specialist should then make occasional rounds to check these boxes and take records.

(5) A good chit system should be set up.

(6) Analytical inspection (see Section 6.8) should be carried out.

Many other ideas are possible, but as long as everyone concerned (including those on the shop floor) is aware of the importance of stratification, and if a little care is taken over the chit system, lot segregation, method of transport, etc., stratification can be achieved relatively easily.

4.3.3 Graphs

Some useful features of graphs are that they show changes over time clearly, enable us to understand things intuitively more readily than a mass of figures does, and they make it easier to spot abnormal changes. Control charts are even easier to understand. However, beginners are advised to get accustomed to using graphs. If possible, control limits should be shown and out-of-control points circled in red to indicate the need for care. Graphs can be interpreted in many different ways, but Figure 4.1 shows some typical ways in which defectives occur.

Figure 4.1(a) shows that defectives increased sharply, beginning on a particular day. In cases like this, the cause is often quickly found by looking carefully to see whether some change took place around the time of the increase.

Figure 4.1(b) shows periodicity. Defectives often increase like this on Mondays, Saturdays, and paydays or the day after.

Figure 4.1(c) shows defectives increasing sporadically and unexpectedly. Such a situation is often misinterpreted to indicate a big problem, but this type of problem is often control-related and can be solved by instituting stricter control.

In Figure 4.1(d) the fraction defective is about 5% to 9% every day and appears to be more or less stable. Defectives of this type are called "chronic defectives," and because they are chronic, people are often resigned to them as natural and inevitable. However, such defectives frequently conceal major problems. It would be a good idea to express things in money terms and do some trial calculations to see how much profit could be made if the average daily fraction defective $\bar{p}$ were reduced from 7.4% to 3% or less (or even if the $\bar{p}$ of 4% in Figure 4.1(c) were reduced to 2%).

The above ways of interpreting graphs are also applicable to control charts.

4.3.4 Exposing Hidden Defectives and Latent Complaints (See Sections 1.4.4 through 1.4.5 and Figure 4.2)

What a company calls "defectives" before it starts quality control is just the tip of the iceberg. In other words, it has more than ten times as many defectives and problems as it thinks it has. As a company starts gradually to uncover defectives and complaints, more hidden defectives and previously unnoticed problems appear in rapid succession. Supervisors and managers who do not feel that rework, adjustment, and minor dissatisfactions constitute defectives or complaints are completely undeserving of their titles. It is all right to start by tackling actual defectives—i.e., those that have appeared on the surface—but when QC has sunk in to some degree, we must turn our attention to uncovering and attacking hidden defectives and latent complaints. If actual defectives and complaints are tackled in earnest, hidden defectives and complaints will automatically come to light.

For example, does your company classify rework as a type of defective, and do you consider adjustment work to be a waste of man-hours? Do you calculate the true go-through rate? Are you in fact achieving your standard yields, standard amounts of material, labor, and mechanical power required for production of a product unit, standard man-hours, and standard operation rates? Are these standards high enough? If you think along these lines, you will find hidden defectives and latent complaints in areas you had simply accepted in the past.

When you start looking for problems in this way, they will come so thick and fast that you may be overwhelmed. You will certainly have your work cut out for you, but once you start looking forward to tackling them, I would say you have a pretty good understanding of QC.

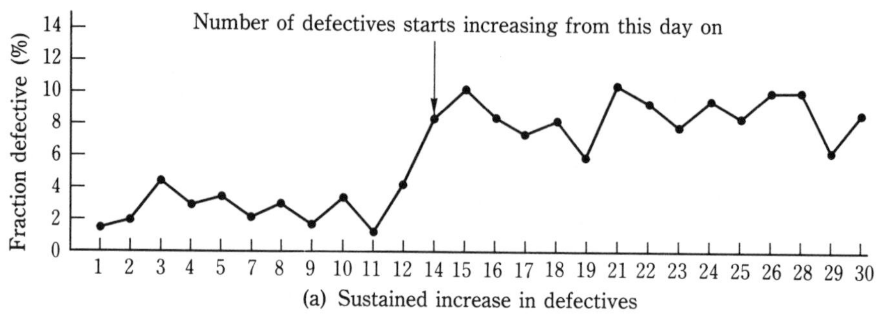

(a) Sustained increase in defectives

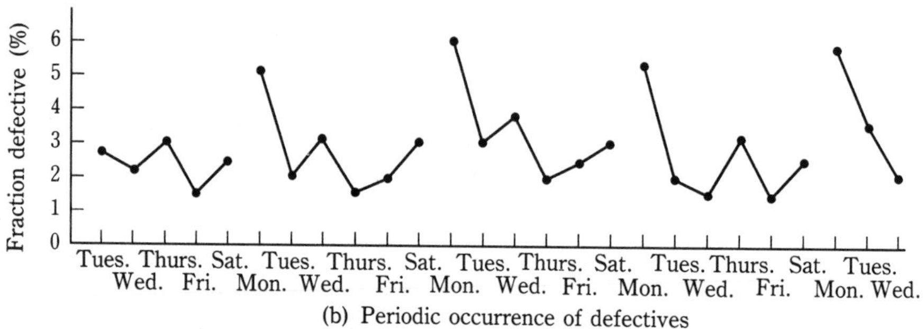

(b) Periodic occurrence of defectives

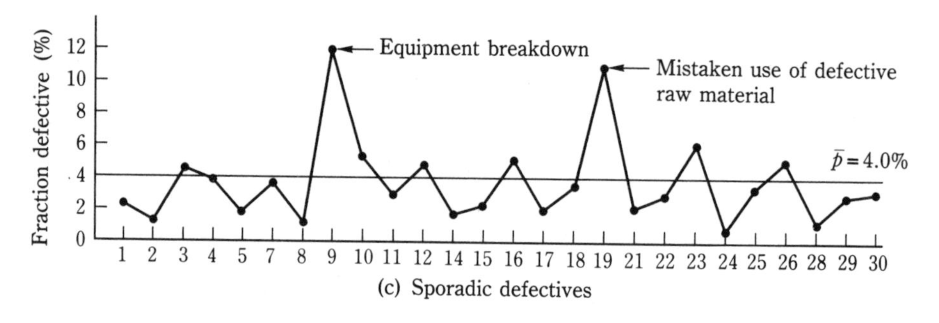

(c) Sporadic defectives

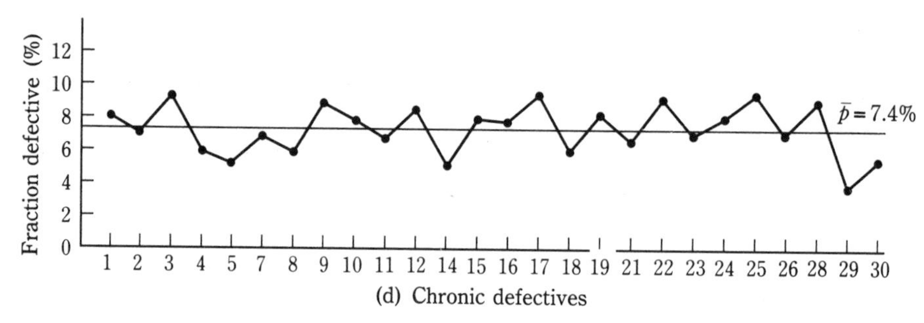

(d) Chronic defectives

Figure 4.1 Graphs Showing Occurrence of Defectives

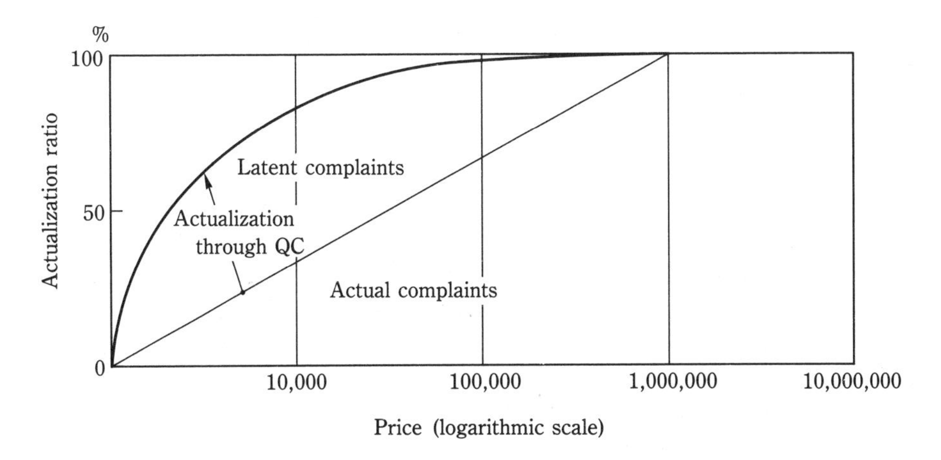

Figure 4.2 Actualization of Latent Complaints

4.4 Deciding what Problems to Tackle

If we know what a particular problem is and what our objectives are, the problem is already half solved; conversely, it is impossible to solve a problem unless we understand its significance and know our objectives.

Some points to bear in mind when deciding what problems to tackle are given below (see Section 1.5.2).

(1) Who should be allowed to have a voice in selecting the problems to tackle? It is important for this to be discussed at QC committee or QC circle meetings. Secrecy is not in order.
(2) The final decision should be in accordance with managerial policy. Everyone in a position of authority should have such a policy.
(3) Decisions should be the result of reasoned argument based on hard facts and data. One must have a firm grasp of the status quo.
(4) Policy should be decided from a broad, comprehensive perspective. One must think about the future and consider how policy meshes with long-term plans.
(5) It is important to think concretely and quantitatively, not in the abstract.
(6) One should always think about breaking out of the status quo.
(7) The focus is on objectives, not methods (see Table 1.3).
(8) Policies must be consistent.

(9) One must be goal-oriented, not organization-oriented.
(10) Chronic problems are serious problems.
(11) The Pareto principle is ever important: Don't waste time carping at trifles.
(12) Worrying too much about what might or might not happen is fruitless. It is said that a product under development is premature if 10% of the company are in favor of it, and has missed the boat if 50% support it. If the probability of success is 50%, take the plunge. Superiors should praise their subordinates' successes and take responsibility for their mistakes. People do not thrive when their successes are taken for granted and their mistakes criticized.
(13) A priority orientation will allow the number of priority problems to be kept to a minimum.
(14) In principle, the problems to be tackled should not be decided bottom-up style but in accordance with policy from above. Since superiors may not have all the facts, however, subordinates should steer them in the right direction using facts and data.
(15) Once a decision has been made, it should be printed out and made known to all employees. All employees should be made equally aware of and informed about the decision, and a consensus of opinion should be obtained.
(16) As well as setting quality, quantity, and cost targets, one must set time targets and deadlines.
(17) One must decide what will constitute an improvement and how it will be appraised, i.e., set targets and decide on evaluation and measurement methods. This will be the result of items 5 and 7 above.
(18) One can never worry about losing or saving face.

The choice of priority problems must be approved by senior management and officially recorded. Progress control must then be handled by having people submit regular written reports, holding briefing sessions, etc. In addition to tackling these priority problems, each department should, of course, if time permits, independently identify the important problems in its own patch and promote steady improvement through activities such as QC circles, starting with the problems that it can handle by itself.

If QC has not yet become well-established or if difficulties are being encountered in getting it across, a suitable process should be selected and analyzed and good results achieved for all to see. In this type of situation, it is best to choose one of the following types of process:

(1) A process overseen by keen middle managers.
(2) A process on which data would be comparatively easy to collect.
(3) A process in which the lot history or data history is already fairly clear or is easily clarified.
(4) A process plagued by more problems than any other in the factory, or a process not previously thought to give trouble or have any problems.
(5) A process shown by control charts to be out of control.

Ultimately, however, all work must be analyzed, trials carried out, and standards created or revised until every operation can be controlled by means of control charts and other statistical tools. This process of analysis and improvement must then be continued for as long as the company survives.

4.5 Organization for Process Analysis and Improvement

Simply ordering people to improve quality or sell more is not enough; things will not improve through willpower alone. When improvement relies only upon motivation, a time always comes when enthusiasm flags. As well as telling their people what they want to do—i.e., giving them policies and objectives—senior managers are responsible for creating a setup that makes these objectives achievable. Such an organization is sometimes established before a problem-finding investigation is started and sometimes after the problems have been determined. The former is generally true if the work is to be performed through the standing organization or through QC circle activities.

From the organizational viewpoint, process analysis and improvement can be carried out in the following ways:

(1) As a part of routine work by the standing organization.
(2) Through QC circle activities.
(3) By QC teams (i.e., task forces or project teams).
(4) Under the project engineer system or project manager system.

A committee system is also a possibility but is not recommended, since it usually blurs responsibility and authority and delays action.

Whichever system is used, the experience of solving a problem and taking control, the memory of how the problem was solved, and the taste of victory are all extremely important. People steadily develop by accumulating such experiences. Both QC circles and QC teams gradually learn to solve more and more difficult problems, and their leaders are the ones who grow the most.

4.5.1 Process Analysis and Improvement through the Standing Organization

If a problem concerns only one particular department, it can be solved by the standing organization as a part of their everyday work. However, problems that can be solved by just one department are usually not very serious.

The standing organization is also used when policies and objectives for defect reduction, cost reduction, etc., are broken down and apportioned among different departments in the form of a business plan. For instance, if a 30% cost reduction were planned, Pareto diagrams could be used to break down the objectives, and the company as a whole or the plant as a whole would set goals for each department. The responsibility for the cost reduction could be divided up so as to give, for example, 15% to the planning department and 5% each to the purchasing, sales, and manufacturing departments, and analysis and improvement would be implemented through the standing organization of each department. When this is done, the manager of each department is responsible and will further break down the objectives and apportion them among the sections and subsections under his or her control.

However, even with such an arrangement, if a problem crops up that involves another department, department and section managers are responsible for establishing good lines of communication to allow the work to proceed. Because sectionalism is often strong and horizontal communication poor, department heads must reorient themselves psychologically to this task and put special effort into it. When a problem closely concerns more than one department or when sectionalism is particularly prevalent, the job should be tackled by either using functional committees or creating QC teams.

4.5.2 Process Analysis and Improvement through QC Circle Activities

Because improvements implemented through QC circle activities (see Section 1.10) generally deal with fairly immediate problems in individual departments or workplaces, circle leaders are usually the point of contact between management and workers. However, since the emphasis in QC circle activities is on independence, it is better for managers not to give them too many detailed instructions. QC circles detect problems and perform analysis and control independently and voluntarily, under policy guidelines from above. However, if a department wants, for example, to lower its fraction defective by 2% or raise its go-through rate by 5%, this can be intimated to QC circles and they can be asked to tackle it as their problem theme. When this is done, each circle should of course be allowed to set its own objectives in terms of which types of defective or rework to reduce after collecting data and analyzing the situation.

Managers attend QC circle meetings as observers, receive QC circle activity reports, support the circle's activities, and check their work. If a problem involving another department crops up during QC circle activities, management can liaise with the other department, or, if the circles are experienced and capable, joint circles can be set up between departments to give them an opportunity to work together to solve the problem.

Whichever method is adopted, the responsibility for encouraging and promoting QC circle improvement activities always lies with management (i.e., factory, branch, department, section, and subsection managers).

4.5.3 QC Team Activities

QC team activities are similar to QC circle activities in that they are both carried out by small groups, but they are intrinsically different and must be considered separately. They are often lumped together as small-group activities, but they are run, controlled, and evaluated in completely different ways (see Table 4.1).

A QC circle is a small group of people from one workplace that continues its activities for as long as the workplace exists. A QC team, on the other hand, is a type of project team made up of the people from different departments needed to perform analysis, improvement, and control activities on a particular project. It is a temporary group that disbands once the problem has been solved. A QC team, which may also be given a title such as "task force," is thus a small group made up of people from several different workplaces formed to address a particular objective. If the project is a big one, the "engineer-in-charge" system or "manager-in-charge" system explained below is used.

For example, if the problem is a particular type of defective, a QC team might be made up of three people: someone in charge of the workplace directly concerned (someone, like a supervisor, who knows the workplace intimately and is responsible for control); someone responsible for engineering and designing the product (not a department or section manager but the person actually responsible); and a QC staff member. Depending on the problem, it is of course acceptable to have four or more people in the group, but because of the nature of group activities, the fewer members the better.

The following are some pointers for running QC teams:

Table 4.1 Differences between QC Circle Activities and QC Team Activities

	QC circles	QC teams
1. Aim	Depends on basic philosophy of QC circle activities	Problem-solving and control
2. Features	Bottom-up style, basic spirit of principles of QC circles	Top-down style, project-team-type management
3. Topics dealt with	Workplace problems selected autonomously with advice from superiors	Mainly directed by superiors in line with policy management
4. Members	10 or less (3—7) from same workplace	20 or more, with 3 to 7 different members changing by topic from different workplaces
5. Formation	Voluntarily, with advice from superiors	As directed by superiors
6. Period of activity	Continuous	Disbanded on completion of topic (improvement and control); if possible, continued for a year or more after control is established
7. Relation with superiors	Some relation (mainly advice)	Close relation (instruction and delegation of authority)
8. Evaluation	Teamwork, cooperation, division of responsibility, ingenuity, use of statistical methods, results, number of meetings and attendance rate, number of problems solved annually	Benefits, subsequent state of control

(1) Selecting project themes and objectives:
A project theme and objective are often decided as the policy of the company, factory or other organizational unit, and a team is then officially formed. This differs from QC circle activities. Naturally, a team or project is sometimes decided as a result of a suggestion from below.

(2) Selecting team members: Nominate a small number (three to seven) of people such as young line supervisors, technical experts, and QC specialists who are well versed in the topic to be tackled. Depending on the scope of the project, the team leader should normally rank lower than section manager and should, if possible, be a foreman or supervisor. The team members' superiors should lighten their daily workload, or if possible, assign them exclusively to the team.

(3) Running the team: It is important to delegate full authority to the team members. This means that all members must be given authority to act without obtaining permission from their superiors, and the job should be left up to them. QC teams are formed in order to smash through sectionalism and solve problems speedily, and they would make little progress if their members had to get approval from their superiors for

everything they did. Some old-fashioned middle managers oppose team activity because it disrupts the standing organization, but they are either mean-spirited people more worried about preserving their own power than about how well the work is going, or are too self-important. In any event, if a department or section manager refuses to be persuaded, a steering team made up of people on the level of department manager can be created to oversee the QC team.

(4) Team duties: The team's work is not over until the problem has been solved, a system of control has been established, and the relevant process is in a stable state of control and can be maintained in that state. A problem cannot be called solved if it is only solved temporarily and things rapidly revert to their original condition. Since there are normally monthly and seasonal variations in many places, we say that a team should remain active for at least a year. At first, the team would of course meet at frequent intervals of, say, once a week; then, when things had settled down to a certain extent, they could meet about once a month to check on results, the state of control of the process, etc. The team members are responsible for the project, but, if necessary, they may link up with QC circles in the workplace and put them in charge of some of the work.

(5) Evaluation: Once a month or at other set intervals, the status of the improvements and the state of control should be evaluated and written up in a report to be studied by the steering team or TQC committee. Managers should occasionally attend meetings to listen to reports and check results.

(6) QC teams should be used particularly extensively during the initial introduction and promotion of QC. They should carry out analysis and improvement in the form of specific projects, create systems enabling control to be implemented using control charts, and build up a control organization by providing each workplace with a succession of usable control charts one by one.

(7) When the problem concerns suppliers or customers, someone in charge of QC should act as an intermediary in helping both of them to create QC teams.

4.5.4 Engineer-in-Charge and Manager-in-Charge Systems

The engineer-in-charge system consists of appointing a full-time engineer to handle a particular project and giving him or her authority to form a task force of the necessary people. In this approach, the work is carried out from beginning to end under the responsibility and authority of one person. This system is used to deal with relatively large projects, such as the development of a new product or technology or the construction of a new plant. The engineer appointed is at least of department or section manager level, and, depending on the situation, could be a company director. In the development of a new product, for example, the engineer-in-charge might start with a task force of five designers, two researchers, one production engineer, and one QC staff member, and reduce the number of designers to three and add two more production engineers, an accountant, someone from sales, someone concerned with subcontracting, etc., as the work progresses. This system works like a QC team, since the engineer-in-charge and the people he or she appoints are taken right away from their original departmental work and other duties.

4.6 Analyzing Problems and Preparing Improvement Plans

4.6.1 Basic Attacking Attitude

Good control is not achieved simply by preparing a bunch of operating standards, regulations, etc., and drawing up a lot of control charts. We must thoroughly analyze past and present data, correctly understand work and process conditions, and obtain real, factual technical information on processes. Without thorough analysis, improvement and standardization are impossible, and we will be unable to achieve good control or prepare control charts usable for control.

(1) Data

The following types of data are used for problem and process analysis:

(i) Routine past data, collected by existing methods.
(ii) Routine data specially collected for easy analysis, e.g., stratified data or data that correspond with other factors.
(iii) Fresh data collected by specially-designed experiments.

Types (i) and (ii) consist mainly of data taken under existing operating and working conditions, while type (iii) consists mostly of data obtained when trying out work under new conditions. Since the first type of data usually contain a great deal of information, it is best to start by analyzing this type thoroughly before moving on to types (ii) and (iii). However, when the first type of data is from places that lack QC awareness, they may be difficult to analyze because such places often fail to stratify or to collect corresponding data; in other words, the history of the data is unclear. In such cases, the second type of data must be obtained. It is usually best to proceed to the third type after the analysis of the first two types has been completed.

This book gives a simple discussion of analyzing data of types (i) and (ii). The collection and analysis of the third type of data should be studied in specialist works on design of experiments.*

(2) Identifying the status quo, actual conditions, and process capabilities

People trying to make improvements and reduce defectives who think like old-fashioned engineers are prone to leap after causes and say things like, "What's the cause of this?. . . Change that and see what happens." It is possible to succeed by a fluke this way, but they are more likely to end up chasing their own tails. When using the QC approach, we should start by preparing cause-and-effect diagrams and QC process control charts, then observe and investigate the site carefully, stratify the relevant data on characteristics or results by various causes, obtain a general understanding of how the changes are manifesting themselves, build up an overall picture of the process capability in its broad sense, and identify the process capability index (see Section 4.6.7). At the same time, we should investigate the variation due to the process (σ_p^2), to sampling (σ_s^2) and to measurement (σ_M^2), as discussed in Section 4A.9.

This basic approach is so important that it cannot be overemphasized. If we grasp the facts about the goal or result we are aiming at, the appropriate countermeasures will automatically become apparent. For example, if we discover that customers are dissatisfied with a particular feature of a product or service, the problem can often be solved instantly.

(3) Breakthroughs and breakthrough tactics

Whether trying to achieve improvements as an organization or individually, all involved should never feel contented with the way things are at the moment. We must work together, take the long-term view, and think about ways of breaking with the status quo. Businesses tend toward sectionalism and maintaining the status quo. One of the fundamental attitudes needed for improvement is that of tackling problems with perseverance, smashing through sectionalism, and breaking out of the straitjacket of the existing situation.

---

* For example, Ishikawa et al.: *Shoto Jikken Keikakuho Tekisuto* (An Elementary Textbook on Design of Experiments), JUSE Press.

The biggest obstacles to improvement and new ways of doing things are within our own companies and groups. Improvement will be impossible if human factors are ignored. We must devise ways of winning over and drawing into the planning process those higher up in the organization, those who disagree, and those with negative attitudes.

(4) Preventing recurrence and removing root causes

It is useless merely to suppress symptoms; one must concentrate on the kinds of improvements that remove causes, especially root causes (see Section 5.3.4).

4.6.2 Items to be Specified in Improvement Plans; Standardization and Control Methods

We must pin down processes into the desired state of control by performing various analyses and preparing and testing improvement plans. To achieve any benefit, improvement plans must therefore specify what standards are to be prepared or revised, and must lay down methods of control, starting with the QC process chart. This means that the following items must be decided:

(i) Measurement method standards, measurement control standards.
(ii) Sampling method standards.
(iii) Quality standards, control levels, inspection standards, quality assurance methods, etc.
(iv) Process capabilities.
(v) Operating standards, technical standards.
(vi) Equipment control standards, raw materials standards, other types of standards.
(vii) Standards for control chart use, process control standards.
(viii) Matters to be studied by the technical department, research department, and other relevant departments.
(ix) Delegation of responsibility and authority. These items concern mainly standardization and control methods.

These goals must always be kept in mind when analysis is carried out; simply analyzing data haphazardly, without a definite objective, amounts to no more than playing around with numbers and is a complete waste of time.

4.7 Examining Process Analysis and Improvement Methods

The various methods of analyzing a problem once it has been identified can be classified as follows:

(i) Analysis and improvement using proprietary technology (see Section 4.7.1).
(ii) Analysis and improvement using pooled knowledge (see Section 4.7.2).
(iii) Analysis and improvement with simultaneous use of statistical methods (see Section 4.7.6).

4.7.1 Analysis and Improvement Using Proprietary Technology

When a problem arises, we probably possess relevant knowledge born of long experience, various theories, or proprietary technology that will enable us to analyze the problem logically. This know-how is extremely valuable; without theories and proprietary technology, problems cannot be solved.

Applying theory and proprietary technology incorrectly, however, can cause serious mistakes and constitute a barrier to rationalization. I would now like to discuss some problems that can arise with experience, theory, and proprietary technology.

(1) Engineers specializing in a particular technology, especially those with experience or who are authorities in the field, are often extraordinarily overconfident and are consequently extremely stubborn and unwilling to listen to the opinions and advice of others. Since they do not know that an engineer who is ignorant of statistical methods is only half an engineer, they find it difficult to accept QC and statistical methods.
(2) Whether or not things will go according to theory is a perennial question. It is also often said that experiments are performed to confirm theories that themselves emerge from the results of experiments. Generally speaking, experiments would be unnecessary if everything went according to theory. However, since theories usually contain certain assumptions and preconditions, and various errors and omissions are also present, data should be collected and analyzed and experiments performed to see how things actually work out in practice. Similarly, good things are not always produced even if made according to the design

drawings when those designs are based solely on theory and past experience. In other words, theory must be taken into account, but relevant analysis should also be carried out for confirmation. In practice, assignable causes not theoretically present and not studied in college often exert a tremendous influence.

(3) Past experience often leads to blind belief; for example, if in a particular process the yield and fraction defective deteriorate at a certain time but return to their original values when someone raises the temperature, people will tend to jump to the conclusion that it is always good to raise the temperature (forgetting that on a different occasion the yield did not improve in the slightest when the temperature was raised). It is human nature to remember the good and forget the bad. In statistics, the effect illustrated by this example is called "interaction."

In the above example, although the yield and fraction defective improved when the temperature was raised, the real cause was the worker's operation. The temperature was raised just when a shift change took over, and workers changed, which was actually why the situation improved. This led to the mistaken conclusion that raising the temperature was the right thing to do. In statistics, this is called "confounding the causes." The possibility of interaction and confounding is often overlooked.

(4) Those who live exclusively by proprietary technology are always too bound up with what happened in the past.

(5) At the same time as they are discussing theory, experience, and proprietary technology, people are often surprisingly confused. This becomes obvious when they are made to draw cause-and-effect diagrams.

(6) When people do not have a feeling for dispersion, they become upset and start leaping up and down at every little change in yield or fraction defective.

(7) Some people are incapable of cooperating with others.

The above points concern the misuse or unskillful use of theory, proprietary technology, and experience, but proprietary technology is essential for improvement. To use it correctly, we must remember the following points:

(1) As I have said over and over again, it is vital to identify the priority problems that everyone must think seriously about and to obtain a firm grasp of the facts.

(2) Good ways must be devised of utilizing experience, proprietary technology, veteran workers, and theoreticians.

(3) People's knowledge must be properly organized. Tools such as cause-and-effect diagrams, histograms, graphs, and control charts are extremely useful for this.
(4) When people make confident pronouncements, it is important to check the grounds for these and to see whether their experience is still relevant. One should also check for interaction and confounding of causes.

All things considered, individual proprietary technology is fallible and liable to bias, so a common body of knowledge should be built up by pooling individual knowledge. Data should then be gathered in line with this, experiments performed (see Sections 4.7.2 and 4.7.6), and statistical tools used to check the theory through facts and data.

4.7.2 Analysis and Improvement Using Pooled Knowledge

Pooled knowledge is the most practical and effective method of analysis and improvement. When a priority problem is actually taken up and people's awareness of it is refreshed, everyone will be found to have some relevant information or opinion, or to have noticed something useful. It is important to start by gathering this information together. QC team and QC circle activities are one way of doing this. The steps in this procedure are delineated below:

(1) Assemble everyone from the foreman and workers of the workplace in charge of the problem to quality control staff, section managers, supervisors, engineers, inspectors, and if necessary, people from design, sales, and materials departments, and hold a QC study meeting, QC team meeting, or QC circle meeting. The person directly confronting the problem (i.e., the foreman of the workplace in charge of the problem) should be the chairman; a foreman who cannot chair a meeting cannot be called a foreman.
(2) Explain the problem very clearly.
(3) Use brainstorming to gather information from everyone present, and draw up a cause-and-effect diagram.
(4) Take the cause-and-effect diagram and go *en masse* to the actual site of the problem. Thoroughly investigate the actual conditions and recheck what kinds of defective and defect are appearing. Next, look at the things regarded as the primary causes, and discuss how the work is being done at the moment, whether the procedure is satisfactory, what should be done to improve things, and how work standards and other standards should be changed.

(5) If no conclusions can be drawn about the causes after the above investigation, analyze the data statistically, as discussed in Section 4A.6, or perform experiments by design of experiment methods.
(6) Execute an improvement plan and check the results by making statistical comparisons using control charts, Pareto diagrams, etc.
(7) Hold study meetings and repeat this process as many times as needed. This type of investigation is rarely successful the first time. There are many possible causes, so the above steps should be repeated for each cause in turn. QC becomes hard work in this situation, but we must apply ourselves tenaciously to the task.

These discussions enable everyone concerned to renew his or her awareness of the problem, its degree of importance, and the various factors that affect it. Foremen who become good chairmen and can guide meetings well are foremen worth their salt.

4.7.3 Creativity and Suggestion Schemes

An important human characteristic is our ability to use our heads to think and generate wisdom. We should be constantly problem-conscious and approach things with a questioning attitude. Creativity and suggestions are a kind of improvement, and the number of suggestions people give is an indication of the desire for improvement in a company. In a company with vigorous QC circle activity, the number of suggestions increases by leaps and bounds, with each employee averaging 12 per year (i.e., one per month). Companies with even more vigorous activity receive up to 50 suggestions per year (i.e., one per week) per employee, and their acceptance rate is 60% to 70%. As long as a company is promoting QC, TQC, and QC circle activities, it must implement a suggestion system. Such systems should be promoted as a part of TQC and should be administered by the QC or TQC promotion office.

The following points should be considered when setting up a suggestion system:

(1) Clear up misunderstandings about the suggestion system. Encourage a pioneering spirit. Turn ideas into reality.
(2) Allow for cyclical variations in the number of suggestions; work to improve the quality of suggestions.
(3) Encourage both individual and group suggestions (from QC circles and QC teams).

(4) Ensure that everyone in the company, regardless of position, feels free to make suggestions.
(5) Make participation easier for people who dislike writing, e.g., by having suggestion system facilitators or counselors help them at first.
(6) Encourage suggestions on specific topics.
(7) Include both suggestions for improvement plans and suggestions of problems that should be tackled.
(8) Process suggestions speedily, implement them promptly, give reasons when suggestions are not adopted, and publish the results.
(9) Consider the connection between suggestion systems and standardization.
(10) Set up evaluation and award systems. Make it fairly easy to have suggestions adopted. Base rewards on cumulative results, giving awards or commendations when suggestions are made, when they are adopted, one year later, five years later, etc.

As the number of suggestions increases, it will become impossible for the suggestion committee to evaluate every single suggestion, so line supervisors should be allowed to evaluate and implement them under their own authority. The suggestion committee is then responsible for monitoring the suggestion system, grasping the trends of the suggestions and giving policy directives, and deciding on high-level awards such as the President's Prize and the Plant Superintendent's Prize.

4.7.4 Cause-and-Effect Diagrams

Cause-and-effect diagrams, an example of which is shown in Figure 4.3, illustrate the relationship between characteristics (the results of a process), and those causes considered for technical reasons to exert an effect on the process. They enable all the cause-and-effect relationships in a process to be summarized. When used in conjunction with other statistical tools such as Pareto diagrams, cause-and-effect diagrams are useful for promoting process improvement on a priority basis, accumulating and organizing knowledge and technology, consolidating the ideas of all employees for control-related activities, and facilitating discussions, education, and a variety of other aspects of human relations. They are also useful for all kinds of quality, quantity, delivery time, and cost control activities in new-product development, research and development, new plant construction, etc. Since they can be easily understood by anyone, they are one of the most important tools for QC promotion and implementation.

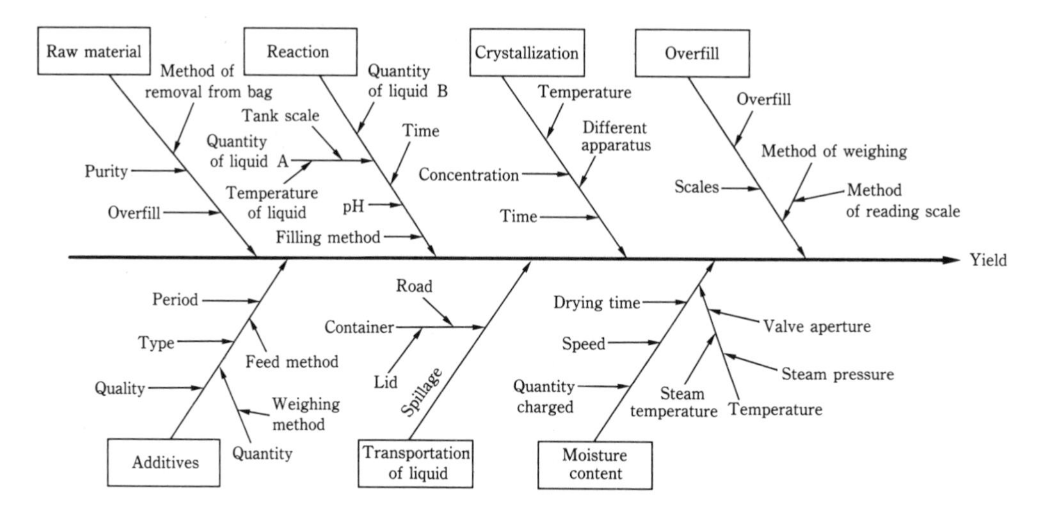

Figure 4.3 Cause-and-Effect Diagram

(1) How to make a cause-and-effect diagram

(a) Decide on the characteristic to be considered.

(b) Draw a horizontal arrow in the middle of a suitable piece of paper, as shown in Figure 4.3, and note the characteristic in question at the right-hand end of the arrow. This arrow, which forms the axis of the diagram, represents the process under consideration.

(c) Choose broad headings for the substitute characteristics or causes, and enter them in the diagram using smaller arrows, starting from the left and following the order of the process. Broad categories such as raw materials, equipment, work methods, people, environmental conditions, sampling methods, and measuring methods should be used. Various methods of classification should be tried, e.g., process order, department, function, etc., to see which is easiest to use.

There are no specific rules for drawing the diagram; the important thing is to break the categories down using the sub-branches, sub-sub-branches, etc., discussed in (d) below, to the point where causes that can be acted on have been identified. The main branches should be labeled with the names of the causes in boxes, as shown in Figure 4.3. Remember the "5Ms" (men, materials, machines, methods and measurements) when deciding on the main branches, as shown in Figure 4.4.

(d) Take the causes and break them down further, using sub-branches and sub-sub-branches. For example, temperature, time, speed, charge, etc.,

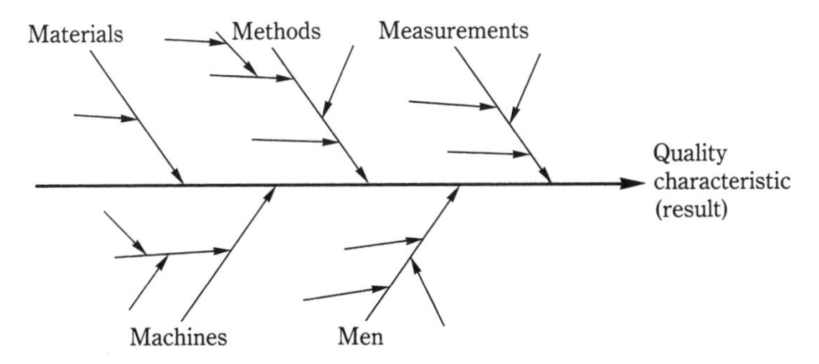

Figure 4.4 For Controlling Processes—The 5 Ms

might be used as sub-branches for moisture content during a drying process. Try to understand the cause-and-effect relationships as fully as possible and keep multiplying the number of sub-branches by repeating the question, "Why? Why? Why?" over and over. Keep writing in sub-branches and sub-sub-branches until a cause that can be acted on is eventually reached. Merely listing causes is not really very useful.

The cause-and-effect diagram is nicknamed the "fishbone diagram," but a diagram with a structure as simple as that term implies is not very useful. It is important to think of the diagram in terms of the branches of a tree or following a river back to its source.

(e) When all the possible causes have been recorded, rank them in order of the influence they exert, based on their technical significance, or as decided by a poll.

(f) Always record the date of preparation when making a diagram, and add the date of any revisions, since these give an indication of progress.

(2) Hints

(a) When preparing a cause-and-effect diagram, as many relevant people as possible should be assembled, e.g., department and section managers, foremen, operators, engineers, designers, QC specialists, etc., and everyone should be allowed to state their opinions freely as the diagram is created. Caution is needed when a diagram is prepared by one person or by a handful of people, since it can easily become biased. If possible, people from other processes should also be included, and brainstorming should be used to stimulate the flow of ideas. The person chairing the meeting should encourage people to talk so that everyone's input can be gathered, and it is particularly important to create an atmosphere in which operators, foremen, and non-specialists feel able to talk freely. In this exercise, there should be no negative comments or discussion after a person has offered an opinion. It is more important to listen to what others have to say than it is to express one's own ideas. Items judged unnecessary can always be rubbed out later. This is not the time to debate whether or not a particular cause affects the process or whether or not it is important.

In principle, the session should be led by someone responsible for the process under consideration, e.g., a section manager, supervisor, or foreman.

(b) Do not neglect management-related causes (those not found in standard textbooks).

(c) Do not forget things like sampling and measurement error and methods of calculation.

(d) Make any number of cause-and-effect diagrams for each characteristic.

(e) Examine how certain causes influence other characteristics (remember the possibility of interaction, confounding, etc.).

(f) Rather than thinking about why a problem has occurred, it is better to concentrate on the best way of solving it.

(g) During the preparation of cause-and-effect diagrams, things that everyone agrees should be done and improvements that everyone agrees should be made should be rapidly standardized and implemented.

(h) For implementing process control, it is useful to choose methods of classification that clearly show the responsibility and authority of departments, foremen, etc., from the control standpoint.

(i) It is useful to differentiate between variable causes and discrete causes.

(j) Be sure to include all causes considered important for technical reasons, regardless of whether or not they are presently being measured or are capable of being measured. Such causes should be distinguished with a special symbol.

(k) Classify causes as sporadic, periodic, or chronic. Mark causes liable to produce abnormalities.

(l) It is also a good idea to use special symbols for causes suspected of interaction.

(m) Classify causes according to whether they are easy, difficult, or impossible to control, taking responsibility and authority into consideration when deciding what can be controlled and what cannot.

(n) If control charts are to be drawn, classify causes according to whether they create dispersion within groups or between groups.

(o) As a process is improved, review meetings should be held to revise the cause-and-effect diagrams monthly, whenever an accident or abnormality has occurred, or whenever Pareto diagrams confirm that the relative effect of the causes on the quality characteristics has changed.

4.7.5 QC Process Charts

To create and control products and services, we must first decide how to create and control the relevant processes, as well as considering factors such as quality, quantity, and cost, and draw up charts to assist with this. Such charts are generally called "control process charts" (or tables). To assure quality, we further prepare QC process charts (or tables) that present, in an easy-to-understand manner, everything needed at each work step, e.g., control items, inspection items, names of those responsible, measurement methods, criteria for judgment, relationship with specifications, control methods, and related standards (see Figure 4.5).

The QC process chart should really be made in two steps: when a new product is being developed, the design department decides at the design stage roughly how to go about making the product and summarizes this in the form of QC Process Chart I; as the work moves into the prototype and pilot production stages, departments such as production engineering flesh this chart out with specific detail and turn it into QC Process Chart II, which can actually be used by the manufacturing department.

QC process charts are also used in process analysis; in this case, the first step is to prepare a cause-and-effect diagram, which is then checked against the QC process chart (newly preparing one if none exists), while one investigates what is happening in practice, and revises the QC process chart on the basis of the results of the analysis so as to facilitate process control.

Figures 4.5 and 4.6 are examples of QC process charts for process analysis and improvement. Forms like these should be used to assist with control, but, since they vary depending upon the type of product or work, they should be custom-made for ease of use. As the charts are developed, it is important to coordinate them with the preparation and revision of standards.

4.7.6 Analysis and Improvement with Simultaneous Use of Statistical Methods

If the statistical methods described in Chapters 2, 3, and 4 are used together with the above methods, the results of these analysis and improvement methods can be clearly checked and the action to be taken will also become obvious. In a debate based on opinion rather than fact, with people saying, "I think this" and, "No, this is what I think," it is often not the correct argument that wins but the

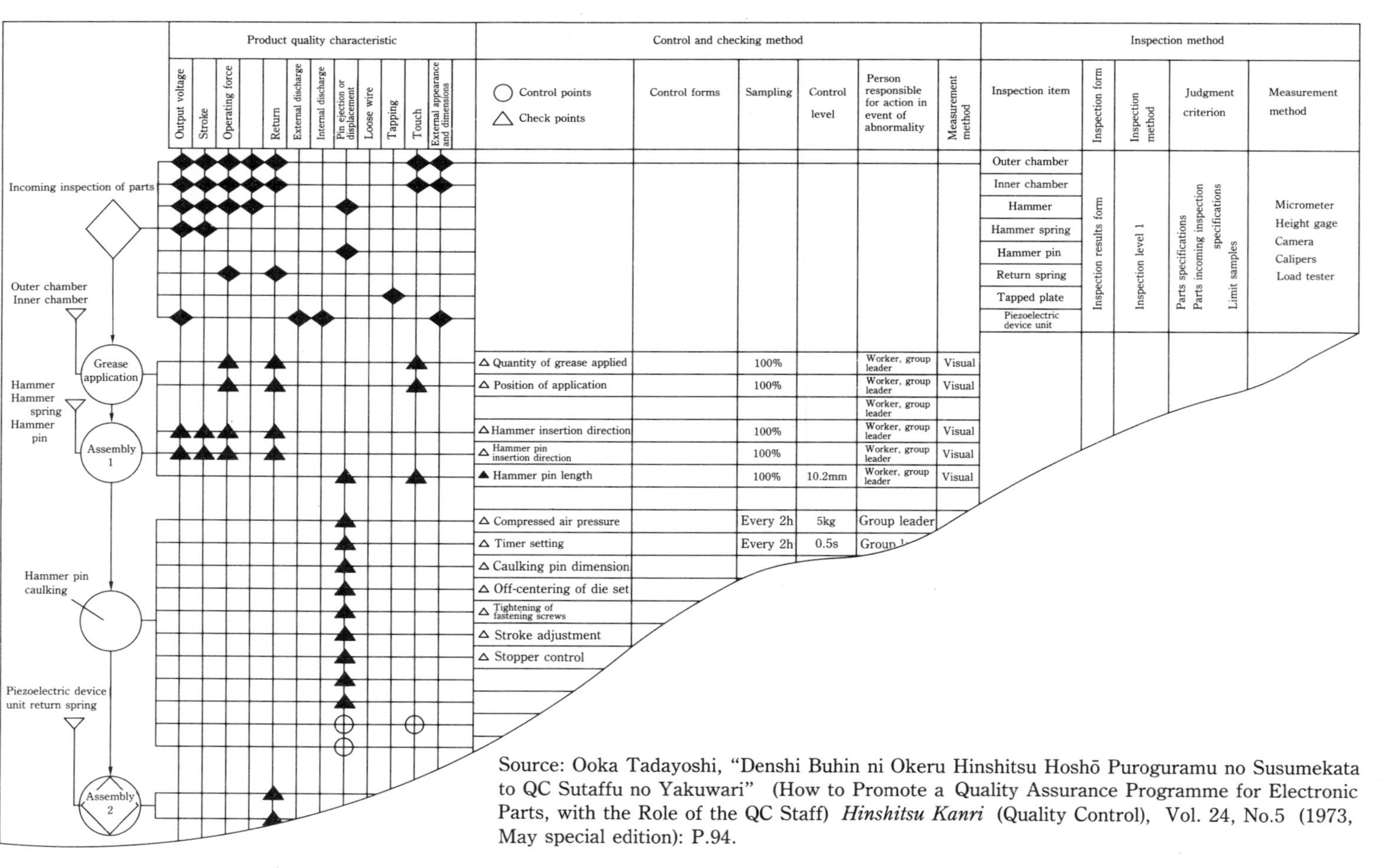

Source: Ooka Tadayoshi, "Denshi Buhin ni Okeru Hinshitsu Hoshō Puroguramu no Susumekata to QC Sutaffu no Yakuwari" (How to Promote a Quality Assurance Programme for Electronic Parts, with the Role of the QC Staff) *Hinshitsu Kanri* (Quality Control), Vol. 24, No.5 (1973, May special edition): P.94.

Figure 4.5 A Quality Control Process Chart

Notes	
Machine drawings:	○To be newly issued ○Existing drawings to be modified and reissued ○Existing drawings to be used without modification ○Not to be issued
Changeover	No.

Process Plan (Draft)
Process Change Plan
Notification of Revision in Process Plan Process

(Changeover No. Modification)

Date: Year Month Day

Department manager	Deputy department manager	Section manager	Super visor	Person in charge

Shop			
Item number			
Item name			
Type of vehicle		Number used	

Process number	Process name	Machine number	Cycle time per machine	Machine: To be newly installed	Machine: As currently in use	Machine: To be transferred from elsewhere	Machine: Existing equipment to be rebuilt	Machine: Setting up required	Ancillary equipment	Machinery requested	Department responsible	Desirable completion	Cutting tools requested: To be newly installed	Cutting tools requested: As currently in use	T·P: Number	T·P: Time	Automation	Compressed air	Power kW	Remarks / Disposal of old equipment

Inspection methods to be specified by inspection department

Process plan to be prepared by production technology department and developed into machine drawings; layout to be planned by production department

(Process preparation) ⇩ (Investigation of operating procedure) ⇨ Job instructions / Job procedure document

QC process chart			Issued: Year Month Day		Item no.														Page	/
Processing location			Process		Item name															
Process no.	Process name	Machinery and equipment	Control item	Normal production standard	Control level	Control method: Initial: Sampling method (number of samples)	Control method: Initial: Measurer / Controller	Control method: Interval	Control method: Regular: Sampling method (number of samples)	Control method: Regular: Measurer / Controller	Measuring instrument	Control chart no.	Check sheet no.	Job instruction no. Job procedure no.	Corrections					

The control plan is summarized on the QC process chart, and the details are settled during the process preparation stage.

This summarizes the control characteristics on the control charts required and gives instructions for implementing these charts.

Control Characteristics Chart (for variables)
Implement the control charts shown below

No.______________

Department manager	Section manager	Supervisor	Person in charge

Control charts for variables, $\bar{x}-R$ and $\bar{x}-R_s$; modified control chart				Location of implementation	Department section group							
File no.	Record no.	Item no.	Item name	Control characteristic	Specified value	Machine no. Processing equipment	Sampling: n	Sampling: Interval	Measuring instrument	Person responsible for control	Date of implementation	Type of control chart

Figure 4.6 An Example of a Process Control Plan for a Machining Process

opinions of the most aggressive, those of higher status, or those which the gift of the gab that carry the day. This can obviously impede future progress and cause hard feelings. When this happens, it is best to replace argument by statistical analysis and arrive at conclusions on the basis of facts—i.e., switch to "control through facts." Carrying out an investigation using proprietary technology and experience alone is like going from Tokyo to Kyoto in a handcart. Using statistical methods as well is like taking the bullet train.

The following points should be borne in mind when using statistical methods to analyze processes, compare conditions before and after improvement, or recognize facts.

(1) Prepare stratified frequency distributions, control charts, etc. to compare conditions before and after improvement.
(2) Draw Pareto diagrams before and after improvement and investigate the absolute and relative values of each item.
(3) Stratify and perform correlation analysis to check for changes in the relationship between causes acted on and characteristics being investigated before and after improvement.

If the above analysis shows statistically that a beneficial change has occurred before and after an improvement, that the improvement was effective becomes a proven fact.

4.7.7 Process Capability Studies

Process capability studies are the cornerstone of quality control. The whole chain of QC activities—that is, quality design, process design, equipment planning and control, process control, and improvement, etc.—is impossible without knowledge of process capabilities. In short, process capability studies play a central role in QC.

The word "process" has many definitions; it could be defined succinctly as "a collection of causes producing a certain result." Some specific examples are:

(a) Individual machines or equipment operated under certain fixed conditions.
(b) Individual machines or equipment operated under conditions (such as operators, materials, times, etc.) with a certain degree of variation, i.e., operated in a particular setting.
(c) A series of operations performed by several interconnected machines or pieces of equipment.

(d) Apart from machinery and equipment, a way of doing a particular job, e.g., the overall work of quality assurance, sales, purchasing, or servicing.

Process capability can be defined as "the performance of a process over a certain period of time while in the statistically controlled state." It is usually expressed in terms of a quality distribution, or by fraction defective, number of defects, etc. More broadly, it expresses the results of a process or a distribution of characteristic values. When the process consists of a single machine, its performance in terms of quality is called "machine capability" or "machine precision." In example (a) above, this would be termed "static machine capability" (or precision), while in example (b) it would be "dynamic machine capability" (or precision).

The generally used term "production capacity" means quantitative performance, and care must be taken not to confuse it with process capability, which means qualitative performance. Japanese industry has traditionally concentrated on investigating production capacity, and even now spends insufficient time on studying qualitative process capability. As long as we are implementing quality control, we must constantly study process capability.

Controlling a process means getting it to deliver its maximum capability in the controlled state, but improving a process means raising its capability, i.e., studying and improving its capability.

The following are some hints on investigating process capability:

(1) When the capability of a process is determined, the process must be in the controlled state, must have been standardized, and must contain no assignable causes. This cannot be done by simply preparing a histogram with, for example, a month's data and using this as the process capability.
(2) Process capability is easier to investigate if illustrated graphically in one of the following ways:
  (i) In the form of an $\bar{x}-R$ control chart or other type of control chart.
  (ii) As a graph with specification values (a process capability chart).
  (iii) In the form of a histogram.
(3) With variables, the process capability can be numerically expressed by $\hat{\sigma}_w$ $(=\bar{R}/d_2)$ or S_H (the standard deviation obtained from a histogram). These must generally be calculated using at least 50 items of data. $\hat{\sigma}_w$ calculated from $\bar{R}$ obtained from rationally subgrouped data expresses the process capability in a more or less satisfactorily controlled state and is also called the "short-term process capability." In contrast, S_H calcu-

lated from data obtained from a process whose control charts show the controlled state includes long-term variation and is therefore sometimes called the "long-term process capability." S_H for a process not in the controlled state is not the process capability. With attributes, the process capability can be expressed as $\bar{p}$, $\bar{c}$, etc.

(4) The process capability index C_p (comparison with specification values) is calculated as follows:

$C_p = (\text{USL} - \text{LSL})/6\hat{\sigma}_w$ (when there are upper and lower specification limits)

$C_p = (\bar{x} - \text{LSL})/3\hat{\sigma}_w$ or $C_p = (\text{USL} - \bar{x})/3\hat{\sigma}_w$ (when there is only one specification limit)

where LSL is the lower specification limit; USL is the upper specification limit; and $\bar{x}$ is the process average.

The following are the categories into which the process capability index is divided, according to values derived by the above formula:

$C_p > 1.67$: (Special class): A process capability of 1.67 or above should be targeted when aiming at ppm control, fraction defectives of the order of millionths, or extra high reliability. Such quality is too high for general purposes.

$1.67 \geqq C_p > 1.33$:(Class A): Very good quality. Inspection can be reduced.

$1.33 \geqq C_p > 1.0$: (Class B): Quite good quality. Sampling inspection is sufficient.

$1.0 \geqq C_p > 0.67$: (Class C): Some defectives will be produced. C_p should be raised to 1.0 or above.

$0.67 \geqq C_p$: (Class D): Very bad.

A few points must be noted. First S_H, which looks on the safe side, may be used in place of $\hat{\sigma}_w$. When S_H is considerably larger than $\hat{\sigma}_w$, further process improvement is needed (see Section 3A.3(6)).

Second, when there are both upper and lower specification limits, a C_p of 1.0 corresponds to 6σ, of 1.33 to 8σ, of 1.67 to 10σ, and of 0.67 to 4σ.

(5) The bias index, D_p (investigation of bias in average) is derived by the following formula:

$D_p = (\bar{x} - \text{LSL})/\hat{\sigma}_w$ or $D_p = (\text{USL} - \bar{x})/\hat{\sigma}_w$

When there are both upper and lower specification limits, it is best if $\bar{x}$ is centrally located between LSL and USL. The categories for D_p values are:

$D_p > 5$	As for $C_p > 1.67$
$5 \geqq D_p > 4$	Pretty much okay. In some cases, $\bar{x}$ should be shifted so as to reduce D_p slightly.
$4 \geqq D_p > 3$	Pretty much okay. If necessary, $\bar{x}$ should be shifted so as to increase D_p slightly.
$3 \geqq D_p$	Since this will result in defectives, either the variation ($\hat{\sigma}_w$) should be reduced, or $\bar{x}$ should be shifted to raise D_p above 3.

(6) Old-fashioned equipment control consisted simply of repairing equipment that had broken down, which then advanced to a second phase, that of controlling equipment so that it does not break down in the first place. For QC purposes, however, the goal of equipment control is to ensure that processes deliver their full capability without any deterioration, and beyond that, to raise process capabilities and increase equipment reliability. In other words, process capability studies and equipment control are two sides of the same coin. When TQC has made a certain amount of progress, TPM (total productive maintenance) should be implemented.

(7) Process capability surveys and studies are normally carried out by departments such as quality control, production engineering, manufacturing, and equipment control, or by QC teams. A competent QC circle, however, can perform them on its own process. In either case, it is important to state clearly who is responsible and make a firm decision about which departments are to perform the investigations and which are to carry out research and improvement.

(8) Process capability studies should not be limited to one's own company; they should cover everything from design through suppliers and the manufacturing process to the distribution system that gets the product to the consumer. Ideally, the process capability of administrative processes, computer systems, information processing, etc., should also be investigated.

(9) The departments that should know about process capability probably include purchasing, manufacturing, design, production engineering, quality control, equipment control, and sales. In the workplace, if supervisors and workers understand process capability, they will gain an interest in controlling and improving it, and tremendous benefits will be obtained if they succeed in doing so.

(10) Do not be content with merely doing process capability studies; they must actually be used for research, control, and improvement.

4.8 Some General Hints on Analysis

The following points should be kept in mind when using statistical tools to analyze data from the workplace with the aim of taking action against assignable causes. The hints given here apply to the statistical analysis of routine data (data types (i) and (ii) in Section 4.6.1(1); see Section 4A):

(1) Data should be stratified by cause during collection.
(2) Data should be collected in such a way as to make causes correspond to the characteristics being investigated.
(3) As far as possible, unprocessed individual values should be used for analysis, while averages and totals should be avoided.
(4) Wherever possible, routine data should be presented in the form of graphs or charts for easy understanding.
(5) Nothing in the records may be ignored; even subjective remarks such as "Things going well" or "Not in good condition" are extremely useful for analysis.
(6) If results contradict past experience or knowledge, the following items should be checked in the order given:
   (i) Use of statistical tools.
   (ii) Method of arriving at conclusion.
   (iii) Correctness of calculations.
   (iv) Authenticity of data and use of sampling and measurement methods.
   (v) Rationale and authenticity of past experience or knowledge.
(7) If analysis of something thought to be an assignable cause shows that it in fact makes no difference to the process, one must not lose heart. Perseverance and tenacity are needed. When a factor is analyzed and found to have no effect, this simply shows that the engineers who suggested it misunderstood the situation. One must be positive; knowing that a particular factor does not affect the end result is still extremely valuable information, since it can allow a cheaper alternative to be adopted.
(8) When the existence of many possible causes makes the situation unclear, each cause must be analyzed one by one, starting with the one thought to have the greatest effect. When there are too many possible causes obscuring the situation, it is best to use multivariate analysis.
(9) It is always best to work in pairs or in groups of three or more.
(10) While performing an analysis, one must keep asking oneself how the variation can be reduced and what kind of action should be taken.

(11) When investigating a particular cause, cause-and-effect diagrams can be prepared for other characteristics and what effect the cause has on these can be determined. Eliminating a cause may make one characteristic better but another worse.
(12) When an assignable cause of variation has been identified, different courses of action should be tried until the variation has been eliminated.
(13) According to the Pareto principle, there are generally two or three major assignable causes; eliminating these will more than halve the number of defectives. Thus, halving the number of defectives (e.g., raising the yield from 60% to 80%, from 80% to 90%, or from 90% to 95%) is usually comparatively simple. This is why I say, "95% of workplace problems can be solved through the use of the Seven QC Tools."
(14) Since data can easily mount up and get out of hand, statistical tools (particularly pictorial ones) should be used at every opportunity.
(15) The results of action should be checked statistically by means of control charts or other methods; the action should be formally standardized if it is beneficial; and control should be established so that the same problem does not occur again.
(16) In many cases, it is best to complete all analysis of past data before starting to collect fresh data and perform planned experiments.
(17) The results of an analysis should always be summarized in the form of a written report for collation and filing by the technical department. Even when unexpected conclusions are reached, such reports will form a valuable store of technical information for the company.

4.9 General Procedures for Statistical Analysis

While the general procedure presented here for carrying out statistical analysis will of course vary according to the process being analyzed and the prevailing conditions, it provides a useful general framework. The fundamental idea is not to analyze causes, but to analyze data on results, obtain a firm grasp of the facts, search out assignable causes, and take action. This method gives ideas and hints on analyzing large amounts of past data that cannot be made to correspond. Nowadays computers are often used to carry out regression analysis, multiple regression analysis, etc. If the data corrections required in Steps 8, 10, 12, etc., are too cumbersome, they may be skipped.

(1) Pool everyone's knowledge to prepare a cause-and-effect diagram for the targeted characteristics (or results), noting effects due to combinations of causes and anything else that might give rise to confounding of causes, and draw up a QC process chart for the process being investigated. Everyone should then work together to check the actual situation and find out what is happening to the characteristics and causes in practice.
(2) Collect at least 100 items of data on each characteristic and cause pertaining to the problem. It is best not to collect all the data over just a one-day or one-hour period, but over a period of several days or one or more months.
(3) Arrange the data on each characteristic in the form of a histogram, then calculate the mean and the standard deviation, and compare these with the targets or specifications. It is very important to stratify the data in various ways when doing this. The same procedure should be followed for causes.
(4) Plot the data on a control chart or graph in sequence according to time, lot, etc., to provide a picture of the actual state of affairs. The within-subgroup and between-subgroup variations should be well separated and carefully investigated for each cause.
(Note: Steps 3 and 4 are the first steps in process capability studies.)
(5) Check the pattern of points on the control chart or graph as follows:
(a) Do any points lie outside the control limits?
(b) Are the points arranged randomly?
(c) Is there any periodicity? If so,
(d) Are there any runs?
(e) Are there any trends?
In short, check for abnormalities in the pattern of points. If an abnormality is found, identify the cause and eliminate it by preparing appropriate written standards. Item (a) covers mainly sporadic causes.

Note: Caution is needed, since data from a workplace usually contain dirty values such as outliers, false data, highly inaccurate data, calculation errors, and incorrectly read or transcribed data.

(6) Prepare modified control charts for characteristics by removing any points or data clearly attributable to assignable causes, subgrouping the data so as to maximize the uniformity within subgroups, and redrawing the charts. It is best to subgroup the data to provide maximum within-subgroup uniformity when performing analysis, but not necessarily when drawing up control charts for process control.

(7) Repeat Step 5, take action, remove any points or data clearly attributable to assignable causes, and recalculate the control limits. Keep repeating this step.

If a situation arises where an assignable cause is discovered but those performing the analysis or involved in the project do not have the authority to take action or standardize, an abnormality report form should be completed and passed to the appropriate party for action. Abnormality charts should also be prepared on which the party contacted can record the action taken or indicate that no action was taken. For the sake of clarity, the cause should also be marked with an " × " or other symbol on the cause-and-effect diagram with an indication of whether or not any action was taken. In short, it is important to organize and assemble the knowledge obtained from the analysis in some specific form.

(8) With discrete assignable causes, the data on characteristics should be stratified by the conditions (in terms of attributes) prevailing at the time the lot was produced, and should be investigated by means of a checklist, histogram, or graph, or by rational subgrouping of the data and drawing up of a separate control chart for each stratum. When there are many discrete causes, the following steps should be taken:

(a) When there are a lot of data, they should be stratified in order of the factors regarded as most technically significant, e.g., first by furnace, and then by work group, material, etc. When doing this, it is best to have at least 100 items of data.

(b) When there is not a lot of data, examine the data by stratifying them in various ways in turn; for example, if different furnaces are found to give different averages, correct the data for this difference only, then stratify further according to work group. If, during this procedure, a particular cause is found to produce no difference in average, return to this cause later and try the same method of stratification again after correcting the data for other causes that do give a difference in average.

If the above analysis reveals points lying outside the control limits, take appropriate action and standardize to eliminate the abnormal difference between strata. If possible, continue stratifying and correcting like this until in the end a more or less satisfactory control chart is obtained.

A few points about stratification procedures should be noted:

First, if two discrete causes may be having a combined effect—that is, if they may be interacting—proceed as in the following example: If there are two pieces of equipment, A_1 and A_2, and two operators, B_1 and B_2, and it is believed that there is interaction, stratify the data by equipment, then further stratify by operator for each piece of equipment, giving a total of four strata, then prepare a control chart or graph for each stratum.

Second, if the same data are corrected too many times, there will be a large error in the final data due to the error in estimating the difference at each correction.

Third, generally, the smaller $\bar{R}$ is, the greater the precision of an estimate. Also, the more subgroups and data there are, the greater the precision.

Fourth, a sign test or other test of differences in means may also be performed without relying only on control charts.

(9) With variable causes, abnormalities can be detected and work standards revised by using measurements themselves to draw up graphs, histograms, or control charts and by performing a technical investigation of the state of the dispersion and the way it changes with time. However, it is also necessary to perform a statistical analysis to see what kind of effect these variable causes have on the process, i.e., on the quality characteristics produced by the process, in the actual operating environment (see Section 4A.5).

To do this, graphs or control charts drawn in time sequence for each variable cause and quality characteristic should be lined up and compared. The median method should be used first to examine whether there is long-pitch correlation between causes and characteristics, between one characteristic and another, or between one cause and another. Next, the segment method should be used to determine whether there is any short-pitch correlation. It is often easier to obtain a picture of long-pitch correlation by drawing a scatter diagram, which can also simplify the analysis.

(a) If the conclusion is that long-pitch, large correlation (see Section 4A.8) exists, the cause has a definite effect on the characteristic regardless of the degree of influence of other causes. The operating standards must therefore be revised so as to reduce the range of dispersion for this cause. When this is done, the histogram for the cause should be narrowed down to about two standard deviations, or to about the range where the operation can be performed satisfactorily from the technical standpoint. It is not good to be too idealistic just because correlation exists, narrowing the range too far and setting standards that are impossible to meet. An alternative solution is to use automatic control.

When there is long-pitch correlation, auto-correlation analysis should also be carried out.

(b) If the conclusion is that short-pitch, small correlation exists, other causes are having a great effect and the present cause is exerting a small, not a large, effect. It is first necessary to control the other causes, but if possible, the operating standards should be revised to control the present cause as well. This type of correlation may be ignored at first, but the cause must of course be suppressed if small fluctuations in characteristics become a problem.

(c) When there are a large number of variable causes, we should start by considering those we think exert a major effect from the technical standpoint and by seeing which causes have a long-pitch correlation to the quality characteristic. At the same time, we should look for correlations between the causes and make up a cause-and-characteristic chart such as that shown in Table 4.2. Information on how operations are performed in the factory or workplace can be obtained by carrying out various technical investigations of this.

A few more points are worth noting here concerning variable causes:

First, variable causes may also be investigated through stratifying by dividing their variation into a number of different ranges.

Second, when results take time to appear, correlations can be

Table 4.2 Cause-and-Characteristic Chart

	Characteristic	Cause I	H	G	F	E	D	C	B
Cause A	** +	⊖	** −	⊖	** −	⊕			
B	⊕		⊕		⊖			** −	
C	⊖								
D	* +		⊖						
E	⊕	** −	⊕		⊕				
F	** −								
G	⊖								
H	⊖								
I	** +								

+: positive correlation
−: negative correlation
*: 5% significance
**: 1% significance
○: no significance
Empty square: not analyzed

obtained by shifting data to different times. In some cases, correlations can be found by taking moving averages.

In Table 4.2, for example, statistical analysis of the existing operation reveals that Causes A, D, and I are positively correlated and Cause F negatively correlated to the characteristic. A look at the correlation between causes, however, shows that Causes A and F are negatively correlated. Further investigation discloses that although F has been taken as an assignable cause, it is difficult to control, and it is technically feasible for F to decrease naturally as A increases. It is therefore sufficient in this case to set clear operating standards for A. If necessary, complementary standards depending on A can also be set for F.

There is correlation between H and A, and it is clear from the technical standpoint that H will decrease by itself if A is controlled well. Cause I is controllable, and operating standards should be set for it, but there is a negative correlation between I and E, while there is no correlation between E and the characteristic. Furthermore, E is controllable. A close examination shows that E is poorly controlled, producing corresponding large changes in I and consequently affecting the characteristic. In this situation, proper standards must therefore be established for E, and the standards for I must be set in accordance with these.

D is an independently controllable factor with little relation to any other cause. Since a different factory is responsible for it, the quality control committee decides to issue an abnormality report and have the other factory narrow the range of the quality standard.

Although B and C are unrelated to the characteristic, there is a negative correlation between them, and C is being adjusted to nullify any changes in B. While this is not directly affecting the characteristic, it is disrupting the process, so the decision is made to set operating standards for B. As for G, the chart reveals that it is sufficient to standardize the existing operation without modification.

As the above example shows, various facts come to light when the chart is carefully investigated from the engineering standpoint; so-called "false correlation" can be seen through, and standardization can be implemented well. If this is done diligently every month or three months at first, the standards can be gradually rationalized in accordance with the existing situation.

It should be noted that when there are several characteristics, they should also be grouped together and their correlation investigated, as in Table 4.2. Caution is needed, since one characteristic might deteriorate as another improves.

Another point to note is that the amount of information provided by the above table will be further increased if small-scale correlations are also shown, using different symbols.

(10) When there are many variable causes, all the long-pitch correlations may be investigated just as described above, but it is sometimes best to use the following devices for obtaining correlations:

(a) Stratify the data according to discrete causes and check the correlation. For example, plot a scatter diagram using different colors for different machines or time periods and find both the overall correlation and the individual correlations for the different strata. Correlation must always be established after stratifying according to the discrete causes, particularly when a combination of discrete causes and variable causes is thought to be exerting an effect (i.e., when interaction is thought to be present).

(b) When correlation has been found between a characteristic y and a cause x, the regression line of y on x should be drawn, and the y values corrected for a standard value of x (e.g., if x is temperature and is set at 600 ± 20°C, the standard value of x will be 600°C).

(c) When two variable causes are thought to be interacting, the analysis should be performed as in the following example: when it is thought best to vary the temperature according to the purity of the raw material, and this varies from about 70% to 90%, the raw material data should be stratified into bands of 70–75%, 75–80%, 80–85%, and 85–90% purity, and the correlation between temperature and the characteristic should be investigated separately for each band. In such a case, the data should be stratified according to the cause that is thought more difficult to control or that has a smaller range of dispersion, and the correlation between the characteristic and the remaining cause should be investigated.

(11) The points discussed above are general principles; in practice, however, a process should be analyzed by trial and error in every possible way using technical knowledge and experience. As this is done and experience in analysis is built up, it will gradually become possible to perform effective analysis with an efficient use of time and effort. At all events, nothing can be found out without analysis, and no technology will be established. The most important thing is to give it a try.

(12) In this way, the data are gradually corrected using regression lines or differences in means, and the control chart is finally redrawn. This re-

vised chart will show an approximate state of control, and the results can be used to set provisional standards if the histogram approximately satisfies the specifications or quality targets. If it does not satisfy these, one or more of the following actions should be taken:

(a) Try performing the operation according to the standards set as described above.
(b) Continue analyzing the process.
(c) Perform new experiments using design of experiment methods, in the factory, on pilot plant, or in the laboratory.

(13) To check whether the results of the above analysis are correct or not, operate the process according to the provisional standards and perform various checks to see whether the targeted characteristics have improved as expected and whether other characteristics are affected. Then revise the standards and continue the analysis as necessary.

(14) When the work has been performed satisfactorily according to the provisional standards for a trial period of 1 to 3 months, the standards should be formalized, and the same kind of analysis continued. As necessary, further improvement should be implemented, analyzing the process by taking data not previously measured and stratifying the product in various ways as it passes through the process.

While proceeding with the above analysis, we should prepare and revise various standards, e.g., technical standards, work standards, control standards for machinery and equipment, sampling and measurement standards, etc., and carry out training so that the assignable causes we have eliminated do not recur. In this way, we steadily take action one step at a time to prevent recurrence and establish firm control. It is of course best to carry out work studies, process studies, and time studies, take snap readings, and hold various education and training programs while doing this. The results of these actions should always be recorded and filed in the form of technical reports.

There may also be times when the process capability is insufficient in spite of everything, and it is impossible to obtain satisfactory products under existing technical and economic conditions. In such cases, discussions should be held with customers, the next process, or company management with a view to revising characteristics, standards, specifications, quality targets, etc. At the same time, of course, it is also essential to take actions such as having the technical and research departments undertake research, planning the installation of measuring instruments and the rebuilding of machinery and equipment into the next financial period's investment schedule, rationalizing contracts, and promoting the implementation of quality control at raw material suppliers.

As people master statistical techniques and gain experience in the type of analysis described above, they will become able to make educated guesses about whether significant differences exist simply by graphing data, without bothering to calculate averages and dispersions every time. In this way, they will become able to proceed with analysis on a priority basis.

4.10 Performing Factory Experiments

Various experimental designs are used for performing experiments on processes, but the details of these will be left to more specialized works. This section gives some brief hints on performing experiments using processes.

(1) When performing experiments in the factory, draw up factory experimental regulations in advance and ensure that experiments are carried out through formal procedures in accordance with these regulations. It is quite wrong to allow people to perform experiments in the workplace as they see fit, reporting good results and keeping quiet about bad ones. Ensure that a formal procedure is followed and written reports submitted.
(2) Perform experiments using the engineer-in-charge system or by setting up a QC team. Since experiments may require the setting of extreme levels, may affect entire processes, and may necessitate far-reaching actions based on their results, someone with the maximum possible authority to take action must be in charge.
(3) Decide on the responsibility and authority for what happens if the experiment produces defectives or rework or disrupts production.
(4) Be sure to arrange and analyze past knowledge and data properly and clearly establish the focus and aims of the experiment. Organize the factors to be tested in order of priority.
(5) As far as possible, experiment with factors capable of being controlled.
(6) Since experiments using processes are often accompanied by danger or possible damage depending on the levels at which the factors being tested are set, start in the first experiment by setting the factors at two or three levels shown by past experience and technology to be fully safe. For example, when a certain factor has a fair amount of variation and it is impossible to determine its best level from past data, but it is considered important from technical considerations, set it at two levels: one thought

to be its standard value, and another at a better value within its range of variation. In unclear cases, set the factor at three levels, high, low and standard. From the second experiment on, use the information from the first experiment. The optimum level is often found to lie outside the previously used levels.

(7) Rather than trying to reach a definitive conclusion from a single experiment, it is better to perform a series of two or three, using the knowledge of each for the next. This method is safer and more economical, and it guarantees the reproducibility of the results. The error variances at each stage should also be compared. In short, experiments should be performed in steps.

(8) When the available technical knowledge is uncertain, double-check by first using a method such as the orthogonal array roughly to identify the particularly problematical factors and finally performing replicated experiments using the two-way layout or similar method.

(9) It can also be advantageous to use the results of analysis of past data for checking. Be particularly careful to check for confounding of factors.

(10) The design of experiment approach is beneficial when the levels of factors can easily be changed randomly step by step without too much time or expense, e.g., in batch-type operations. This approach is sometimes difficult or dangerous in continuous operations, but factories with well-maintained automatic control equipment are suited to this method.

(11) Assign the factors and levels of the experiments in the form of work standards, and always prepare detailed experimental work orders. It is also good to prepare data collection sheets in advance and to take supplementary measurements.

(12) Since reproducibility is so important, replicated experiments should be carried out at different times, to check whether the R control chart shows a state of control between repetitions. Experiments should be planned to allow detection of differences between blocks and interaction between blocks and factors. When people, machines, raw materials, etc., are taken as factors, experiments should be repeated at least twice within the same stratum.

(13) The experimental sequence should be as random as possible. To achieve this, the sequence should be clearly specified in experimental plans and instructions, and the experiment itself must be closely controlled. When randomization is impossible, use the split-plot approach, make a thorough technical study of the experimental sequence and the factors possibly subject to confounding, and state these clearly in the experimental report and conclusions. Operations performed according to existing standards should also be included randomly for comparison.

(14) Uncontrollable factors must of course also be tested, with stratification by, for example, equipment or raw material. Experiments should be planned to detect interaction between uncontrollable and controllable factors, since it is important to set operating standards for the interaction, i.e., for each stratum.
(15) When a conclusion has been reached, the conditions should actually be tried out in the process and studied by means of control charts. When this is done, it is helpful to compare the square root of the experimental error variance with $\bar{R}/d_2$ from the R control chart.

4.11 Process Analysis with Scant Data

Statistical methods of analysis are easy to use in continuous processes or when plenty of data are available. Statistical analysis is often also possible with scarce data (e.g., when there are only 20 to 30 items of past data), but in some cases good analysis may prove difficult. Some points to note in such cases are given below:

(1) Consider the reason for the lack of data. When people have been using the old "world of averages" approach and have reduced the data to a few monthly averages, it is sometimes better to use quick methods to analyze the original data. Precise methods must of course be used when data really are scarce, but thought should be given to switching to control experiments in the future so as to provide plentiful, easily analyzed data.
(2) When workplace data are used for analysis and data are scarce, there is a danger that the assignable causes may be confounded with other, non-randomized causes. It is thus necessary to carry out a technical review to see whether other causes are confounded with the results of the analysis. Attention must of course also be paid to this when data are plentiful, but particular caution is needed when data are scarce because of the poor randomization of other causes.
(3) When data are scarce, changes in other causes are often small, and it is dangerous to extend the conclusions beyond the range of the analysis.
(4) Advanced statistical tools such as the design of experiment approach and analysis of variance should also be used for analysis under these conditions.

4.12 PREPARING AND IMPLEMENTING IMPROVEMENT PLANS

When deciding on an improvement plan, attention should be paid to the following points:

(1) Be sure to have all those who will actually implement the plan take part, educate and train them, and be sure that they accept the plan. Liaise with other departments affected.
(2) Use practicable methods, but remember that approaches regarded as impossible surprisingly often succeed when actually tried.
(3) Delegate decision-making authority as far down the hierarchy as possible, e.g., to QC circle or QC team leaders.
(4) Prepare a definitive plan with the intention of creating draft standards such as operating standards, technical standards, raw materials specifications, etc.
(5) What is decided at this stage is strictly a trial plan and provisional standards. They should only be adopted as formal standards after they have been tried out and the results have been checked and found good.
(6) Thoroughly check a second time what effect the improvement will have on other characteristics, conditions, and departments. The optimum conditions for one characteristic or department are not necessarily the best for other characteristics or for the company as a whole.
(7) As mentioned before, there will always be opponents within a company whenever anyone tries to implement an improvement plan or to do anything new. Rout these opponents and carry out the plan courageously.
(8) Before starting to implement the plan, prepare standards showing who is responsible for measuring and evaluating the improvement and how and when this should be done, and how the improvement plan is to be controlled.

Proceed in this way to the control and steps for further improvement described in the next section.

4.13 CHECKING RESULTS: CONTROL AND FURTHER IMPROVEMENT

However much proprietary technology we have and however much statistical analysis we have performed, an improvement plan is still only a plan, and we do not know how it will work out until it is actually tried. An extraordinary number of plans thought technically sound have failed when put to the test. If the brilliant technical ideas we have really were so good, we would be able to produce far better products than we do. Some people even go so far as to say that things improve when we do the opposite of what engineers suggest.

This means that once an improvement plan has been carried out, we must *always* check the results. For example, when we make a design change, we must only ship the product after thoroughly checking the consequences. We must also set up customer liaison systems to check that everything is all right after the product has reached the consumer. This was not done in the past because people blithely assumed that they were making their products well, but omitting this step and failing to rotate the control cycle has led to a great many failures.

The following are some hints on checking:

(1) Make a habit of checking everything. This means making it standard practice to record on improvement report forms the methods of checking (i.e., the control methods) used and the results and benefits obtained, i.e., always ensuring that reports on the results of implementation are submitted. This does not always happen: for example, expected results are often included on draft proposals and memoranda for equipment investment, but no reports are submitted on the results obtained after the investments have been made.
(2) Check not only the characteristics and causes that are the subject of the improvement but other related ones as well.
(3) Continue the process of checking and controlling for a fairly long time, e.g., at least one year. Industrial processes are highly susceptible to seasonal changes and the reliability of the results also needs to be confirmed.
(4) Check how the process capability and state of control have changed as a result of the improvement.
(5) In summary, an improved process should be controlled and checked using control charts for at least one year. If the state of control is then stable, the QC team has fulfilled its responsibilities and is disbanded.

The results obtained by checking should be examined to see whether further improvement is needed, and the procedure repeated. By going through the steps of analysis, improvement, control, further analysis, and further improvement diligently and perseveringly, it is usually possible in about six months, even with existing equipment, to halve the number of defectives, halve the variation in process capability, achieve a 50% increase in production volume, reduce the number of man-hours by 30%, and raise productivity by 30%.

It is no good relaxing just because a single improvement has produced good results. We must keep on making one improvement after another, and continually strive to reach higher and higher targets.

4.14 Preparing Reports

Whenever process analysis is carried out and an improvement is implemented, a written report must be prepared, even if the improvement was a complete failure or repeated mistakes were made on the way to eventual success. This clarification of the facts on failures and successes is very important not for the benefit of individuals, but for the sake of collecting technology for the business or organization. If it is not done, the same mistakes will be repeated in the future, because people tend to think along the same lines. For the same reason, it is also very important to prepare a report so that the significance of the improvements and the technology used can be communicated to less experienced employees and those coming after.

The goal of preparing reports is twofold:

(1) To enable superiors and other interested parties to understand thoroughly the purpose of the analysis, the process, and the results so that they can take action if necessary.
(2) To accumulate technology for the business or organization.

To meet these goals, the report must be written to be easily understood by someone unfamiliar with the subject. Like QC, it should be written not just for the writer, but for the writer's customer, i.e., the reader.

It is usually unwise to try to start writing a report from scratch after the analyses and experiments have been concluded, because this is time-consuming, and in extreme cases, the report is never prepared at all and the information gets locked away in the writer's head or private notes. It is better to start with the intention of writing the report right from the outset, by thinking about the procedure for writing it and preparing appropriate charts and diagrams.

In general, the contents of a report are best organized according to the scheme described below, commonly known as the "QC Story." The QC Story or QC-style report uses the headings listed below and differs from the standard, old-fashioned type of business report, which was concerned only with results, i.e., whether the objectives have been accomplished or not; such reports are based on the philosophy that "if the results are good, then all is well," and top management and superiors used to concur in this. In QC, we are concerned with the process by which the results are accomplished, or in other words, items 2 through 7 below. QC-style reports focus on the methods, means, and processes by which goals are attained. If the process by which the goals are attained can be made absolutely clear, then experience and technology are built up and the same process can be repeated in the future.

The old-fashioned approach depended on motivation and effort, and good results may sometimes have been obtained even on the basis of false data, the business environment and luck permitting. Old-fashioned "management by objectives" achieves only objectives limited to a particular situation and that cannot be reproduced elsewhere, while QC focuses on the improvement process itself and aims at preventing the recurrence of problems.

The chairman of a certain company noticed that the company president often visited factories and branch offices to conduct presidential QC audits. When asked why it was necessary to make so many such trips, the president replied, "You believe that everything is fine as long as the results are good, but I am more interested in the process by which they were achieved. That's what I go to find out." That was the last time the chairman broached this subject.

A QC Story or QC-style report must be written in such a way that anyone reading it, whether a superior, an engineer, or someone taking over a job, can clearly understand each of the items below:

1. The topic selected.
2. The reasons for choosing this topic (the Pareto principle).
3. Identification of the current situation (facts and stratification).
4. Analysis of results and processes (investigation of assignable causes).
5. Countermeasures and their execution.
6. Confirmation of countermeasures.
7. Standardization (preventing backsliding), recurrence prevention.
8. Establishment of control.
9. Review of improvement and consideration of remaining problems.
10. Future plans.

This type of internal company report is different from old-fashioned academic-style reports. I recommend that internal reports and QC audit reports be divided into the following three parts:

The first part succinctly describes the problem and the requisite action in one to three pages for the benefit of top management and other busy individuals.

The second part briefly states the main data and conclusions in the order of the above list of items in four or five pages for the benefit of fairly busy people such as section and department managers.

Parts 1 and 2 should be provided with references to page numbers in Part 3 to allow easy access to additional information in areas of particular interest to the reader.

The third part includes detailed data and a full discussion of both failures and successes, written for easy understanding by those coming after. Since the raw, original data are the most important type of data in QC, it is best to include this here. If this makes the report too bulky, the data can be compiled in a separate volume or stored on computer. In my experience, the original, untreated, basic data contain a great deal of information, much of which can be revealed through subsequent analysis. When reports are filed on computer disk, an index of key words should be prepared to allow easy retrieval of information.

At all events, although preparing reports is a troublesome and time-consuming chore, it is extremely important for organizing our ideas so that others, including our superiors, can understand them, and for accumulating technical know-how for our businesses or organizations.

4A.1 Investigating Measurement Methods *

It is impossible to obtain data without taking measurements; this means that every item of data we obtain contains some measurement error. Moreover, measurement is so important, we could almost say that any advance in quality control depends on the progress we make in measurement methods. It is therefore obvious that, before analyzing a process itself, we must first review our measurement methods from both the statistical and the engineering standpoints. In investigating measurement methods, the following points should be covered:

---

* Sections 4A.1-4A.9 deal with some very simple statistical methods for analysis that anybody can soon learn to use. For frequency distributions, histograms, Pareto diagrams, and check sheets, see Chapter 2. For analysis using control charts, see Sections 3.9.2 and 3A. For statistical tools and design of experiment methods that require a certain amount of calculation, please refer to other works.

(1) Devising ways of quantifying: whatever the situation, quality assurance and improvement are facilitated by numerical evaluation. Whenever possible, we should devise methods of quantification and exercise control using figures.
(2) Reconsidering whether measurements are being taken for inspection and assurance, for process control, or for process improvement: these purposes are often confused; for instance, measurements taken for inspection purposes are sometimes used for analysis or control despite their unsuitability. It would be better to use improvement or control measurements for inspection.
(3) Investigating which is more appropriate, measurement by variables or measurement by attributes: measurement by variables usually allows us to get by with smaller samples and provides more information, making it easier to decide what action to take, but collecting and organizing the information is expensive and time-consuming. Measurement by attributes often simplifies data collection and organization, enables large numbers of units to be checked, and is easily understood by people in the workplace, but yields less information. Since it is impossible to give general rules as to which type of measurement should be selected, I will offer illustrative guidelines:
    (a) When the variation between sample units is extremely large and the sampling error is great, it is necessary to take a large number of samples and perform many measurements in order to control a process. In such a case, it may be advantageous to measure by attributes.
    (b) When one wants to perform quality assurance together with process control, it may be necessary to take a fairly large number of samples. In this case, measurement by attributes is sometimes advantageous.
    (c) Measurement by attributes can be better for showing the overall state of control of a factory to senior management and may be more easily understood by workers and line foremen.
    (d) Measurement by variables provides more information and is therefore better for process analysis by QC circles and QC teams.
    (e) It is sometimes helpful to use measurement by variables until a process is under control and measurement by attributes thereafter. The opposite is also sometimes true.
    (f) It is best to use variables gages for measurements of the order of 1/100 mm and below (or recently 1μ) and attributes gages for measurements above this.
    (g) Items to be measured by sensory testing should be ranked in three to five classes.

(4) Investigating measurement errors: in process improvement and control, what we are concerned with in measurement error is the reliability and reproducibility (particularly that of measurements taken in the same laboratory but on different days, by different people, or with different instruments) of the measurements. We must of course also be concerned with bias when the measurements for test for control are used to guarantee the results, but it is good enough if the amount of bias and correlation are known and if the measurements are reliable. During process analysis, out-of-control points may be due to measurement abnormalities, i.e., to unreliable measurements, and $\bar{R}$ is often large as a result of poor measurement reproducibility.

In general, if we call the variation in the process itself σ_P, the variation due to sampling σ_S, and the measurement reproducibility σ_M, the variation in the data, σ, will appear in the form:

$$\sigma^2 = \sigma_P{}^2 + \sigma_S{}^2 + \sigma_M{}^2$$

Thus, if other variations such as $\sigma_P{}^2$ or $\sigma_S{}^2$ are ten or more times $\sigma_M{}^2$, the measurement method is more or less ideal for controlling the process, but if $\sigma_P{}^2 \approx \sigma_M{}^2$ or $\sigma_P{}^2 < \sigma_M{}^2$, it becomes impossible to tell whether the control chart is being used to control the process or the measurement. I sometimes come across such control charts. In such cases, we must either repeat the measurements and take averages or revise the measurement work standards so as to reduce σ_M. The opposite case—when the measurement precision is good—sometimes occurs, and we have $\sigma_P{}^2 \gg \sigma_M{}^2$ (e.g.,$\sigma_P{}^2 \geqq 100\sigma_M{}^2$). This kind of precision is needed for assurance tests but is unnecessarily high for test for control experiments, and it is better to switch to a cheaper and simpler form of control experiment if the measurements are too time-consuming or costly. As process analysis and control progress, σ_P generally decreases gradually, and this may cause σ_M to become large in comparison with σ_P even though it may have been small at first. It is thus necessary to check the measurement method occasionally to make sure that it is still suitable.

As can be seen from the above, it is important to check the reliability and reproducibility of measurement in process analysis. If this is not done, much time and effort will be wasted analyzing processes and searching for assignable causes when many of the out-of-control points may be due to measurement error or unknown causes. It is of course the duty of inspection, testing, analysis, measurement control, and jig and tool sections to control the measuring instruments and methods for which they are responsible and to quantify their reproducibility and bias.

(5) Preparing and controlling measurement work standards: when measurement errors are reviewed, measurements are surprisingly often found to be unreliable or to have extremely poor reproducibility. This is because existing standard measurement methods and test methods often leave much to be desired as work standards. In fact, I occasionally come across measurements and tests that are almost completely uncontrolled. Measurements and tests should be regarded as another type of process, and work standards should be prepared for them with this in mind. In addition, those responsible for taking measurements should be thoroughly trained, and methods of controlling the measurement process should be devised, e.g., by quietly passing standard samples through the process as a check from time to time.

(6) Speeding up measurement times: for good control, it is essential to speed up the feedback of data and information. To achieve this, we should investigate whether the methods of analysis and testing currently in use are satisfactory from the time standpoint and consider using quicker, easier methods. Feedback methods should also be improved.

(7) Preparing measurement control instructions: although many different instruments and gages are used in the workplace for measuring quantities such as temperature and weight in relation to both characteristics and causes, it is necessary at the analysis stage to make people responsible for controlling them and to prepare appropriate instructions. These instructions should include the following:
   (a) Instrument purchasing specifications, standards for incoming inspection, and installation checks.
   (b) Method of use for each instrument, assignment of people responsible for control, standard control methods.
   (c) Work standards for cleaning, maintenance, calibration, and inspection of instruments.
   (d) Items concerning repair of instruments.

4A.2 Investigating Sampling Methods

Sampling is an important topic forming the foundation of mathematical statistics and statistical quality control. Sampling methods cannot be treated separately from process analysis and the preparatory stage of process control. I will discuss the general theory of sampling here, but the sampling method must be chosen for each particular situation in consideration of knowledge obtained from process, analysis, existing technical knowledge, the purpose of controlling the process, and the state of control of the process.

While sampling is one of the chief buttresses of quality control, when it comes to implementing quality control, it is only part of the problem. From the viewpoint of process analysis and control, we regard a process as a population, and a lot resulting from a process is obviously a sample from that process. Since the data we obtain are either data on a lot or data on a sample taken from a lot, when we think of sampling in process analysis or control, we must consider how to sample in order to control the process (i.e., the population; see Fig. 2.1). When selecting a sampling method, we must consider the following points:

(i) the sampling methods presently used
(ii) the purpose of sampling
(iii) sampling locations
(iv) sampling errors
(v) sampling and subgrouping methods
(vi) the control status of the process and sampling interval
(vii) sampling method standards.

(1) Investigating the sampling methods presently used

Some of the sampling methods used in the past may have been rationalized empirically, but the following irrational aspects often exist, and all sampling methods should therefore be reviewed:

(a) The purpose may be unclear: it may be unclear what the data obtained from sampling are to be used for: process control, process analysis, quality assurance, or inspection.
(b) The method may be unsuitable for the purpose:
  (i) Sampling for inspection purposes may be in use for process control.
  (ii) The method may be unreliable.
  (iii) The method may not have the appropriate precision.
  (iv) Unnoticed bias may be present.
(c) The sampling method may not be controlled.

There are not a few examples of factories in which process analysis and control have foundered because of irrational sampling methods.

(2) What is the purpose of sampling?

Sampling can often be rationalized simply by clarifying the technical purpose of taking samples. This may be done by considering quality standards, the requirements of the next process, the results of analyzing the process in question and the state of control of the previous process, and initially selecting what is thought to be a rational method in light of technical, empirical, and statistical knowledge. It is even better if the method can be used jointly for lot assurance and process control. The method can then be gradually improved later by carrying out more detailed statistical studies and implementing process control.

Generally, when we are trying to control a certain dispersion, we should take samples, collect and subgroup data so as to reveal the underlying variations. Thus, the purpose of sampling is closely related to subgrouping in control charts.

(3) Deciding on sampling locations

If the quality characteristics in question are selected and the check points, control characteristics, and objectives are decided, the sampling locations will decide themselves. However, the following points should be noted:

(a) One of the most important principles of process control is stratification; sampling should therefore in principle be carried out after stratification. In other words, sampling locations should be chosen from technical considerations to enable sampling with stratification by raw material, machine, process route, time of day, work group, etc., to be carried out and to provide as much information as possible. It is usually unsatisfactory to take samples after items from different sources have become mixed together.

(b) It is best to choose locations that facilitate random sampling or sampling at fixed intervals (systematic sampling). The simplest way of doing this is to take samples while a lot is passing through the process, i.e., while it is actually moving. Random sampling may occasionally be easier to perform on stationary lots but it is usually more difficult.

The above points should be considered when designing or reorganizing factory layouts.

(4) Sampling errors

As mentioned in the section on measurement error, we are particularly concerned in process control with precision and reliability. Bias is of course also a problem in quality assurance. Whether one is considering existing sampling methods or planning new ones, it is the responsible department's duty to control the sampling procedures, ensure that reliable sampling is performed, and clarify its precision and bias statistically.

Reliability must be controlled so that when abnormal data appear, the workplace is unable to dodge the issue by blaming poor sampling.

If we call the variation in the process average σ_P^2, the sampling reproducibility σ_S^2, and the measurement reproducibility σ_M^2, the variation in the data, σ^2, is given in the case of discrete units by:

$$\sigma^2 = \sigma_P{}^2 + \sigma_S{}^2 + \sigma_M{}^2 \qquad (4A.1)$$

(See Section 4A.9.)

The variation in $\bar{x}$, $\sigma_{\bar{x}}^2$, when the subgroup size is n, is given by:

$$\sigma_{\bar{x}}{}^2 = \sigma_P{}^2 + \frac{1}{n}(\sigma_s{}^2 + \sigma_M{}^2) \qquad (4A.2)$$

In the case of bulk materials, particularly when forming compound samples, it facilitates understanding to split the process variation σ_P^2 into the between-subgroup variation σ_b^2 and the within-subgroup variation σ_w^2. If we do this, we have:

$$\sigma^2 = \sigma_b{}^2 + \sigma_w{}^2 + \sigma_S{}^2 + \sigma_M{}^2 \qquad (4A.3)$$

And the variation in $\bar{x}$ is given by:

$$\sigma_{\bar{x}}{}^2 = \sigma_b{}^2 + \frac{1}{n}(\sigma_w{}^2 + \sigma_S{}^2 + \sigma_M{}^2) \qquad (4A.4)$$

In formulas 4A.2 and 4A.4, the first term on the right-hand side is the between-subgroup variation and the second term is the within-subgroup variation. Of the different components of σ^2, the more immediate, short-term, within-subgroup variation is often the largest, so it is best to choose a sampling method that will reveal this.

With discrete units, the sampling precision is not so much of a problem, since the within-subgroup variation indicates the within-lot variation, and the subgroup size and method of subgrouping are therefore more important.

With bulk materials, when measurements are performed on every sample unit and subgroups are formed from a number of these measurements, $\sigma_w{}^2$ and $\sigma_s{}^2$ become identical and the formulas are the same as formulas 4A.1 and 4A.2 for discrete units. However, $\sigma_S{}^2$ depends on the way in which the sampling unit is chosen. In contrast, when compound samples or average samples are taken, the applicable formulas are 4A.3 and 4A.4 (where $\sigma_S{}^2$ also includes the sample reduction error $\sigma_D{}^2$).

Since what we want to control with the *R* chart is mainly $\sigma_w{}^2$, it is desirable to keep $\sigma_S{}^2$ to about 1/10 of $\sigma_w{}^2$ if possible. When $\sigma_S{}^2$ is large, it is sometimes advantageous to take two or more separate samples by the same sampling method and use these as a subgroup.

In the $\bar{x}$ chart, the variation in $\bar{x}$ is given by formula 4A.4 and we wish to control $\sigma_b{}^2$ within control limits drawn by the second term in the formula. Since $\sigma_S{}^2$ is multiplied by $1/n$, it is not really a problem. However, even in this case, if $(1/n)\sigma_S{}^2$ is much larger than $\sigma_b{}^2$, or if the sampling method is not controlled and the sampling is unusually unreliable, people in the workplace sometimes draw up control charts in the fond belief that they are controlling the process, when what they are in fact controlling is the sampling. When this happens, the causes of out-of-control points often remain obscure.

As explained above, when carrying out process control, we must thoroughly check our sampling methods statistically for each process and ensure that they are reliable and of reasonable precision.

(5) Sampling method and subgrouping

Sampling methods can be classified in various ways, but here I would like to discuss only what is necessary for process control in relation to subgrouping. It is best to stratify in any situation whenever possible, since random sampling from the whole process without stratification loses information. However, when we reach the stage of actually sampling from something in front of us, we must remember to observe the principle of random sampling. In other words, we take random samples from groups whose internal variation we do not need to know about; when we do need this information, we stratify as much as possible before sampling. I would like to start by briefly discussing the sampling of discrete items.

(a) It is naturally advantageous to stratify the material into lots as it passes through the process and to take samples from each stratum. There are three ways of doing this (see Fig. 4A.1):

(i) Take a lot manufactured in a certain time period, stratify it according to, for example, machine, etc., take random samples of *n* items from each stratum, and use these as subgroups. This

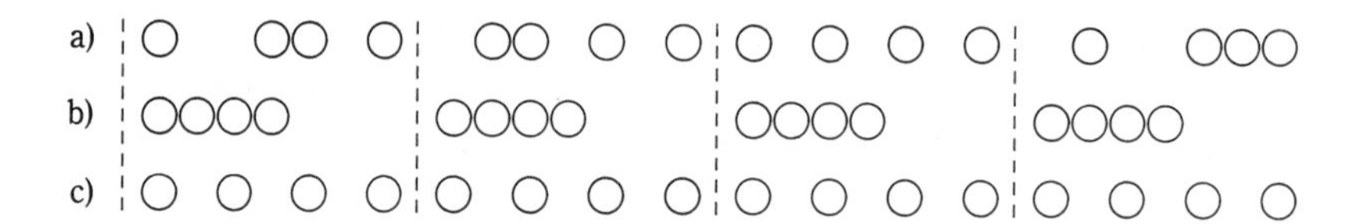

Figure 4A.1 Sampling Methods

method treats the products made in a certain time period as a lot and attempts to control the process using the variation within lots as a datum. This is satisfactory when the variation within lots satisfies the quality standards.

(ii) At set intervals, sample n successive product units produced by the process and use them as a subgroup. This method is suitable when the process is relatively well controlled and its operators understand the philosophy of quality control. If this method is used before this stage is reached, there is a danger of too many out-of-control points making the control charts unsuitable for controlling the process, or of missing process variations or slipshod work between samples.

(iii) Take n samples at fixed intervals within a certain time period and use these as a subgroup. This method enables the within-subgroup variation within a certain time period to be detected quickly, but the within-subgroup variation will be larger than in method (ii). This method is often used in ordinary process control, particularly in its initial stages.

Each of the above three methods has its advantages and disadvantages, and it is impossible to make any sweeping statements about which is best, since this depends on the state of the process. The third method provides the most information about the process, but there is a risk of overlooking variations between sampling intervals. There is less chance of missing this variation with the first method, but information on time variation within subgroups is lost. The second method minimizes the within-subgroup variation but neglects variation occurring after one sample and before the next.

The most important thing with all three methods is to stratify the materials into lots for passage through the process and to make sure the lot history and data history are clear, as well as to decide on the sampling interval and the time period to be treated as subgroups.

Expressed in the abstract, the sampling interval should be such that no assignable causes can slip through the sampling net. When relatively uniform raw material lots are supplied and processed stably, machinery is infrequently adjusted, and there are technical grounds for believing that abnormalities are more likely to occur systematically than sporadically, it is relatively easy to decide on the time at which samples should be taken. For example, it might be every time a new raw material lot is introduced, an adjustment is made, or a shift is changed. However, when raw material lots are fed to the process haphazardly and machinery is constantly being adjusted, it is not so easy to decide on the sampling interval. In such a case, the first consideration is radically to reorganize and standardize the way in which raw material lots are supplied and to standardize equipment adjustment.

Nevertheless, it is better at least to do something until such a reorganization can be achieved, so sampling should be performed in the meantime at fixed intervals, the data subgrouped, control charts drawn, and reliable recurrence-prevention measures taken every time lack of control is indicated. If this is done, even if assignable causes arising between samples are at first overlooked, they will be spotted sooner or later and the process will gradually come under control.

(b) With bulk materials, data may be produced in the form of either composite samples or individual sampling units. Either case will be identical to (a) above if the unit from which the data are obtained is treated as a discrete unit. However, I will leave this point to specialist works on sampling, since it is a fairly complex matter involving questions of how to prepare composite samples, how to carry out sample reduction, and how to decide the size of the sampling unit.

Taking composite samples also involves the questions of whether snap sampling is satisfactory and at what interval the samples should be averaged. It is generally disadvantageous to average them over one day, since information on the variation within a day will be lost, and a sampling method that will reveal the average over 1 hour, 4 hours or one shift is often better. Whatever interval is used, it is best to install automatic sampling equipment.

When individual sampling units are measured—for instance, in sampling textiles and paper—the sampling unit should in principle be decided according to quality standards and customer requirements. Existing test methods should be reviewed, since many are unsuitable for these purposes or for customer requirements.

(6) The control status of the process and the sampling interval
As stated above, the sampling interval should in principle be chosen so as not to overlook process abnormalities. However, when a process is out of control and changing rapidly and it is impossible to tell when an abnormality may occur, we cannot be constantly sampling, measuring, and looking out for abnormalities. In such a case, we must standardize work procedures and actively suppress abnormalities. Meanwhile, since long-period cyclic and trend-like changes are often present along with rapid short-term fluctuations when the above type of variation occurs, we must find the average over a certain time period and either adopt some method of controlling these long-period cyclic changes and trends or install automatic recording and control equipment. However, this is sometimes a difficult technical problem, and it is questionable whether we should rush to install such devices just because a process is unstable. We should only do this if it can be objectively proven beneficial after a full statistical and technical study using control charts.

It is also not necessarily profitable to reduce the sampling interval immediately on the grounds that the process is unstable. In fact, even if the interval is rather long and some abnormalities occurring between samples are overlooked, the state of control will gradually improve if we patiently work to bring the process under control, revising work standards or taking other action to prevent assignable causes from recurring whenever an abnormality is discovered. The sampling interval thus also depends on whether or not the process is acted on without fail to prevent assignable causes from recurring. The sampling interval must be short if control charts are being used for taking prompt action to suppress abnormalities, but it is permissible to lengthen it slightly if reliable recurrence-prevention measures are always taken.

As the state of control of the process gradually improves in this way, the sampling interval can be lengthened, the lot size increased, or the sample size decreased. Lengthening sampling intervals and increasing lot sizes is a great economic benefit of implementing quality control, but this depends on the state of control of the process, the level of awareness of quality control within the company, and the progress of standardization. These items must therefore be promoted.

In general, if progress has been made with the above points, if there are technical grounds for confidence, and if at least 100 successive in-control points have appeared on the control charts, steps such as the following can be taken:

(i) Lengthen the sampling interval by a factor of 2–5.
(ii) Increase the period over which samples are averaged by a factor of 2–5.
(iii) Decrease the sample size by a factor of 1/2–1/5.
(iv) Sample without stratifying.

(7) Formulating sampling method standards

Since sampling is also a type of work, standard methods must be stipulated in the same way as for ordinary processes. Standards may be formulated using the same approach as for ordinary work standards. Items to be decided include the following:

(i) Who is responsible for sampling and in charge of witnessing it and carrying it out.
(ii) When and how sampling should be carried out, by what method, and using what equipment.
(iii) How samples taken should be disposed of, numbered, reduced, stored, transported, etc.

In addition, attention should be paid to points such as the following in the same way as for ordinary standardization:

(i) Always put standards in writing.
(ii) Constantly strive to rationalize.
(iii) Ensure that the methods specified are technically feasible in the workplace.
(iv) Give specific criteria for action and use random sampling cards.
(v) Clarify the authority and responsibility of everyone involved.
(vi) Ensure that everyone concerned approves the standards.
(vii) Select reliable, easily controlled methods.
(viii) Use methods of known precision and reliability.
(ix) Select methods that are not susceptible to bias.

(8) Summary

The items discussed above should always be borne in mind during process analysis and must always be reviewed and specified at least once when analyzing a process and moving toward control. In the present state of quality control in Japan, sampling is being investigated in the following three types of factory:

(a) Factories where process analysis and control cannot be attempted without studying sampling methods, owing to the irrationality of the methods used in the past.
(b) Factories that have carried out analysis and control and have improved the state of their processes, and where sampling has now become a problem.
(c) Factories that are experiencing difficulty in introducing quality control and where quality control engineers are enthusiastic about studying sam-

pling because it is the easiest area to attack, i.e., they are escaping into sampling.

That the number of factories of the second type has recently increased is gratifying proof of the progress of quality control.

4A.3 The Concept of Statistical Testing

Some general ideas on statistics were described in Section 2.2, but I would like to treat the procedure for statistical testing in a slightly more theoretical way here, elaborating on the example (of throwing a die) given in that section; thus, each step is here described in specific detail.

(1) Decide on the purpose of the analysis or experiment (i.e., the purpose of making a judgment): to complain if cheating is taking place.
(2) Make assumptions: we will assume that the die is true, not loaded.
(3) Form a hypothesis (a null hypothesis): "The die is being thrown fairly, without cheating, and even and odd numbers appear with an equal probability of 1/2."
(4) Calculate probabilities from the data in view of the distribution of statistics, assuming the hypothesis to be correct: calculate the probability of obtaining five even numbers in succession when the die is thrown fairly. In this case, the probability is:
$$(1/2)^5 = 1/32 \approx 0.03 = 3\%$$
(5) Compare the probability with the significance level or confidence level and pass judgment: the hypothesis is normally definitely wrong if the probability is 1% or less, and we should decide that the hypothesis is rejectable in this case. In the example of the die, the probability of 7 successive even numbers appearing is 1/128, which is less than 0.01, so the hypothesis that cheating is not taking place can definitely be rejected if this happens. In other words, we can say, "Cheating is taking place." If the probability is 5% or less, we should decide that the hypothesis appears rejectable, i.e., if 5 successive even numbers are obtained, we can say, "Cheating appears to be taking place." If the probability is 10% or less, the hypothesis is a little dubious, i.e., we can only say, "Cheating may possibly be taking place."

    These probabilities are called "confidence levels" or "significance levels." As explained in Section 2.2, a confidence level is the probabil-

Table 4A.1 Standard Significance Levels

	Significance level (α%)		
	Possible difference	Apparent difference	Definite difference
Laboratory	10–30	5–10	1–5
Pilot plant	5–20	1– 5	1
Workplace	5–10	1– 5	0.1–1

This table shows one example; in practice, the values depend on the action to be taken.

ity of committing a type I error, e.g., deciding that cheating is taking place when in fact it is not (i.e., rejecting the hypothesis when in fact it should not be rejected). In control charts, this significance level is set at around 0.3% (equivalent to approximately 8 even numbers in a row in the case of the die), so it is acceptable to decide that there is definitely something wrong with the process if a point falls outside the control limits.

(6) Take action according to the decision: complain that you are being cheated.

We must note that steps 4 and 5 are often combined and a comparison is made with the significance level in step 5 without calculating the probability in step 4. For example, the t (10, 0.05) value may be compared with the t_0 value as in the following example.

(1) Test of difference in means

The ideas described above will now be illustrated by an actual example in which we test whether there is a difference in population means, i.e., whether the mean of a distribution has changed.

The average yield of a certain product appears to differ depending on the type of raw material used. To check this, the yield was measured six times for each raw material in random order and the values shown in Table 4A.2 were obtained.

(a) Purpose of experiment: to buy the better raw material if there is a difference and the cheaper if there is none.

(b) Assumptions:

(i) That the tests are carried out with the process in the controlled state.

(ii) That the measurements are performed in random order.

(iii) That the experimental error variance is the same for both raw materials.

(c) State hypothesis: "There is no difference in yield between raw material A and raw material B, i.e., the yield is the same for each." This is shown symbolically as:

H_0: $\mu_A = \mu_B$ (null hypothesis)

(d) Calculate probability based on hypothesis: at this stage, a calculation is performed using statistical formulas. This is the only difference between this method and old-fashioned common-sense judgment.

The manual method of calculation is described below, but it can be done more easily with a calculator or computer.

Step 1: Simplify the calculation by coding the data as shown in the table. In this example, $X = x - 80$.

Step 2: Calculate the mean and the sum of the squares of the deviations, S (see Section 2A.2).

$\bar{x}_A = 83.83$ $\bar{x}_B = 80.83$

$S_A = 52.83$ $S_B = 46.83$

$n_A = 6$ $n_B = 6$

Step 3: Calculate t_0 from the following formula:

$$t_0 = \frac{\bar{x}_A - \bar{x}_B}{\sqrt{\dfrac{S_A + S_B}{n_A + n_B - 2}\left(\dfrac{1}{n_A} + \dfrac{1}{n_B}\right)}} = \frac{83.83 - 80.83}{\sqrt{\dfrac{52.83 + 46.83}{6 + 6 - 2}\left(\dfrac{1}{6} + \dfrac{1}{6}\right)}} = 1.65$$

It has been proven statistically that, if $\mu_A = \mu_B$, t_0 follows the t-distribution with degree of freedom $\phi = n_A + n_B - 2$. The test is therefore performed using the t-distribution.

Note: Formulas of this type have been obtained statistically for various situations.

(e) Compare probabilities and make a decision:

Step 4: Find the value of t at a probability (i.e., a significance level) of α (e.g., 0.05, 0.01) from t-distribution tables at $6 + 6 - 2 = 10$ degrees of freedom. This is generally expressed as $t\ (n_A + n_B - 2, \alpha)$.

$t\ (10, 0.05) = 2.228$ $t\ (10, 0.01) = 3.169$

Thus, $t_0 = 1.65 < t\ (10, 0.05) = 2.228$, and the hypothesis that $\mu_A = \mu_B$

Table 4A.2 Difference in Yields with Raw Materials A and B

Raw material A	Raw material B	$x_A - 80$	$x_B - 80$	$(x_A - 80)^2$	$(x_B - 80)^2$
83%	80	3	0	9	0
79	85	−1	5	1	25
83	83	3	3	9	9
87	80	7	0	49	0
88	76	8	−4	64	16
83	81	3	1	9	1
Total –	–	23	5	141	51
Average 83.83	80.83	3.83	0.83		

cannot be rejected. In other words, we cannot say that there is any difference between the two sets of data. A difference of $\bar{x}_A - \bar{x}_B = 3.00\%$ between the means can occur by chance when there is this degree of dispersion in the data.

(f) Take action according to the decision: since difference could not be proven, it was decided to use the cheaper raw material, A, in future.

In this case, even if a difference were found, the decision as to which raw material to use should only be made after estimating the difference and taking account of technical and economic considerations.

(2) Test of difference in variance

This test is performed using F distribution tables. The value of $F_0 = V_1/V_2$ (where V_i is an unbiased estimator of variance and $V_1 > V_2$) is compared with the value of $F(\phi_1, \phi_2, \alpha/2)$ from F distribution tables, and a difference can be said to exist at a confidence level of $\alpha\%$ if $F_0 \geqq F(\phi_1, \phi_2, \alpha/2)$.

4A.4 The Concept of Statistical Estimation

As described in Sections 2.2 and 2.10, data always contain dispersion and various types of error, making it impossible to estimate the true population values (population parameters, e.g., population mean and variance) exactly. We therefore estimate them at a certain precision or within certain limits. The distributions of statistics mentioned in Section 2A.3 may be used for this. The following points are advantageous in making statistical estimates:

(i) There should be no bias, i.e., the estimates should be unbiased.
(ii) The precision or confidence limits of the estimates should be known to a definite probability.
(iii) Statistics should be used that minimize the sample size and give good precision (in other words, statistics with good efficiency).

Population mean is estimated by:

$$E(\bar{x}) = \mu \quad \therefore \quad \hat{\mu} = \bar{x}$$

If β is set at a precision of 95% probability,

$$D(\bar{x}) = \sqrt{\frac{N-n}{N-1}} \frac{\sigma}{\sqrt{n}} \approx \frac{\sigma}{\sqrt{n}} \qquad \text{(when } 1/10 \geqq n/N\text{)}$$

$$\therefore \quad \beta = 1.96(\bar{x}) \approx 2\mathrm{D}(\bar{x})$$

The confidence interval at a probability (confidence level) of 95% is given by:

$$\bar{x} - 1.96\mathrm{D}(\bar{x}) \leqq \mu \leqq \bar{x} + 1.96\mathrm{D}(\bar{x})$$

When σ is not known,

$$\bar{x} - t\,(\phi, 0.05)\sqrt{\frac{V}{n}} \leqq \mu \leqq \bar{x} + t\,(\phi, 0.05)\sqrt{\frac{V}{n}}$$

where ϕ (the number of degrees of freedom) is the number of degrees of freedom of V.

The main feature of estimating statistically in the above way is that the error and confidence limits are clear. For example, when a mean of 74.3% is obtained, the data will probably be used differently depending on whether the precision at a probability of 95% is $\pm 1\%$ or $\pm 10\%$.

4A.5 Difference in Means of Two Sets of Corresponding Continuous Data—Simple Method

Some data on product yield have been obtained by splitting a raw material lot into two every day and feeding each half to a different reactor (Reactors 1 and 2; see Table 4A.3). Is there a difference in yield between Reactors 1 and 2?

In this example, we can say that the two sets of data correspond to each other, since they originate in the same raw material lot used on the same day.

Table 4A.3 Yield Data (%)

No.	Machine 1	Machine 2	Sign	No.	Machine 1	Machine 2	Sign
1	85	64	+	21	80	85	−
2	73	82	−	22	92	88	+
3	88	76	+	23	70	56	+
4	90	72	+	24	82	83	−
5	99	79	+	25	64	78	−
6	63	64	−	26	84	60	+
7	95	56	+	27	70	80	−
8	97	61	+	28	80	71	+
9	88	56	+	29	70	78	−
10	59	89	−	30	73	71	+
11	75	74	+	31	81	78	+
12	89	74	+	32	94	60	+
13	75	87	−	33	73	75	−
14	85	99	−	34	81	57	+
15	87	83	+	35	89	78	+
16	92	72	+	36	88	71	+
17	75	57	+	37	81	80	+
18	66	90	−	38	73	89	−
19	94	81	+	39	91	77	+
20	89	72	+	40	75	56	+

Step 1: Compare corresponding values from Reactors 1 and 2 and record a plus sign in the +/− column if the Reactor 1 value is higher and a minus sign if it is lower. If the two values are the same, record a zero. Step 2: Count up the pluses and minuses. In this example, there are 27 pluses and 13 minuses. Step 3: Compare the lower total (13 in this case) with the values in Table 4A.4. Here, since the number of data values (k) is 40, we look along the $k=40$ row. The figure in the 0.05 column in this row is also 13. Step 4: If the lower total is higher than the value in the 0.05 column (if it is 14 or more in this case), there is no difference in the means of the two sets of data. If it is higher than the value in the 0.01 column but equal to or lower than the value in the 0.05 column (i.e., if it is 12 or 13 in this example), we can say that there appears to be a difference in the means. If it is equal to or lower than the value in the 0.01 column (i.e., if it is 11 or less in this example), we can say that there is definitely a difference in the means. Since the lower total is 13 in this case, there appears to be a difference between Reactor 1 and Reactor 2. Reactor 1 appears to give a better yield.

The following additional points should be noted concerning this procedure:

First, when this kind of data is analyzed, it is best to have at least 30 pairs, or if possible 50 or more.

Second, in this kind of example, the difference in the means can be ana-

Table 4A.4 Sign Test Chart

k	0.01	0.05	k	0.01	0.05	k	0.01	0.05
20	3	5						
21	4	5	46	13	15	71	24	26
22	4	5	47	14	16	72	24	27
23	4	6	48	14	16	73	25	27
24	5	6	49	15	17	74	25	28
25	5	7	50	15	17	75	25	28
26	6	7	51	15	18	76	26	28
27	6	7	52	16	18	77	26	29
28	6	8	53	16	18	78	27	29
29	7	8	54	17	19	79	27	30
30	7	9	55	17	19	80	28	30
31	7	9	56	17	20	81	28	31
32	8	9	57	18	20	82	28	31
33	8	10	58	18	21	83	29	32
34	9	10	59	19	21	84	29	32
35	9	11	60	19	21	85	30	32
36	9	11	61	20	22	86	30	33
37	10	12	62	20	22	87	31	33
38	10	12	63	20	23	88	31	34
39	11	12	64	21	23	89	31	34
40	11	13	65	21	24	90	32	35
41	11	13	66	22	24	100	36	39
42	12	14	67	22	25			
43	12	14	68	22	25			
44	13	15	69	23	25			
45	13	15	70	23	26			

Note: The figures in the table represent the total of pluses or minuses, whichever is the lower. If the total is higher than the figure in the chart, there is no significant difference.

When $k > 100$, use the value calculated from the following formula, rounded down to the nearest integer:

$$(k-1)/2 - K\sqrt{k+1}$$

Example: When $k = 100$,

$$(100-1)/2 - (1.29\sqrt{100+1}) = 36.6 = 36$$

Probability	K
0.01	1.29
0.05	0.98

lyzed by the usual two-way layout method or by taking the differences between corresponding data values and performing numerical calculation.

Third, this kind of data can be tested in the way described in note 1 of Section 4A.8, using a 50:50 dividing line on binomial probability paper.

Table 4A.5 Numbers of Defectives Produced by Machines 1 and 2

Machine \ Day		1	2	3	4	5	6	7	8	9	10	Total
Machine 1	Number of products	58	63	65	57	70	62	60	52	72	65	624
	Number of defectives	5	7	5	7	3	4	6	11	4	6	58
Machine 2	Number of products	55	65	63	60	65	53	68	50	70	59	608
	Number of defectives	7	6	7	6	6	9	6	12	7	6	72

Total: number of products $n = 1{,}232$, number of defectives $r = 130$

4A.6 Difference in Fractions Defective of Two Sets of Data–Method Using Binomial Probability Paper*

In Section 4A.5, we were concerned with variables. Data on fraction defective can be dealt with in the same way if the sample sizes are approximately the same and correspondence exists, but fraction defective is usually investigated by the following method:

The same parts are being made on two different machines using the same kind of materials every day. Data taken over 10 days are shown in Table 4A.5. Can we say that there is a difference in the fractions defective of the two machines?

Step 1: Find the total number of products and defectives for each machine. In this example, the number of products produced by Machine 1 (n_1) is 624 and the number of defectives (r_1) is 58. The values for Machine 2 (n_2 and r_2) are 608 and 72, respectively.

Step 2: Find the total number of products (n), defectives (r) and non defectives (a) for both machines. In this example, $n = n_1 + n_2 = 1{,}232$, $r = r_1 + r_2 = 130$, and $a = n - r = 1{,}102$.

Step 3: Plot this point on binomial probability paper as shown in Figure 4A.2 with a on the horizontal axis and r on the vertical axis and label it P. When doing this, divide both values by 10 if a is more than 600 or r is more than 300. Here, since a is 1,102 and is therefore more than 600, a and r are plotted as 110.2 and 13.0, respectively.

Step 4: Join P to the origin (O) with a straight line (see Fig. 4A.2). This is called the "dividing line."

---

* See Nakasato and Takeda: *Nikō Kakuritsushi no Tsukaikata* (How to Use Binomial Probability Paper) (revised edition), pub. JUSE Press, 1965.

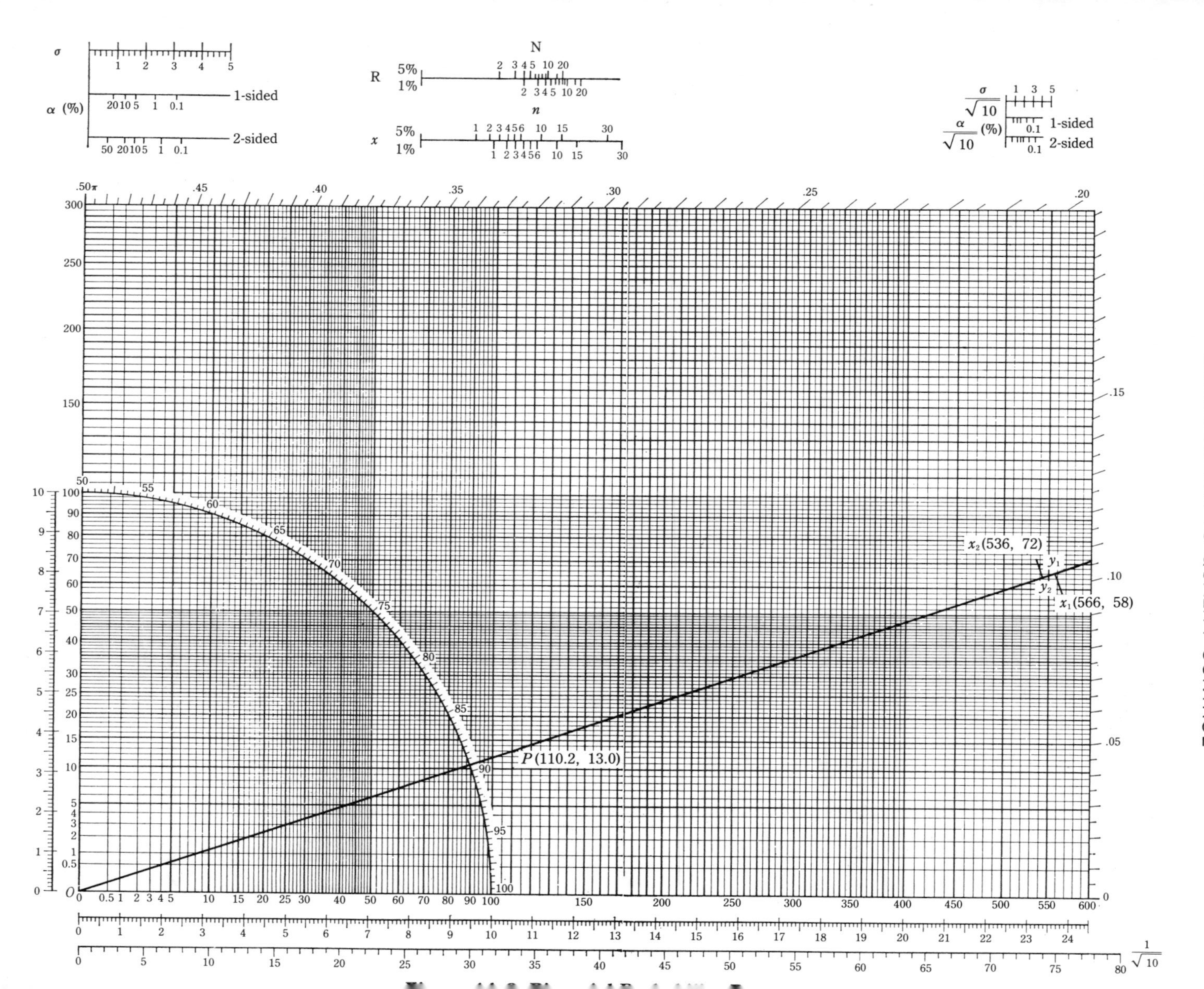
σ
α (%)
1-sided
2-sided
N
R
5%
1%
n
x
$\frac{\sigma}{\sqrt{10}}$
$\frac{\alpha}{\sqrt{10}}$ (%)
1-sided
2-sided
x_2(536, 72)
y_1
y_2
x_1(566, 58)
P(110.2, 13.0)
O
$\frac{1}{\sqrt{10}}$

Step 5: Find the total number of non-defectives for each machine. Here, $a_1 = n_1 - r_1 = 566$ and $a_2 = n_2 - r_2 = 536$.

Step 6: Plot the points $x_1 = (a_1, r_1)$ and $x_2 = (a_2, r_2)$ using the number of non-defectives and defectives for each machine. Here, $x_1 = (566, 58)$ and $x_2 = (536, 72)$.

Step 7: Drop perpendiculars from x_1 and x_2 to OP and label the points of intersection y_1 and y_2, respectively.

Step 8: Measure the distances x_1y_1 and x_2y_2 and compare the sum of these two lengths with the 5% and 1% values on the R scale at the top of the diagram for $N=2$. If the distance on the R scale is longer, we cannot conclude that a difference exists; if it is shorter, we can. In this example,

$$\begin{array}{r} x_1y_1 \approx 5.0 \text{ mm} \\ \underline{x_2y_2 \approx 5.0 \text{ mm}} \\ 10.0 \text{ mm} \end{array}$$

and the total length, 10.0 mm, is less than the 5% distance on the R scale for $N=2$.

Step 9: Conclusion: Since the total length is less than the 5% value, we cannot say that there is any difference in the fractions defective of the two machines.

4A.7 Difference in Means of Two Sets of Data (Variables) when there is No Correspondence

When, for example, machines are assembled from parts supplied by Company A on one day and Company B on the next, the difference in the day or other factors may affect the machines' characteristics, and we cannot claim correspondence in the data. In cases like this, we proceed as follows:

Five samples were taken from each product lot, a certain characteristic was measured, and the averages were plotted in Figure 4A.3. Data on machines made with parts supplied by Company A are shown by crosses, while data on machines made with parts from Company B are shown by circles. Can we say that the performance of machines made with Company A parts differs from that of machines made with Company B parts?

Step 1: Plot the data in time order using different symbols for Company A and Company B, as shown in Figure 4A.3.

Step 2: Draw a horizontal line with half the points above it and half below it (the median line; see notes).

Note 1: The points may also be divided horizontally by another line close in value to the median line. If this is done, the dividing line drawn in step 4 should not be the 25:25 line but a line corresponding to the numbers obtained, e.g., 27:23.

Note 2: Points lying on the median line are not included in the calculation in step 3.

Step 3: Add up the number of points above and below the median line separately for Companies A and B (i.e., add up the number of crosses and circles above and below the line) and arrange them as shown in Table 4A.6. This table is called a 2×2 contingency table.

Table 4A.6 2×2 Contingency Table

	Company A	Company B	Total
Above	18	7	25
Below	5	20	25
Total	23	27	50

Step 4: On binomial probability paper, connect the 25:25 (= 50:50) point Q (because the median line was used) to the origin O with a straight line (the dividing line; see Fig. 4A.4).

Step 5: Plot the points $x_A = (18, 5)$ and $x_B = (7, 20)$ for A and B.

Note: Binomial probability paper uses an approximation to the binomial distribution. When the number of data points is small, it is best to test by constructing "observational triangles." In this example, two right-angled triangles are constructed, one for A from the points (18, 5), $(18+1=19, 5)$, and $(18, 5+1=6)$, and one for B from the points (7, 20), $(7+1=8, 20)$, and $(7, 20+1=21)$. Testing is then performed according to steps 6 and 7, taking the shortest distance to the dividing line. In this example, the total of the distances to the dividing line from points (18, 6) and (8, 20) is measured according to step 7. Since the total, 24 mm, is longer than 18.5 mm, the conclusion remains the same. When the total is close to the distance on the R scale, precautionary testing should be performed using observational triangles.

Step 6: Drop perpendiculars $x_A y_A$ and $x_B y_B$ to OQ from x_A and x_B, respectively.

Step 7: Test in the same way as in Section 4A.6 by calculating the total length $x_A y_A + x_B y_B$ and comparing it with the value on the R scale for $N=2$.

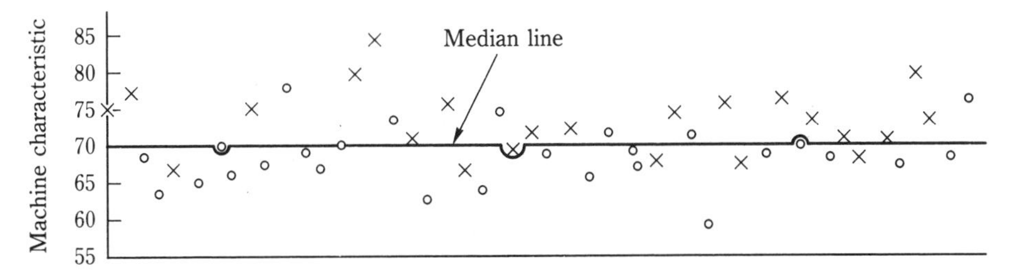

In order of product lots

×: assembled from company A parts
○: assembled from company B parts

Figure 4A.3 Graph of a Certain Characteristic ($n=50$)

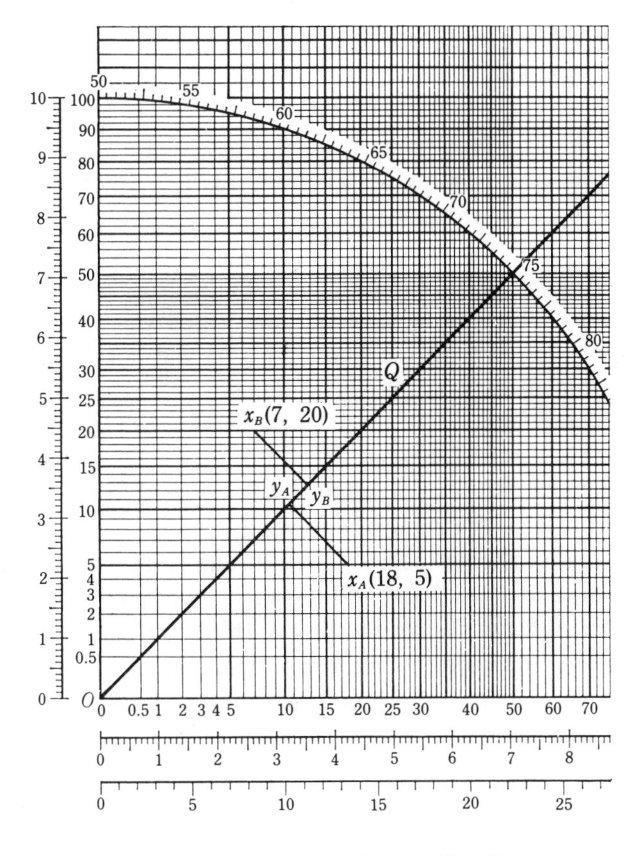

Figure 4A.4 Binomial Probability Paper

Here,

$x_A y_A$ = 14.5 mm
$x_B y_B$ = 13.0 mm
Total: 27.5 mm

The distances on the 5% and 1% R scales for $N=2$ are 14 mm and 18.5mm, respectively and the total length is greater than both of these.

Step 8: Conclusion: In this case, since the total length is greater even than the distance on the 1% R scale, we can say that the performances of the machines made with Company A and Company B parts are definitely different.

4A.8 Relation between Corresponding Sets of Data—Correlation

As an example, we have some corresponding data on atmospheric humidity and the percentage moisture content of a certain textile product. Can we say that the humidity affects the moisture content?

Step 1: Plot the data on a correlation chart with the cause on the horizontal axis and the effect on the vertical axis (see Fig. 4A.5). In this example, the data are plotted with the humidity x on the horizontal axis and the moisture content y on the vertical axis.

Step 2: Draw a horizontal line at a certain value of y with half the points above and half below (the median line XX').

Step 3: Draw a vertical line at a certain value of x with half the points on the left and half on the right (the median line YY').

Step 4: Count the number of points n_1, n_2, n_3 and n_4 in each quadrant (usually, $n_1 = n_3$ and $n_2 = n_4$) and add the number of points in the first quadrant (top right) to the number in the third quadrant (bottom left), i.e., find $n_1 + n_3$. Also add the number of points in the second quadrant (top left) to the number in the fourth quadrant (bottom right), i.e., find $n_2 + n_4$. Here,

$$n_1 + n_3 = 34 + 34 = 68$$
$$n_2 + n_4 = 16 + 16 = 32$$

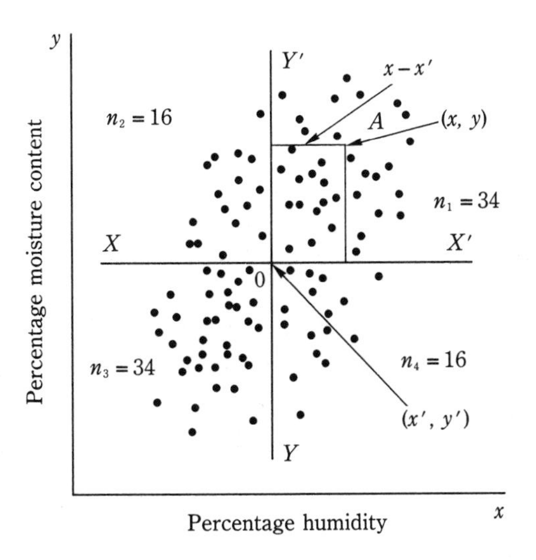

Figure 4A.5 Relation between Humidity and Moisture Content

When this is done, the total number of points should equal the original number. Do not include any points lying on the median lines in the calculations.

Step 5: Compare the smaller of the two totals with the values in Table 4A.4. In this example, the smaller total of 32 is less than the 1% value of 36 for $k = 100$.

It should be noted that if $n_1 + n_3 > n_2 + n_4$, the correlation is positive, but if $n_1 + n_3 < n_2 + n_4$, the correlation is negative.

Step 6: Conclusion: In this example, we can definitely say that a positive correlation exists. In other words, we can say that the moisture content of the textile will certainly be high if the atmospheric humidity is high.

There are various other methods of testing and making decisions apart from the above, but these will not be covered here. The method we have discussed is one of the most commonly used.

Note 1: This test may also be performed using binomial probability paper, as described below:

Step 1: Plot point A $(n_1 + n_3, n_2 + n_4)$ on binomial probability paper as shown in Figure 4A.6. For example, let us take A as the point (110, 40). When doing this, plot the larger value (110 in this case) on the horizontal axis.

Step 2: If A lies outside the 5% or 1% limits of the dividing line, a correlation

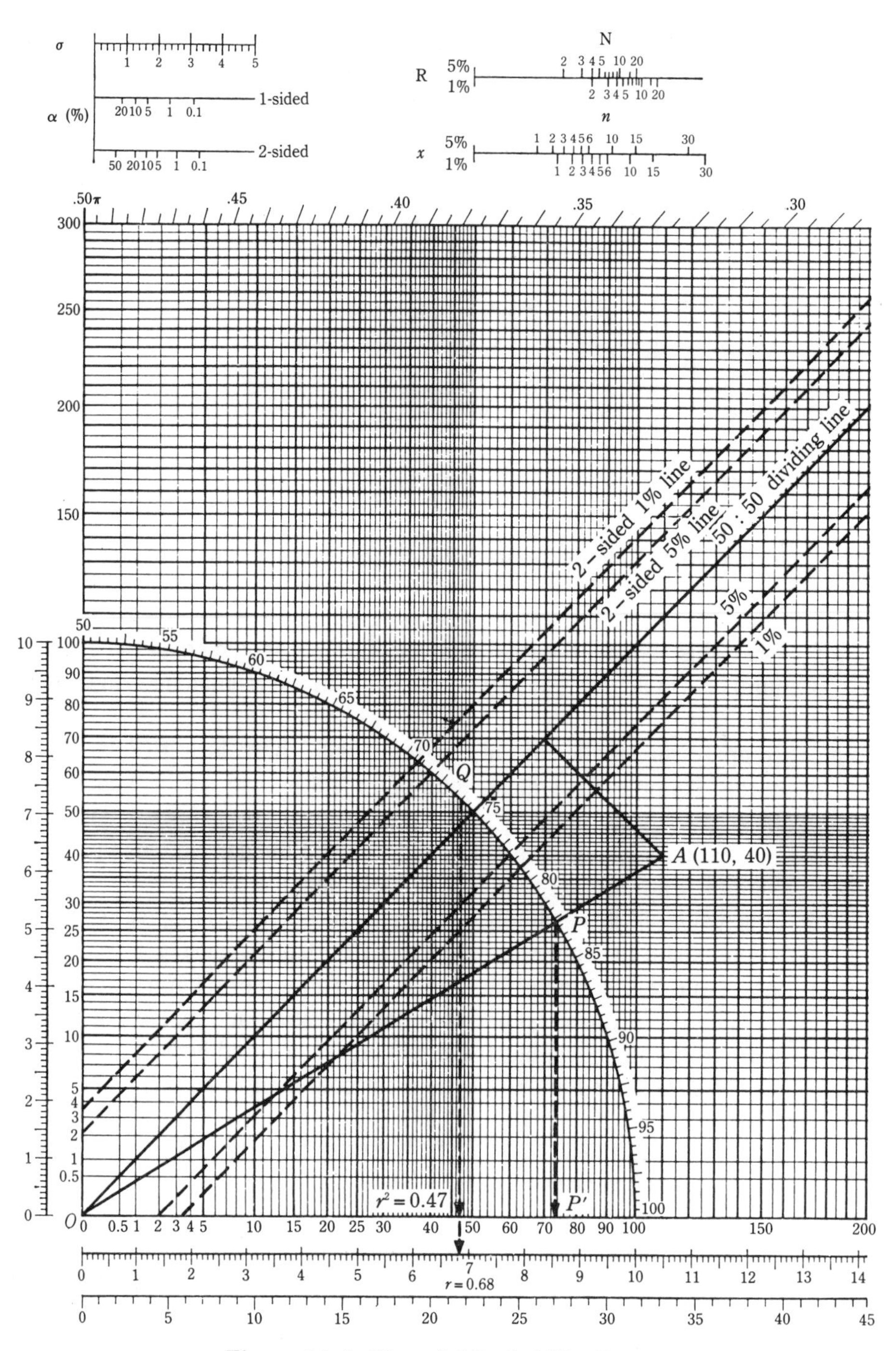

Figure 4A.6 Binomial Probability Paper

can be said to exist. Since the point (110, 40) does lie outside these lines, a correlation definitely exists.

Step 3: The degree of correlation between x and y is expressed by a quantity called the "correlation coefficient," r. If the correlation coefficient is zero, there is no correlation. If it is unity, there is a very close correlation between the two sets of data and they may, for example, lie on a straight line. The correlation coefficient can be obtained easily from the data by using binomial probability paper, as follows:

Find the point of intersection P between the arc of the circle and the straight line joining O to A, drop a perpendicular from this point to the horizontal axis, and read off the value at P', the point where the perpendicular intersects the axis. In this example, it is 74. Then find the point Q (the value of 74 on the arc scale), drop a perpendicular from this point to the horizontal centimeter scale, and read off the value. A correlation coefficient of 0.68 is obtained by multiplying this value by 1/10. When this is done, 1/100 of the value on the horizontal axis gives a coefficient of determination (r^2) of 0.47.

Note 2: Calculating a regression line by the median method: If correlation is tested by a method such as the above and is judged to exist, a straight line estimating y from x (a regression line estimating y from x, i.e., the regression line of y on x) can be obtained as follows:

Step 1: Draw one median line through the points on the right of YY' dividing them equally top-to-bottom and another dividing them equally left-to-right, and find the point of intersection of the two lines.

Step 2: Repeat step 1 for the points on the left of YY'.

Step 3: Draw a straight line joining the two points of intersection found in steps 1 and 2. This is the regression line estimating y from x.

Note 3: The regression line estimating x from y can be found by carrying out steps 1 to 3 for the points above and below XX'.

Note 4: The correlation coefficient can be calculated from the following formula:

$$r = \frac{\Sigma(x_i - \bar{x})(y_i - \bar{y})}{\sqrt{\Sigma(x_i - \bar{x})^2 \Sigma(y_i - \bar{y})^2}} = \frac{\Sigma x_i y_i - (\Sigma x_i \Sigma y_i)/n}{\sqrt{\{\Sigma x_i^2 - (\Sigma x_i)^2/n\}\{\Sigma y_i^2 - (\Sigma y_i^2/n\}}}$$

Note 5: The straight line estimating y from x can be calculated from the following formula:

$$y = \bar{y} + b(x - \bar{x})$$

where $b = \Sigma(x_i - \bar{x})(y_i - \bar{y})/\Sigma(x_i - \bar{x})^2$ and the straight line estimating x from y can be calculated from the following formula:

$$x = \bar{x} + b'(y - \bar{y})$$

where $b' = \Sigma(x_i - \bar{x})(y_i - \bar{y})/\Sigma(y_i - \bar{y})^2$

4A.9 Additivity of Variances

The principle of additivity of variances tells us what happens to the net dispersion resulting from adding together two items of data x_1 and x_2, each of which contains its own individual dispersion. For example, some people think that if parts with a variation of ± 0.5 mm are joined randomly to parts with a variation of ± 0.4 mm, the resulting dispersion will be ± 0.9 mm. This is wrong. In cases like this, we must base our thinking on the laws described below.

(1) When there is no correlation between x_1 and x_2 and their respective distributions are $(\mu_1, V(x_1))$ and $(\mu_2, V(x_2))$, and if both x_1 and x_2 are taken at random, the distribution of y, the sum of x_1 and x_2, can be calculated in the following way:

$$y = x_1 \pm x_2$$

$$\mu_y = \mu_1 \pm \mu_2$$

$$V(y) = V(x_1) + V(x_2)$$

(2) Slightly more generally, when y can be expressed as a linear function of several variables x_i, i.e.,

$$y = a + bx_1 + cx_2 + \cdots$$

where a, b, c, etc. are constants, if the x_i terms are independent (i.e., there is no correlation between them) and are sampled randomly, the variance of y can be calculated as follows:

$$V(y) = b^2 V(x_1) + c^2 V(x_2) + \cdots$$

(3) When y can be expressed as the functional relation $y = f(x_1, x_2, \ldots x_n)$, we have, under the same conditions as (2) (independence and random sampling),

$$V(y) \approx \left(\frac{\partial f}{\partial x_1}\right)^2 V(x_1) + \left(\frac{\partial f}{\partial x_2}\right)^2 V(x_2) + \cdots + \left(\frac{\partial f}{\partial x_n}\right)^2 V(x_n)$$

where the variation in each x_i term is within approximately 20% of the corresponding μ_i.

(4) In (2), if there is a correlation between x_1 and x_2 expressed by the population correlation coefficient ϱ, we have

$$V(y) = b^2 V(x_1) + c^2 V(x_2) + 2\varrho bc\sqrt{V(x_1) V(x_2)}$$

The above property of variances is extremely important and is widely used when discussing tolerances and errors or when an effect y is the result of many causes x_i, i.e., when discussing errors in functional relations.

Note: The formula for error variance given in Section 4A.1 is an example of the practical application of the theory of additivity of variances.

Chapter 5
PROCESS CONTROL

5.1 WHAT IS PROCESS CONTROL?

The process control I am talking about here means the control of all kinds of processes. In Japan, the term was often used in the past to describe production control or progress control, but we should really make a distinction between the terms.

The process referred to here is, as stated in Section 4.7.7, "A collection of causes producing a certain result." In other words, it includes not only manufacturing processes, but also work performed in various ways. I would therefore like the principles described here to be used for controlling not only production processes but all types of work process.

The control methods described in this chapter may be implemented basically by applying the approach described in Section 1.5, which readers might want to review carefully before studying the present chapter. At all events, good process control first requires proper process design and analysis (see Chapter 4). Some further requirements are as follows:

(i) Enthusiasm, leadership, and determination on the part of top management; clear policy and delegation of authority.
(ii) Understanding of the QC approach and real keenness on the part of all employees; instruction, dissemination, education, and training.
(iii) Analysis of work and processes; identification of relationships between characteristics (effects) and causes.
(iv) Promotion of standardization.
(v) Establishment of statistical checking systems; decisive action. Recurrence-prevention countermeasures are particularly important.

Because they are often misunderstood, I would like once again to explain the difference between "control" and a number of terms with which it may be confused:

"Control" and "Improvement"

"*Control*" means maintaining a process in its existing state for the time being. However, if an abnormality occurs, its cause is eliminated and prevented from recurring, and this action results in a slight improvement. In this sense, control is a passive kind of improvement. "*Improvement*" consists of actively seeking out problems, solving them, effecting a permanent fix to prevent any backsliding, and controlling the new situation. It means actively continuing to improve things (see Figure 1.18).

"Control" and "Inspection"

"*Control*" consists of making comparisons with control standards or control limits on control charts, seeking assignable causes in the process or work if the standards are violated or points fall outside the limits, and taking action on the process. In contrast, "*Inspection*" consists of making comparisons with inspection standards, and, if these are not met, taking action on the products, e.g., designating individual products as defective, deeming lots unacceptable, or instituting 100% inspection (screening-type inspection). The idea that 100% inspection should be brought in if a control chart shows that a process has gone out of control is mistaken. Deciding whether or not to perform 100% inspection should be based on the standards for screening-type inspection.

"Control" (elimination of assignable causes) and "Adjustment"

"*Control*" means searching out and removing the causes of abnormalities whenever the points on a control chart fall outside the control limits, i.e., the chart shows lack of control. "*Adjustment*" means performing some regulatory action such as altering the temperature when certain adjustment limits are exceeded. It is generally wrong to perform adjustments such as altering the temperature or changing the bite of a cutting tool when points fall outside the control limits on a control chart. Such adjustments may be made as a stopgap measure, but this does not remove the true causes of the abnormality. Control limits and adjustment limits are usually inherently different.

This chapter discusses process control methods based mainly on control charts, but processes and work can be controlled in more or less the same way without actually using control charts provided that we base our judgment on the control-chart approach.

5.2 Quality Design and Process Design

Since this book is about quality control, the goal of the processes we are talking about is quality. Thus, although this section discusses quality design and process design, it can be applied in more or less the same way to other types of control, e.g., cost control, by replacing the word "quality" with the appropriate word, e.g., "cost."

Sections 1.4 and 1.6 discuss the problem of what to manufacture. Top management must make this decision in view of consumer requirements and business policy; the important thing from the control standpoint is process analysis (discussed in Chapter 4), particularly the in-house and external process capability studies mentioned in Section 4.5.7. Although we may have specified the quality of what we want to produce, it is generally impossible to design it unless we know the relevant process capability and other facts. What is more, the process capability is often insufficient.

In many businesses that do not exercise proper control, process capabilities are often not fully exercised. If process capabilities are improved, either by QC circles formed from shop foremen and workers or by QC teams formed by the QC staff or those responsible for particular processes, it is often found that products of satisfactory quality can be made efficiently with existing equipment. Thus, in quality design, it is best to start by identifying what the process capability would be if it were fully exercised. Conversely, giving people firm targets as a matter of policy often results in dramatic improvements in process capability through process capability studies without any great investment in equipment.

5.2.1 Quality Standards

(1) The distinction between quality standards and control characteristics or control levels

As stated before, a "*quality standard*" is the level of quality that can be obtained in view of costs and policies relating to consumer requirements, process capability, and quality, if the process is actually controlled and its capability is fully exercised. It is usually expressed in the form of a distribution with a mean and standard deviation or with a mean and range. If, in spite of everything, the process capability is deficient and is less than 1.0, there is no choice but to carry out process control and start production as things stand, subjecting the product to 100% screening. Depending on the way work standards are formulated, quality standards are not always control characteristics.

"*Control characteristics*" are characteristics that express the result of a process, and examining these characteristics shows the state of control of the process. In other words, they are the characteristics plotted on control charts and graphs. Control characteristics are therefore not limited to quality and may be anything from production volume to unit cost, necessary amount for a product unit, sales volume, attendance rate, or amount of overtime. If the work standards all deal with assignable causes, some of those factors that can as a matter of common sense be considered to be the results of the process will be control characteristics. However, if the work standards are poorly thought out and are the type that try to shut the gate after the horse has already bolted, causes will be plotted as control characteristics on the control charts and results will be thought of as work standards. This is a confusion of causes and results.

A "*control level*" is the level of a control characteristic; it is normally expressed by $\bar{\bar{x}}$, $\bar{R}$, $\bar{p}$, $n\bar{p}$, $\bar{c}$, $\bar{u}$, etc. The process capability is the level obtained when a process is in the controlled state and is usually taken as the control level, but values corresponding to targets and plans may also be used.

(2) General guidelines for setting quality standards

(a) The quality characteristics we are concerned with in process control are those that customers or the next process are interested in, or their substitute characteristics as identified by quality analysis.

(b) Some means of measuring quality must be provided, as well as suitable sampling and measurement methods; moreover, the size of the errors they introduce must be checked and controlled (see Section 4A.1).

(c) The quality characteristics should be ranked in order of importance and categorized as major, minor, and slight, or major and minor (see Section 1.4.4(4) and Figure 1.4).

(d) Since quality standards are different from quality targets, reasonable ones must be selected in full consultation with customers and the next process. People usually try to purchase goods of as high a quality as possible, but they should really consider the economic aspect and demand products of the lowest permissible quality level. This should also always be borne in mind when purchasing raw materials, since it will lead to reasonable prices for your products. In general, technology consists of making good products with raw materials of poor quality. However, it is advantageous to set quality levels slightly higher than inspection levels and do away with inspection. Analyzing this statistically and technically is a kind of value analysis (VA).

(e) Quality standards must be set for each complete and every intermediate process, and care should be taken not to confuse quality standards with inspection standards.

(f) Quality standards must have a certain tolerance.
(g) Quality standards must be constantly revised and rationalized.
(h) What quality standards are to be decided under whose authority and responsibility should be spelt out for each quality characteristic. The technical department generally investigates them, and draft plans are then drawn up by the quality control or quality standards committees or their subcommittees. Quality standards for the final product should be decided in line with top management policy, and raw materials standards or standards for intermediate processes should either be decided by the technical department or at the managerial level one stage higher than the process. For example, a department head might set the standards for the final products from the sections in his department, the factory manager for the final products from each department, and the company president for the final products from each factory.
(i) It is desirable for technical standards and work standards to be laid down for the purpose of achieving the quality standards.

(3) Comparison with specifications, quality targets, etc.

As discussed in Chapter 4, if a process is analyzed thoroughly, if the process capability is identified, and if the process is controlled in accordance with the work standards, the state of the process and the dispersion in the products that will be produced in the future will be known. We should consider separately whether this satisfies customer requirements, specifications, and standards, or whether it meets the quality levels required by the next process. We can only be satisfied with the process capability and quality standards when the quality levels actually emerging from the process satisfy our customers' standards (see Sections 2.8 and 4.7.7).

We can normally assume that the specifications are satisfied and hardly any out-of-specification product is produced if the control chart shows that the process is substantially in a state of control, there are at least 100 data points, and the raw data lie within the specification limits, and the value of $\bar{x} \pm 4s$ calculated from histograms lies within the specification limits. If $\bar{x} \pm 5s$ lies within the limits, we can claim a fraction defective of ppm order.

When the actual process capability is compared with specifications and control targets using histograms, etc., and is found not to match them, the following actions should be taken:

(a) When the process capability exceeds the specifications—
 (1) Narrow the specification limits to about $\pm 4s$ or $\pm 5s$.
 (2) When the specifications can amply satisfy consumer requirements

as they are, increase the process dispersion or change the process average when this is economically advantageous.

(b) When the process capability does not meet the specifications—
   (1) When the process average is out of place, alter it, if this is a simple technical matter.
   (2) Since the variation and $\bar{R}$ cannot be changed at will, special action must be taken to "exterminate R" when these are too large (see Section 3.9.2).
   (3) If there is no way of changing the process capability to produce products that meet the specifications even after the various technical and statistical aspects have been investigated, the following steps should be taken:
      (i) Consider relaxing the specifications; existing specifications and standards are often statistically and technically irrational.
      (ii) Change the process capability by making radical technical improvements to the process.
      (iii) If possible, carry out 100% screening to remove defective products.
      (iv) When the process capability falls far short of the specifications, segregation may make the product usable. For example, selective fitting or assembly may be used. Measurement errors must be taken into account when this is done.

(4) Control and revision of quality standards

Since quality standards are like living things, they must be controlled and revised so that they never become out of date. To achieve this, standard procedures for setting and revising quality standards must be laid down, and regulations for controlling them must be formulated.

Quality standards should be revised on the following occasions:

(i) When consumer requirements change (e.g., for different countries and different customers).
(ii) When the company's business policy changes.
(iii) When process capabilities change, technical modifications are introduced, or work standards are revised.
(iv) When the product changes as a result of a change in raw materials.
(v) When the economic climate changes.
(vi) After a certain period has elapsed since the last quality standards were issued.

5.2.2 Process Design, Process Analysis, and Preparation of Quality Control Process Charts

"*Process design*" here means the type of design that uses QC flow charts (quality control process charts, or "QC process charts") for the manufacture of a particular product to specify how to control the various assignable causes and build quality in the process. It could also be called the quality control planning of the product (see Section 4.7.5 and Figures 4.5 and 4.6).

The difference between quality control process charts and ordinary flow charts is that the former are more detailed and go as far as showing relations with work standards, sampling method standards, measurement standards, raw material standards, and other standards for controlling assignable causes, together with the check points (the causes, i.e., check sheets) and checks to be carried out through control characteristics (the results, i.e., control charts) needed for controlling the process and assuring quality. The charts specify where and when these checks are to be done and who is to do them, as well as stating what quality characteristics are to be inspected by whom. The existence or absence of related standards, together with their numbers, conditions, etc., should be briefly noted on these flow charts. Also, when the flow charts are being drawn up, cause-and-effect diagrams should be prepared for each characteristic to ensure that no assignable causes are overlooked.

(1) When work is already proceeding or the factory is already operating

In this case, as described in Section 4.7, the work or process should be analyzed and the QC process chart prepared while considering how to proceed in order to achieve a good result or a good-quality product, bearing the above points in mind. The chart is then tried out in practice and revised in light of the results. It is then further revised while listening to the requirements of customers and the next process, or taking action to prevent the recurrence of abnormalities. Revising the QC process chart improves quality, cuts costs, boosts productivity, and raises technical levels.

(2) When starting new work or developing new products or technology

As described in Section 4.7.5, QC Process Chart I is prepared at the design and planning stage and is gradually perfected during prototype fabrication and pilot production. QC Process Chart II is then prepared to ensure that good process control can be carried out during initial and full-scale production and that production startup proceeds smoothly. The technique for doing this will improve at each experience of new product startup.

QC process charts and cause-and-effect diagrams are pivotal for process control. They may also be used for reviewing process control methods, performing quality control audits, etc. If factors such as efficiency, cost, and time are also recorded on them, they can become what are generally known as production standards.

5.3 Action

5.3.1 Types of Action

Although taking action with respect to a process on the basis of control charts is fundamental to process control, it is often misunderstood, and control charts are therefore often not used properly, *quality standards are confused with work standards, inspection is confused with control,* and *adjustment is confused with eliminating assignable causes.* People then mutter that control charts are useless, and quality control fails to get under way.

Here I would like to discuss how to interpret action when using control charts (see also Section 5.1).

The action normally taken in processes can be classified into the following two types:

(I) Action on the process

(a) Prompt action taken with regard to the process:
   (i) Action in line with work standards — mainly adjustment and automatic control.
   (ii) Action based on control charts — investigating and eliminating assignable causes or performing temporary emergency adjustments.

(b) Action taken to prevent assignable causes from recurring in the future: investigating and eliminating assignable causes, i.e., revision of various standards, education and training, personnel reshuffling, etc.

(II) Action on the product

(a) When defectives are present, screening of individual products by 100% inspection.

(b) Passing, scrapping, or screening lots, or discounting their price, through sampling inspection and statistical estimation.

Action on individual products or product lots should be based on inspection standards, and in principle, control charts should not be used for this purpose. When people try to use control charts as a basis for taking action on individual products or product lots, their approach is often mistaken. Process control and inspection are different, and this is a confusion of inspection with control.

5.3.2 Adjustment Graphs

Item (I), (a), (i) above means items concerning assignable causes and operations laid down in work standards. They should really be called adjustments and are completely different from item (I), (a), (ii). For example, they might consist of instructions such as "Adjust the temperature to 700 ± 5°C," or "When the temperature reaches 703°C, increase the airflow by $10m^2$/hr." In situations such as measuring pH or oxygen percentage to find out when to terminate a chemical reaction, or measuring a dimension to find out when to stop machining, the characteristics should be specified in the form of work standards. To enable such operations to go smoothly, it is sometimes very helpful to draw graphs that show limit lines, but these are strictly speaking *adjustment graphs,* not control charts. Automatic control consists of automating this type of procedure.

It must be noted, as discussed in this book, that any graph showing statistically calculated limits may generally be called a control chart, and in that sense it can still be termed a control chart even when used for adjustment purposes. However, it is better to call it an adjustment graph and distinguish it from a control chart in this case. If the various errors involved are not properly considered, it is easy to overadjust and overreact, trying to force the process to stay within the limits. Professor Genichi Taguchi has researched methods of calculating adjustment limits.

Items (I), (a), (ii) and (I), (b) above cover action using control charts or action to eliminate abnormalities. The characteristics that should be plotted on control charts are therefore those control characteristics that are the result of the process.

The control charts we use for process control can be classified under the following two headings from the standpoint of taking action:

(1) Control charts aimed chiefly at prompt action: item (I), (a), (ii) above.
(2) Control charts aimed chiefly at action for preventing abnormalities from recurring in future: item (I), (b) above.

These two types of action are, of course, not entirely independent. For example, even with control charts aimed mainly at prompt action, technology will not improve and the process will not advance if recurrence-prevention action is not considered. And even with control charts aimed chiefly at recurrence prevention, action must be taken as quickly as possible.

Whether the chief purpose of a control chart is prompt action or recurrence prevention, it should in principle be used to plot the results of a process, discover the causes of abnormalities, and eliminate them. Control charts should not be used for adjustment.

5.3.3 Control Charts Aimed Mainly at Taking Prompt Action

In principle, control charts should be used for promptly detecting and immediately eliminating assignable causes whenever points fall outside the control limits. However, if the cause is not known, cannot be eliminated, or cannot be eliminated quickly enough, it is sometimes necessary to perform an interim adjustment by some other factor. If this is done, the real cause of the abnormality must be rapidly eliminated and the temporarily adjusted process restored to its original condition.

In well-standardized factories, workers and supervisors use control charts in the above way. Particularly with the recent increase in automation and the use of robots, workers are more and more becoming supervisors and controllers of machinery and equipment, and control charts are gradually being used more by lower-level supervisors and workers. For control charts to be used like this, a process must satisfy the following conditions:

(1) Sampling must be simple and measurements must be capable of being taken immediately. Rapid feedback must be possible.
(2) Workplace supervisors must be able to plot control charts and must have a good understanding of how to interpret them.
(3) All assignable causes must be fully understood in technical terms.
(4) Assignable causes must be capable of being eliminated immediately.
(5) The above four items must be standardized.
(6) It is desirable for the product quality characteristics and control characteristics to satisfy specifications and targets based on the above conditions.

These conditions must be quickly fulfilled and control charts that allow prompt action to be taken must be prepared. To enable control charts to be used in this way, the technology for the process must be firmly established, and activities such as those of QC circles must have brought those in the workplace to the level of being able to use these charts. If it is impossible to meet the conditions at the moment, efforts must be made to do so by promoting standardization while using control charts that focus on recurrence prevention.

5.3.4 Control Charts Focusing on Recurrence Prevention

When the conditions for using control charts that focus on taking prompt action are not satisfied, i.e., in factories that are not well standardized, priority must perforce be given to the use of control charts that focus mainly on recurrence prevention, i.e., action for preventing the causes of abnormalities from recurring in future.

This type of action includes the following:

(1) Preparation and revision of standards: revision of existing operating, technical and control standards, materials specifications, storage and control methods, equipment standards (including standards for the modification and installation of equipment and measuring instruments), and organizational rationalization.
(2) Education and training in standards, implementation of standards, personnel reshuffling as needed.

When using control charts for recurrence prevention, we are concerned mainly with whether or not action is definitely being taken, problems are being properly solved, and results are being checked. In other words, we want to know whether standard problem-solving procedures are being followed and whether process abnormality reports are being properly prepared and utilized.

Without this approach, not only will it be impossible to improve the process or the technology, but it will also be impossible to move on to the use of control charts for prompt action. In other words, if effective recurrence-prevention action is taken, the points on the control chart will naturally start to fall between the control limits as a result of standardization and training. As the above discussion implies, this way of using control charts consists of using them to control the process while promoting standardization and rationalizing the organization.

Recurrence-prevention action can be divided into three main types:

(1) Eliminating symptoms (stopgap measures and adjustments).
(2) Eliminating immediate causes.
(3) Eliminating basic underlying causes (see Section 1.5.3).

People often used to consider recurrence prevention a matter of simply eliminating symptoms, but this is a misuse of the term. We have to go back to the causes, and even the root causes, and take action to change our working methods, procedures, standards, and regulations.

5.3.5 Process Abnormality Reports

It is the responsibility of managers to know right away whenever something out of the ordinary happens in their processes and to check quickly whether appropriate steps are being taken to deal with it. When any kind of abnormality occurs in a process, a process abnormality report should be prepared and used.

(1) The purpose of preparing process abnormality reports

Process abnormality reports are prepared with the following aims in mind:

(a) For rapid reporting of process abnormalities.
(b) To confirm that correct action is being taken.
(c) To expedite analysis of abnormalities and prompt action, particularly recurrence prevention.
(d) For classifying abnormalities and systematizing the investigation of countermeasures, and for reference when deciding on equipment investment priorities.

(2) Contents of process abnormality reports

Process abnormality reports should be presented on standard forms with space for the following items (see Table 5.1):

(a) Reference number.
(b) State of process: control chart number, title of process, name of product and control characteristics, lot number, status of lot, name of operator, control lines, and other control chart details.
(c) Details of abnormality: date and time of occurrence, description of phenomenon, name of person detecting abnormality.
(d) Cause: whether known or not known; if known, give details of cause and opinions of those responsible.
(e) Action: interim countermeasures, action with respect to cause and process, details of immediate action, whether elimination of cause or simple adjustment, date and time, state of liaison with other departments, details of disposition of lot if necessary. Whether problem is completely solved, half-solved or not yet tackled.
(f) Investigation: details of investigation into recurrence-prevention countermeasures.

Table 5.1 Process Abnormality Report

Ref. No. UA-009 | Process Abnormality Report Form | Issued 5 February 1974

Occurrence of abnormality

Name of machine	ENT-86814	Control chart ledger number	20-2-Tuu-A3-2	Data and time of occurrence
Name of process	PRE · TEST	Lot number		15 February a.m. 5:00 p.m.
Quality characteristic	Electrical performance (waveform)	Operator (Inspector)	Akemi Yoshikawa	
3% on stratified control chart showing pre-test waveform chart errors				Detected by: Tabuchi

3.0
UCL = 1.12
CL = 0.3
9 10 11 14 15

Investigation of causes

In the past, the position of the cam was determined in relation to the rotor shaft groove (i.e., by dimension C). To improve efficiency, the soldering jig for the cam was altered so that the cam position was determined by dimension B. Because of burrs and other irregularities on the end of the rotor shaft, this increased the variation in dimension C, causing the cam to foul the chassis and altering the waveform.

Dimension B
Cam
Dimension C
Solder
Rotor shaft
Burr
Interference
Chassis
Dimension A

It is desirable to use the present efficient jig in order to cope with future production increases.

	Investigation of cause	
1	When ?	16 February
	Who ?	Tabuchi
2	When ?	Day Month
	Who ?	
3	When ?	Day Month
	Who ?	

Emergency action

When soldering earth spring, check whether cam is interfering with chassis

During rotor assembly process, correct rotor shaft cam soldering jig

Liaison with related departments
17 February
Investigation request sent to technical department (UTU-014)

	Emergency action	
1	Who ?	Tabuchi
2	When ?	17 February
	Checked by	Tokunō

Recurrence-prevention action

During the rotor assembly process, control the shaft / cam soldering dimension (dimension B) with an $\bar{x}-R$ control chart (from 17 February)

Change the dimension of the part where the chassis contacts the cam (dimension A) from 5.5 to 6.5 mm

Recurrence-prevention	
When ?	28 February
Who ?	Tokunō
Confirmation of details of action:	Aoki

Confirmation of effect of recurrence-prevention action

After altering dimension A, no waveform errors occurred. Since the p control chart for waveform errors continued to show zero defects, it was discontinued.

The $\bar{x}-R$ control chart for the rotor / cam soldering dimension was also discontinued.

Check	
When?	8 March
Who?	Tokunō

Storage period 3 years	Tuner division	Section Manager	Foreman	Group Leader
Format number TG-Q001	MP Shop Production Section UHF Assembly Group	Aoki	Tokunō	Tabuchi

Note 1: This report clearly shows the date and time of each step from the discovery of the abnormality to the implementation of recurrence-prevention action and the confirmation of the results.

Note 2: It enables progress to be traced right up to the confirmation of the results of action.

Note 3: A column is provided for noting details of contact with related departments.

(g) Recurrence-prevention countermeasures: radical countermeasures for recurrence prevention, opinions regarding the future, dates of initiation and completion of countermeasures, effects of countermeasures, etc.
(h) Confirmation of countermeasures: confirmation of countermeasures and future control methods.
(i) Other: e.g., person responsible, contact person, circulation list, filing location, etc.

(3) Report handling procedures

Standard procedures should be laid down for the handling of process abnormality reports. These should include the following:

(a) Details of who should write the reports, when they should be written, what they should include, and how many copies should be issued.
(b) The method of circulation.
(c) The procedure up to the final solution of the problem described in the report, particularly how recurrence-prevention action should be expedited and its effect confirmed.
(d) General methods of analyzing and utilizing the reports.

Even when a process abnormality report has been issued, an *abnormality list* should also be prepared to ensure that the matter is not shelved until recurrence-prevention countermeasures have been firmly established. This chart should be used to keep a check on progress and ensure that the same problem does not recur. The problem must be pursued doggedly until action to eliminate the fundamental causes has been taken.

5.4 Work Standards and Technical Standards

5.4.1 What are Work Standards and Technical Standards?

One of the first problems encountered when quality control was introduced to Japanese industry was that many companies had no rational work standards (including technical standards). This made it necessary to raise people's quality consciousness; everyone had to become aware of their responsibility for performing process capability studies, analyzing processes, and establishing production technology, while technical departments formulated, revised, and perfected technical standards and investigated ways of improving technology to achieve quality targets, and manufacturing departments and inspection departments worked according to work standards to produce products that were in accord with the process standards. The construction industry, service industries, and sales industries are still like this; they have only recently introduced TQC, and they still lag behind in standardization.

In the past, Japanese work standards were riddled with defects; some were so useless that they were simply filed away and forgotten. Some of these defects were as follows:

(1) Many standards, especially work standards, were inept because people were not used to formulating them.
(2) Some were entirely of the old-fashioned IE type, work-study type, or work-efficiency type, while others omitted these elements completely.
(3) Many were no more than documentation for its own sake.
(4) Some people became infected with a mania for regulation and standardization and spent all their time making rules because they thought that control meant tying people down. Like petty bureaucrats, they made the rules so strict that there was no hope of their being obeyed.

The upshot of this was that obeying the standards produced defectives rather than preventing them.

One of the basic preconditions for promoting quality control, which is at the same time a means of establishing real technology in any industry, is to strengthen the workplace through QC circles, carry out extensive process studies and analysis, and formulate rational work standards.

If the concept of control is widely understood and management policy and organization have become firmly established, quality control can be implemented without preparing control charts if quality standards and work standards are available and are put to use. However, if there are no rational quality standards or work standards (though there may be work standards and regulations on paper), if the organization has not been rationalized and if management policy is hazy, any control charts that are prepared will merely be viewed as graphs, and quality standards and quality assurance will be non-existent.

With recent advances in automation, in the use of robots and in computer control, processes have speeded up and floods of defectives will be produced if work standards and technical standards have not been established, process capabilities are inadequate, or the system is such that process control is not properly exercised. Before promoting automation, sound process analysis must be carried out and a process control system put firmly in place. If this is done, the equipment will take over the work, and the people in the workplace will become responsible mainly for monitoring, as in the process industries. Work standards will therefore change, and equipment and measurement control will become more important.

5.4.2 Quality Characteristics, Control Characteristics and Work Standards

Work standards should really only be decided after quality standards and process targets have been set. In practice, however, quality standards and work standards are often determined by mutual trade-off. Quality standards take precedence when technology is established and process capabilities have been fully analyzed.

In many Japanese factories, there is confusion about whether to give instructions in the form of quality standards or work standards. This is because causes and effects are not clearly distinguished and authority and responsibility are ill-defined. The control characteristics of a particular process or job manifest themselves as the results of that process or job, and work standards state specifically what to do about the assignable causes producing those characteristics. Giving workers instructions in the form of quality characteristics or results is a poor way of doing things, since it is like telling someone to shut the gate after the horse has bolted. If preemptive standards of the type described in Section 1.5 have been formulated, quality characteristics will become control characteristics.

From the point of view of using control charts to eliminate the causes of abnormalities, the measured values should be given as work standards if the data are to be used for adjustment purposes. If the data are to be plotted on control charts and the causes of abnormalities are to be sought out and eliminated when points fall outside the control limits, the measured values should be given in the form of control characteristics. For example, if the temperature (the result) can be changed simply by turning a valve slightly (the cause), it may be specified in a work standard. However, when several operations are needed (i.e., when a number of assignable causes must be changed) in order to alter the temperature, this should be treated as a control characteristic and each of the operations should be specified in work standards.

In controlling a process that has been automated or that uses robots, many of the assignable causes will be automatically controlled, and it is necessary to think hard about what kind of results should be plotted on control charts as control characteristics and how the automated process should be checked. Even when there is a high degree of automation or robotization, it is good to select quality characteristics and use control charts for controlling the process as a complete system.

5.4.3 Purposes and Types of Work Standard

(1) Purposes

There are various aims of preparing work standards, depending on the particular standard concerned. Here I would like to talk mainly about standards relating to process control, from the following aspects:

(i) Quality } from the QC viewpoint
(ii) Control }
(iii) Standard motions, work efficiency, production—from the old-style IE viewpoint
(iv) Cost
(v) Safety

Work standards, of course, incorporate all of the above. Control charts are used to check whether or not the work is being performed according to work standards prepared with the above aims in mind.

The aims of preparing work standards can also be classified as follows:

(i) For education (for new entrants, employees of three months', one year's, or ten years' standing, etc.).
(ii) For workers and supervisors.
(iii) For accumulating technology (for the organization rather than for the individual).
(iv) To create a historical record (not a very meaningful aim).
(v) To qualify for the JIS mark or other marks of distinction (this can easily lead to standardization for its own sake).

The more general aim of standardization is to make it easier for people to do their jobs, to enable authority to be extensively delegated, to capture and preserve technology, and to prevent mistakes from being repeated.

(2) Types of standard

I would now like to discuss the best way of categorizing standards, focusing on the manufacturing process. Since the method of classification will depend on the type of process and the type of product being produced, I will confine myself here to a general classification that I believe should apply in every case. The basic philosophy is the same for the service sector, although there will be differences between different industries. Selling services through personal relationships necessitates standards of judgment that can be adapted to meet customers' needs and preferences.

Technical standards are used mainly by engineers and managers of middle level or above. They deal with matters thought important from the technical viewpoint, together with their historical background; their aim is to build up a body of technology for the organization. Depending on the situation, they may include manufacturing process charts, QC process charts, quality standards, process capabilities, cause-and-effect diagrams, process control standards, inspection technology standards, etc.

Design standards and design technology standards are used mainly by design departments. Design standards are used for standardizing and unifying the design process, while design technology standards are used for standardizing important technical items relating to design and are aimed at building up a body of design technology.

Work standards have various purposes and are known by various names, but I would like to define them here simply as "specified ways of how to perform work." Ultimately, they lead to automation and robotization. In the service sector, a great variety is needed to suit different customer requirements.

Work instructions set out how a job should be done in the form of commands.

Work procedures are a type of work standard that give just the key points for carrying out a job.

Such work standards, work instructions, and work procedures are categorized and named in different ways depending on the history of the organization, but the method of classification should be studied and adapted to suit each individual workplace.

Processes can be divided into the following three main types:

(1) Contract-based customized manufacturing processes or high-variety, low-volume production processes in which the same kind of work is repeated in spite of the variety of products.
(2) Automatic lathes, transfer machines, processes using industrial robots, and other process-industry-type processes where each machine or plant has its own individual method of operation. Many processes in the process industry fall under this heading, as does much work that consists of the monitoring of equipment, i.e., in which workers are in the position of controlling plant and equipment.
(3) Assembly processes, i.e., processes in which products such as TV sets and cars are assembled by fitting together and adjusting a variety of parts. Packaging processes also fall under this heading. The use of industrial robots in these types of processes has increased recently and they are beginning to become more like the process industry. This type of process is also common in the service sector.

There are of course many other processes that fall in between these categories or combine elements from more than one.

In processes of the first type, when the same kind of operation is repeated in various forms and each operation can be broken down into standard motions, common work standards can be prepared for each standard operation and combined with technical standards and design standards to create work instructions showing, for example, a machining process together with the process sequence and methods. This is also applicable to many service industries. An example might be as follows:

"Produce X items of product Y by such and such a date, using Work Standard No. S-10547 and processing in accordance with Drawing No. ABC-18247 and Technical Standard No. E-35764."

With the second type of process, work is mainly of the process type. Those work standards covering important technical conditions are selected to form the technical standards, and authority and responsibility are clearly stated in the work standards. The work standards are used without modification as work instructions.

With processes of the third type, assembly methods, adjustment methods, and autonomous control and inspection of parts should be decided for each process.

5.4.4 Conditions to be Included in Work Standards

The following points must be considered when deciding on the contents of work standards:

(1) They should be formulated with a view to achieving specific objectives.

(2) They should deal with assignable causes, i.e., they should be preemptive. They should show which causes affect which characteristics and should make use of cause-and-effect diagrams.

(3) They must be user-friendly. Work standards should be such that the job can be performed well even by workers who are not fully trained and may be slightly inattentive. Use should be made of jigs and tools and measuring instruments.

A note on foolproofing is warranted here: since it is human nature to make mistakes, safety devices, jigs and tools, and methods of checking should be devised to ensure that things proceed smoothly even if workers' attention wanders or they commit errors. This is called foolproofing. When QC circle activities get in the groove, the circles themselves will start to think about foolproofing. Rather than getting angry about mistakes and carelessness, superiors should cooperate with the workers to think up foolproofing methods.

(4) Work standards should not be abstract: they should indicate specific criteria for action. To make this possible, suitable measuring instruments

and scales must be provided. As quantification advances, work standards become more specific and easier to set.

(5) Work standards must suit existing conditions and must be actually capable of application with the equipment and skills available. Idealistic standards are useless. In other words, standards must take process capabilities into account. One must never give people standards that cannot be followed. When work standards are first issued, conditions and tolerances should not be too strict; it is better to set conditions that can be adhered to without too much difficulty and make sure that they are faithfully obeyed.

(6) One should not aim for perfection at first. Work standards are living entities and are always imperfect. It is better to take the line that standards need constant revision. If standards are not being revised, it is proof that they are not being used, and that the progress of technology has come to a halt.

(7) Priorities must be identified; processes are only seriously affected by one to three causal factors. If standards dealing with these are prepared, the number of defectives can be halved.

(8) Standards should clearly indicate where people's responsibilities lie.

(9) Standards should state people's scope of authority clearly and specifically, and authority should be delegated.

(10) Standards should be accepted by everyone they affect. For example, they should be prepared with the agreement of QC teams, QC circles, workplace QC study groups, etc. They should be revised using ideas from the workplace adopted through the proper channels via a suggestion system.

(11) Standards should be formulated in such a way as to ensure that technology and skills are amassed in written form.

(12) Work standards often become complex when raw materials or assignable causes that are the responsibility of other processes are not properly controlled. Relatively simple work standards can be produced if they are skillfully assigned.

(13) One must decide what action should be taken under whose authority when a process goes out of control. This may be published separately in the form of a control standard.

(14) Standards should avoid negative instructions of the "should not" and "must not" type.

(15) Everything must be taken into consideration when preparing standards. The worst thing that can happen is for standards or regulations to be mutually inconsistent.

(16) As a result of the above, it should become easier for everyone to do their work.

5.4.5 Preparing Work Standards

When preparing work standards for the first time, the following procedure should be used:

(I) Preparation methods

(a) The sketch method: this is a method of standardizing a workplace which is already operating. In it, the work presently being done is sketched as it stands. Its advantages are that it gives a clear picture of how work is currently being carried out, it clearly shows operations being performed in hit-or-miss fashion and hazy or ill-defined areas, and the whole of the workplace is standardized in one fell swoop. Its disadvantages are that people tend to relax and become complacent once the sketch has been completed, the standards are bulky, they take time to write and their subsequent control is cumbersome, and if the quality control and process control approach is not well understood in the workplace, the standards are liable to be extremely inept and to miss out the important points.

A few helpful hints for using this method are:

(i) Give those responsible for drawing up the standards a good grounding in QC before they start.

(ii) Based on QC circles, make use of the process of drawing up the standards to give workers and supervisors a thorough education in matters such as the importance of adhering to standards.

(iii) Since it is so important to revise the standards once they have been prepared, link this up later with analysis, improvement, and QC circle activities, and make people aware that they themselves are responsible for improving standards.

(b) The priority method: The word "priority" here has two meanings: one is to start by standardizing the most important characteristics from the most important processes, while the other is to start by standardizing the most important causal factors of the most important characteristics in each process.

The advantages of this method are that good results will always be achieved if the really important causal factors are identified statistically and standardized; the workplace will form a good opinion of standardization and will come to trust the results of process analysis; this type of standardization is less time-consuming; and since standards prepared by this method are simpler, documentation, revision, and control are easier.

Its disadvantages are that it is difficult to identify the really important causal factors; it is impossible to standardize simultaneously for all characteristics and processes; and the purpose of the standards easily becomes limited to such a type of process control as merely reducing variation; there will then be insufficient connection with cost control, production volume control, efficiency wages, etc.

Some hints on this method include:

(i) Make use of QC teams and QC circles.

(ii) Give workplace supervisors and technicians thorough training in process analysis and process control methods.

(iii) Flesh out the standards after their initial preparation:

(1) Incorporate IE, VE, safety aspects, etc.

(2) Whenever a control chart shows the presence of a cause of abnormality, always revise the standards so as to prevent it from recurring.

(iv) Start by standardizing the most important causal factors and proceed with unrelenting effort, working down to the smaller details.

(c) The orthodox method: In this method, engineers are mobilized and the technical department takes the lead in cooperation with the workplace and with departments in charge of quality control and efficiency to form QC teams or QC circles under the engineer-in-charge system. Processes are selected for analysis in order of priority, experiments are conducted if necessary, and comparatively rational standards are drawn up from the outset. With this approach, the detailed areas are dealt with through discussion with QC circles. It is important to form the habit of using this method when *constructing new plant*, newly installing or rebuilding machinery and equipment, or before *the pilot production of new products.* The startup of new work is an ideal opportunity for promoting standardization.

Some hints on the orthodox method are:

(i) Take care that engineers do not get carried away by desk theory and forget about actual conditions in the workplace.

(ii) The technical department must be staffed by top-flight engineers.

(iii) Engineers must study QC thoroughly. Which of these methods should be used for preparing standards depends on conditions in the company and the workplace. Whichever method is chosen, the draft standards should be prepared by *the people who know the process and the work best.*

(II) Preparation procedure

Good, usable standards cannot be prepared without a clear idea of their aims and necessity.

Standardization in already-operating companies and businesses requires a different approach from that in new plant or for new products. Here, I would like to discuss the procedure for standardization in already-operating companies and branches. The following procedure is generally suitable for conditions in Japan:

(1) Set up a standardization committee and sub-committees.
(2) Decide on a system of standardization, a policy for preparing work standards, and the forms to be used. Have the standardization committee draw up the rules (work-standard preparation procedures, work-standard handling procedures, etc.), classification scheme, and forms that will constitute the backbone of the work standards.
(3) Decide on the organizations that will formulate the standards: teams, the technical department, the workplace, QC circles, etc.
(4) Specify the causal factors and work that should be written up in the form of work standards.
(5) Devise means of quantifying the causal factors selected.
(6) Lay down specific work methods and tolerances for the selected causal factors. Use cause-and-effect diagrams, statistical analysis, the knowledge of veteran workers, and QC circles extensively for this.
(7) If necessary as a result of the above, carry out factory experiments.
(8) Draw up draft standards. Gather together as many of those concerned with the workplace as possible and use QC circles or other forums to investigate whether or not the standards prepared will be practicable in the workplace. At this juncture, educate people in the significance of standards and the necessity of working in accordance with them.
(9) Establish a preliminary trial period of one to three months and try out the standards in practice.
(10) Prepare a standards ledger and record the standards in it.
(11) Revise the standards.

Once a standard has been officially recorded in the ledger, any revisions should always be carried out according to official procedures. The workplace must not be allowed to change standards at its own discretion without obtaining permission. This means that it is important to lay down regulations for the revision of standards. However, authority for doing this should be delegated as far as possible, and it should be made easy for individual workplaces to carry out revisions.

It is necessary to specify clearly responsibilities and procedures for matters such as drafting revisions and the forms to be used for this, the circulation route for the draft revisions and who is to approve them, entering revisions in the standards ledger and checking the entries, the decision-making body, and the procedures for ensuring that old work instructions are collected and new instructions are issued to all.

5.4.6 Implementing and Controlling Work Standards

Implementing standards and giving appropriate education and training are just as important as preparing the standards in the first place.

Making sure that standards are properly implemented is the responsibility of line supervisors, section managers, group supervisors, foremen, and others in positions of leadership. The importance of education for this purpose was discussed in Section 1.5. Various *methods of education* may be used, including (1) group education (e.g., lectures); (2) on-the-job training by superiors; (3) discussion in QC study meetings, QC teams, QC circles, etc.; (4) having people participate in the preparation of standards; (5) delegating authority and allowing people to educate themselves; (6) holding QC audits; and (7) pamphlets, posters, slogans, QC circle symposiums, etc. It is best to use these methods in combination.

If standards are not always usable, they will end up not being followed. Thus, as mentioned before, they should be regarded as living entities, and the standards ledger should be rapidly updated as needed, with rationality and practicality in mind. Control of standards is extremely important.

Since standards, once set, cannot be changed at the discretion of the workplace without going through formal procedures, the following items must be taken into account:

(1) Revision and control of standards

(a) Those responsible for controlling standards must decide what items can be changed to what extent under whose authority. In making this decision, the authority for making revisions should be delegated as far as possible.

(b) A set procedure for revising standards must be laid down.

(c) One must ensure that people in the workplace are able to offer their suggestions and opinions easily through the proper channels. Naturally enough, it is the group leaders, supervisors, and workers who have the best grasp of the conditions in their workplaces. They can be expected to come up with a variety of improvement suggestions since they are the ones who are constantly in touch with the workplace. To make full use of this, QC circle activities, creativity and ingenuity campaigns, and invention and originality campaigns should be carried out in conjunction with the revision of standards.

(d) The details of revisions should always be recorded in the standards ledger, together with the reasons for them, their dates, and the people responsible. Particular care must be taken to organize technical and design standards revisions so that anybody reading them can easily understand the history of the changes and the reasons for them.

(e) Whenever a revision is made, all the old standards and drawings must be recovered and replaced with the new versions or corrected and reissued after being signed or stamped.

(2) Situations in which standards should be revised

(a) When points fall outside the control limits on control charts: if standards have been formulated, even though they may be imperfect, points may fall outside the control limits on control charts for quality characteristics in the following situations:
   (i) When the standards are not being followed.
   (ii) When the standards do not give sufficient guidance.
   (iii) When the standards do not cover that area of work, i.e., when the standards are incomplete.
   (iv) Through the fault of the raw material or another process.
   (v) For unknown reasons.

Note: I would like to mention here the idea of *operator-controllable problems* and *management-controllable problems.* Of the things that go wrong in the workplace, only about one-third to one-fifth are the responsibility of the people on the shop floor, i.e., are operator-controllable. This corresponds to situation (i) above. Situations (ii)-(v) are principally management-controllable, and these account for two-thirds to four-fifths of workplace problems. In both situation (i) (to find out why the standards could not be followed) and situations (ii)-(v), management must get together with shop-floor personnel to search out the causes and take action. This means that *management must on no account show anger at mistakes by people in the workplace.*

(b) On suggestions from the workplace.

(c) When mistakes or gaps are discovered in the standards.

(d) When quality standards change.

(e) When technical improvements are made to machinery, equipment, or methods.

(f) When new measuring equipment has been installed or old equipment has been improved.

(g) When raw materials or other causal factors (work standards) have changed.

(h) At regular intervals after the initial adoption of work standards.

5.5. Control Levels

5.5.1 Selecting Control Items

The importance of checking in the practice of control has already been mentioned in Section 1.5. However, when those in positions of authority actually try to create a systematic, companywide checking network, build control systems, and implement control, they will encounter many problems. Different people and different companies currently use a great variety of classifications and terminology in selecting control characteristics. It is common knowledge that *there is great confusion in business management terminology worldwide;* this is because control items are categorized in rather too much detail. When categorizing control items, it is best to focus on the underlying ideas rather than the actual words used, and to choose categories that are adaptable for use from different standpoints.

(1) Classification of control items

(a) Classification according to cause and effect
As already discussed in Sections 1.5, 5.2.1(1) and 5.4.2, we must differentiate clearly between causes and effects as a method of checking when implementing control. Here, we will call assignable causes to be checked "*check points,*" and characteristics to be checked when checking through results "*control characteristics*" or "*control points.*" All of these will be referred to in general terms as "control items."

Control items { (i) Causes . . . "check points"
(ii) Results . . . "control characteristics" or "control points"

Causes and results are relative terms; depending on a supervisor's rank and how the work standards are allocated, what is a cause for one person may be a result for another and vice versa. For example, in Figure 5.1, moisture content is obviously a control point for the supervisor of that particular work area because it is a result of the process. Temperature is a cause of the moisture content and is therefore a check point. However, as far as the worker responsible for adjusting the temperature of the drying machine is concerned, pressure and valve aperture are causes and are therefore check points, while temperature is a result of these and is therefore a control point.

If the work standards are written in such a way that they try to "shut the stable door after the horse has bolted," they will instruct the worker to adjust the valve in response to changes in temperature. This will make

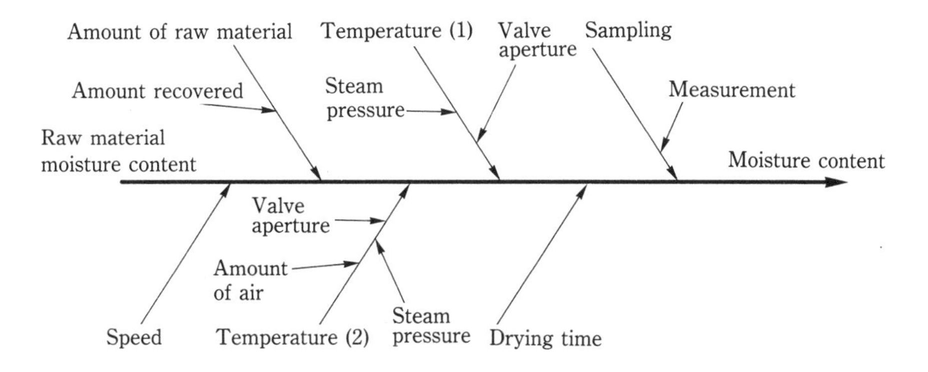

Figure 5.1 Discrimination of Causes and Effects

temperature a check point and drying time a control point. Also, if the work standards are even more ineptly formulated and instruct the worker to adjust the valve aperture in response to changes in the moisture content, the moisture content will be the check point and the drying time will be the result or control point.

The above relationship between supervisor and worker applies equally between department manager and section manager and between section manager and supervisor.

Generally speaking, the number of check points increases and the number of control points decreases as we proceed down an organization. Causes should be checked by lower-level managers, and it is not good for upper-level managers to become too closely involved in this. Upper-level managers should monitor through control points and exercise control from a broader perspective with an eye to the future. People who still want to inspect and verify data on detailed systems of causes even after becoming department managers or directors are what I have called "*artisan department managers*" and "*artisan directors.*" It is sufficient for upper-level managers to check the cause, if necessary, when a control point (a result) indicates an abnormality. And yet, the control system should be such that a report from lower-level management on the cause of the abnormality will already have arrived on their desk by the time they get around to doing this.

Since check points are monitored mainly through comparisons with work standards and regulations, it is usually a good idea to do this using check lists or the like. Control points, as results, will contain dispersion and should therefore in principle be monitored using control charts; at the very least, they should be graphed. Since such graphs are to be used for monitoring and detecting abnormalities and exceptions, they must of course include statistically calculated control limits. If this is difficult, they should at least include lines showing standard values, specification values, mandatory and desirable targets, or plans.

It should be noted that department and section managers may sometimes use the same control points but will then have a different interval between checking (for example, section managers may check daily and department managers once a month) and will take different action.

(b) Classification in terms of action
   (i) Adjustment and regulation.
   (ii) Elimination of phenomenon.
   (iii) Elimination of immediate cause.
   (iv) Elimination of root cause.

   This classification is also rather fuzzy, but (i) and (ii) are mainly check points (including adjustment criteria), while (iii) and (iv) are control points. However, (i) and (ii) may also be control points, and actions (iii) and (iv) may be taken in accordance with check points.

(c) Classification in terms of responsibility and authority
   (i) Items for which a person's subordinates have responsibility and authority for taking action.
   (ii) Items for which a person himself has responsibility and authority for taking action.
   (iii) Items for which a person's superiors have responsibility and authority for taking action.
   (iv) Items for which another department has responsibility and authority for taking action.

This classification is also constantly changing according to responsibility and authority and individual capabilities and experience. Items (iii) and (iv) could also be said to be included in item (ii), since in these cases the person must get his or her superior or another department to take action. Anyone who cannot do this cannot be called a manager.

Even before I became involved in QC, it was my pet theory that *managers or engineers who cannot use their subordinates skillfully are less than halfway to becoming true managers or engineers. In addition, they can only become proper managers or engineers when they become able to use their superiors and other departments well, i.e., when they become able to make people act in the way they want them to.*

There are various other ways to classify control items, apart from the above three, but combining them will produce dozens of different types of control item and will lead to confusion. I recommend that they be classified simply as either check points or control points.

(2) Some general advice about control items

The following are some common points to note in the case of both causes and results:

(a) Since there will always be repetition even with high-variety, low-volume production and customized manufacturing, workers should be given standard measures and scales for checking in the workplace.
(b) Criteria for judgment (whether adjustment criteria or control limits) and the action to be taken if the limits are exceeded should be decided in advance.
(c) One must decide on the responsibility, authority, and method of reporting for when abnormalities occur.
(d) The above points should be standardized and ample education and training should be provided to all concerned.
(e) Managers are responsible for ensuring that their immediate subordinates are given appropriate check points and control points.
(f) The control points should be reviewed at regular intervals or whenever a change or abnormality occurs in the process, and revised whenever necessary.
(g) One should choose items (these may be substitute characteristics) for which data can be obtained rapidly and prompt feedback is possible.
(h) When deciding on control items, each person's responsibility and degree of delegation of authority should be considered, as well as how work standards and regulations are to be allocated.
(i) *The aim is not to prepare a table of control items but to implement specific control.*
(j) One must decide on the checking cycle—e.g., every hour, once in the morning and once in the afternoon, once a day, every week, every month, every financial period, every year, etc. This cycle will generally be shorter for lower-level managers and longer for higher-level managers. Lower-level managers must submit reports of any abnormalities to their superiors whenever they occur.
(k) If the control items are skillfully selected and organized for use as control tools in the form of checklists, control charts, or graphs, poring over daily or monthly reports will become a thing of the past. Put the other way round, the control items should be chosen in such a way as to make it unnecessary to see such reports.

(3) Selecting check points

Apart from the items mentioned in (1) above, the following matters must be considered when choosing check points, as I have already repeated many times:

(a) Select causes as check points.
(b) Select check points that can be adjusted or regulated while monitoring some result.
(c) Assign check points mainly to workers, foremen, supervisors, and others directly responsible for overseeing the workplace. Arrange things so that, as far as possible, managers at the section level and above do not usually have to check causes directly. In other words, these managers should normally have no check points, although of course they have many control points.
(d) It is not necessary to make check points of all the causes; those causes considered most likely to give trouble should be given priority and selected as check points. This means that they will change with time. The people in the front line of the workplace will have many check points.
(e) Workplace foremen or those responsible for the check points can select the check points themselves. There are several reasons for doing this:
  (i) Thinking about the best things to check is good training for workplace supervisors.
  (ii) Those responsible for a particular work area know the conditions and risks of that area best.
  (iii) People's immediate superiors can look at the check points selected and use their observations for deciding on the need for education and training.
(f) Checking should generally be carried out by preparing check lists or graphs.
(g) As well as constantly updating check lists, the results should be periodically collated and used as a basis for recurrence-prevention action according to the Pareto principle.

(4) Selecting control points

The following points should be considered in addition to the items mentioned in (1):

(a) Control points should be selected from among the results of one's subordinates' work.
(b) Lower-level managers should use more stratified data as control points, while higher-level managers should use more aggregated or averaged data.

(c) Every aspect of a person's routine duties—e.g., personnel, quality, cost (profit), quantities, delivery times, safety, environmental factors, etc.—should be taken into account when selecting control points. If only one aspect (e.g., quality) is checked, another control point (e.g., efficiency) will generally suffer. When one characteristic is forcibly controlled, others will deteriorate.
(d) As far as possible, one should choose as control points not final results but substitute characteristics or intermediate characteristics that give an early indication of results. This makes it important to exercise ingenuity in selecting control points and devising measurement methods.
(e) Some measurement error is acceptable provided that it is smaller than the process dispersion.
(f) It is better to use actual measured values and raw data rather than composite values calculated from many different measurements. Errors build up in calculated values and it is possible to lose sight of what it is desired to control.
(g) As leaders, we should select as control points those characteristics that indicate our policies and targets.
(h) Control points should be selected from among those characteristics (results) that can be plotted on control charts and for which action can be taken. In other words, the control points should be selected after thorough process analysis has been performed.
(i) In processes where mistakes and lapses of attention are likely to occur, characteristics should be selected that provide a check on the results of such errors.
(j) Considering items (b), (c), and (g) above, the number of control points (i.e., the number of control charts and graphs to be monitored) should normally be as follows:
—Workers: *1-3*
—Foremen and supervisors: *5-20*
—From section manager up to company president and director: *15-50.*

5.5.2 Setting Control Levels

As discussed in Sections 1.5 and 5.5.1, in setting control levels, the guidelines are to decide what each person in authority should check (i.e., the control points), to quantify as much as possible, and to set the levels of the control characteristics (i.e., the control levels and control limits) as described below.

It should be noted that control levels and control limits are different from target values. Control levels and control limits are guidelines and limits for control purposes, while target values are values to be aimed at when making improvements. Target values may be plotted on control charts and graphs, but care should be taken not to confuse control with improvement. As mentioned previously, target values should be divided into mandatory values and desirable values.

To determine control levels (i.e., control limits), it is generally necessary to prepare control charts for preparing to implement control. The general method of statistically determining control levels is as follows (see Section 3.9.2):

(1) When past data have been analyzed and standards have been prepared up to a certain extent, one should try implementing those standards for a while. When this is done, measurement and subgrouping must be carried out by the sampling and measurement methods specified in the standards.
(2) When 100 or more data values have been collected by this method (it is acceptable to use as few as 20–50 values if obtaining each value is time-consuming, but the precision with which the control lines are estimated will suffer if this is done), a control chart is prepared using these values.
(3) One should check whether this control chart shows a state of control or not, and whether or not the standard values and target values are being met.
(4) If the control chart shows a state of control, the control lines can be used as control levels for controlling the process in the future. In practice, control lines can be extended in this way and used for controlling processes and other operations without too many problems when the number of points falling outside the control limits is no more than 0 out of 25, 1 out of 35, or 2 out of 100.

One must be aware, however, that when the data obtained in preparation for control show that the standard values or specifications are not being met, further analysis is needed. However, the existing control lines are generally adopted as the control levels for the time being, tentative process control is initiated using these, and analysis and improvement are carried out separately.

As can be seen from the above, we take the following steps when preparing to control a process:

(1) Analyze the process.
(2) Set standards designed to create a state of control.
(3) Calculate the control limit lines.

If, when doing this, past data are subgrouped simply and control charts vaguely showing a state of control are obtained, the charts are often not very useful for good process control in the future. It is necessary to clarify the meaning of the control levels by taking steps such as the following:

(i) Clearly state the aims of controlling the process.
(ii) Identify the meaning of the subgroups.
(iii) Formulate work standards.
(iv) Clarify where the responsibility lies.
(v) Set standards for the use of control charts.

The above are the principles for using control charts seriously, and it is best to follow them as far as possible. Nevertheless, temporary control levels can in fact be set, processes controlled, and results obtained even when, as is often the case, no work standards exist or it is impossible to analyze past data and draw up control charts showing a state of control that satisfies the conditions for extending the control limits into the future. However, doing this for long will lead to poor control and bring the process up against a brick wall. It is therefore better to move on to using control charts in line with the above principles as soon as possible. When doing this, the following points should be observed:

(1) Even if no work standards exist, control charts can be used to check whether the work is being done in the same way as in the past. At the very least, graphing the work in this way will have some motivational effect.
(2) When past data are analyzed and the number of points falling outside the control limits for unknown reasons does not meet the conditions (0 out of 25, etc.) given above, more such points will probably appear in the future if the control limits are extended as they are, and the future distribution of the product cannot be assured with confidence. However, if the cause is painstakingly investigated whenever a point falls outside the extrapolated control limits, and work standards are set one by one to prevent the causes from recurring, the number of points falling outside the limits will gradually decrease, the number of unknown causes will also decrease, and progress can be made towards proper control.
(3) When control levels have not been established, we should in principle take data from the previous month or recent data thought for technical reasons to approximate current conditions, remove abnormal data for which the causes are clearly known, calculate the control levels, and use these for controlling the process in the present month.

5.5.3 Control and Revision of Control Levels

Since control levels are living entities, they must be revised whenever necessary. This is particularly important at the stage of preparing work standards and revising them while controlling a process, since the control charts will not be able to fulfill their role as a process control tool and will become mere graphs if the levels are not properly revised.

Control levels should generally be revised (see Section 3.9.3) at approximately the same time as quality levels (see Section 5.2.1(4)). Who is responsible for revising the control levels should be decided for each control chart and recorded in the control chart ledger. The method of revision should also be decided in advance if possible.

(1) If there is no change in the process or policy, control levels should be recalculated every month or every 100 data points for some time after process control has been initiated.
(2) When the process has remained in control for a long period and all the points fall within the control limits, the control levels may be revised once every three months, or 500 data points, or at longer intervals. The estimates will also become more precise when the control levels are estimated from large amounts of data like this.
(3) If the process does not show a state of control and some points occasionally fall outside the limits, it is best to review the control levels once a month or every 100 data points. In this case, we naturally search out the causes of abnormalities and recalculate the control levels after eliminating abnormal values whose causes have been found and acted upon. In principle, we should leave in abnormal values whose causes are known but cannot be acted upon and values whose causes remain unknown.

A few points must be noted here. First, it is not good to use control levels that contain many values falling outside the control limits whose causes cannot be acted upon within the control responsibilities of that process, even if those causes are known, since this blurs the responsibility for controlling the process. In such cases, we should either recalculate the limits by estimating the effect of those causes and modifying the data, or calculate stratified control levels and use these as the control limits for the process.

Second, the following examples of inadequate control levels should prove helpful:

(a) When the control charts show points forming runs over relatively long periods or falling outside the control limits daily.
(b) When points are plotted on the control chart every month without drawing control lines, and these are calculated and drawn in at the end of the month.
(c) When specification values are plotted and action (mainly adjustment) is taken when the points fall outside these values. This is confusing process control with inspection or the removal of the causes of abnormalities with adjustment.
(d) When strict limits are set without reference to past data or the actual capabilities of the process.

5.6 Causes of Abnormalities and Control Standards

5.6.1 Causes of Abnormalities

The causes of process abnormalities can be categorized in various ways, as described below:

(A) Classification of causes of abnormalities in terms of standards

(1) Arising through not following standards
(2) Arising because standards cannot be followed
(3) Arising because of mistakes in standards
(4) Arising because of gaps in standards.

(B) Classification of causes of abnormalities by type

(1) Causes arising from inadequate control:
 (i) human causes
 (ii) mechanical causes
 (iii) raw materials causes
 (iv) metrological causes (errors incidental to sampling, measurement, calculation, etc.)
(2) Causes requiring technical investigation
(3) Unavoidable causes due to external conditions
(4) Unknown causes.

In factories practicing quality control, causes of type 1 are generally the most common, followed by types 2, 3 and 4. If control charts are drawn skillfully, the proportion of type 1 causes increases. When there are many unknown causes (type 4), it is often because process analysis and therefore standardization are inadequate, technology has deteriorated, or the philosophy of control is not well understood in the workplace (particularly when control charts are badly drawn or subgrouping is inadequate). Unavoidable causes (type 3) are extremely rare. Since someone in the company must be responsible for taking action, it is important to be persistent. For example, when there are problems with subcontracting, the company ordering the work is responsible and is in a position to act 60–70% of the time.

(C) Classification of causes of abnormalities according to mode of appearance

(1) Systematically occurring causes of abnormalities (mainly the responsibility of technical staff):
   (i) causes that appear systematically and instantaneously, i.e., those that surface suddenly at regular intervals
   (ii) causes that appear systematically and produce a succession of abnormalities
(2) Sporadically occurring causes of abnormalities (mainly the responsibility of line control staff)
(3) Chronically occurring causes of abnormalities (mainly the responsibility of engineers or management).

Type 1 causes are easily discovered through stratification or correlation analysis. Type 2 causes can be revealed by searching diligently whenever points fall outside the control limits. These causes are easier to trace than type 1 causes if the control charts are skillfully drawn. The occurrence of a type 3 cause is proof that recurrence-prevention action is not being taken. Examples include inattention by workers, defective installation of jigs, tools, measuring instruments, and other equipment, and low-quality raw materials.

(D) Classification of causes of abnormalities by statistical type

The above causes can be further classified into the following two statistical categories:

(1) Fixed-effect model causes
(2) Random-effect model causes

(E) Causes of abnormalities outside the process

Causes of abnormalities are found outside the process surprisingly often. They include:

(1) Lots: contamination by lots of different quality
(2) Sampling:
  (i) biased sampling
  (ii) unreliable sampling
  (iii) errors in sampling method standards
(3) Samples: mistaken samples, wrong handling of samples
(4) Measurements and tests:
  (i) uncontrolled measurements and testing: measurement errors, reading errors, wrong use or wrong installation of measurement and testing apparatus
  (ii) poor measurement method standards
(5) Data: recording and calculation errors, plotting errors, processing errors, contamination by different data.

(F) Summarizing causes of abnormalities

The causes of abnormalities can be categorized in various ways as described above. When formulating criteria for recurrence prevention and process control, it is best to use the following procedure to organize them:

(1) Check the frequencies of causes and draw Pareto diagrams to show which crop up most often.
(2) Check which characteristics are affected by which causes and how abnormalities are indicated, e.g., whether $\bar{x}$ or R changes, or whether points fall outside the control limits or form runs, trends, or other patterns.
(3) Check the long-term state of control by calculating the numbers or percentages of points falling outside the control limits every week or month and plotting them on a graph or control chart. Then standardize methods of searching out the causes and taking action according to the situation.

5.6.2 Control Standards

To perform control skillfully, authority and responsibility for effecting control, making judgments and taking action should be clarified, standardized, and delegated as far as possible. The rules specifying these matters are called control standards.

The following discussion of control standards focuses mainly on the situation in which control charts are to be used, but control standards may be formulated by the same approach even when control charts are not in use.

To expedite process control smoothly as a form of routine work through the use of control charts, we must prepare work standards (which can be called one type of control standard). To start with, even if it is only one for each workplace, *we should prepare usable control charts and acquire the taste of using them.*

When doing this, the following points should be observed:

(1) Number each control chart and keep a record in the control chart ledger.
(2) Decide who is to draw in the control limits, when they are to be calculated and recalculated, and whose permission must be obtained before they are drawn in.
(3) Decide who is to perform sampling and measurement, and when and how it is to be done. Decide what forms are to be used for recording the data, whom it should be reported to, and by what date the report should be submitted.
(4) Plot the points on the control charts. Decide who is to do this, when it should be done, and how the calculations should be performed.
(5) Decide who is to see each control chart, how often it is to be checked, and whose control point it is.
(6) Use the points plotted to assess whether or not the process is in a state of control. Lay down judgment criteria for deciding when a process is out of control.
(7) Decide on the action to be taken whenever the control chart shows that a process is out of control. Decide who is to search for the causes and how this is to be done, and what sort of action is to be taken if the cause is discovered or if it cannot be discovered. In other words, delegate and standardize as much as possible according to rank, from the lowest level up. As mentioned in Section 1.5, standardization for the purpose of delegating authority is one of the key points in management. In doing this, it is important to start at the bottom and make the standards as specific as possible.
(8) Decide when and how reports should be made to superiors or other related departments.
(9) Determine the action that may be taken by each manager on his or her own responsibility.
(10) If necessary, decide on a procedure for taking action to ensure that a cause does not recur in the future. Specify how this is to be linked with the standards.
(11) Decide on how action is to be taken after a process has continued in the controlled state for a long time. Some hints on this are:
    (a) With control points that are vital for the process, continue using the control charts in the same way as before.

(b) With control points that are not very important for the process, gradually lengthen the sampling interval or decrease the number of samples taken. In certain situations, economize on control activities by continuing to take measurements but no longer plotting the data on control charts or by suspending measurements altogether.
(c) In stratified control charts, when the process has continued in a state of control for a long time and no difference is indicated between the different strata, pool the data from the different strata, decide on a single control level, and plot the data on a single control chart. Continue to stratify lots and data even if they are pooled together for plotting on a single control chart, so that they can still be used for future process analysis.
(d) Lengthen the intervals at which control levels, quality standards, and work standards are revised.
(e) If a process has continued in the controlled state for a long time, inspection shows no unacceptable product, and there are no customer complaints about the product's characteristics, lengthen the product inspection interval, decrease the number of products inspected, and switch to check inspection. Ultimately, switch to zero-inspection shipping.
(f) If the state of control of the process still does not satisfy the quality targets, carry out further process analysis and try to raise technical levels.

5.7 How to Check whether Good Process Control is being Implemented

(1) Checking through overall results

The results of process control should ultimately be judged in terms of overall results such as quality, cost, necessary amount for a product unit, efficiency, safety, profit, sales, and market share, or by graphing improvements in process control, decreases in dispersion, or increases in technical capabilities. However, for reasons such as the following, these may be desirable but impractical measures for evaluating the results of process control:

(a) They include the effects of extraneous factors such as raw materials, components, equipment, and workers.
(b) They are difficult to obtain in data form; for example, they may rely

on sensory testing, data may take a long time to collect, or the data obtained may be impossible to stratify.

(c) They lack clear criteria for judgment.

(2) QC diagnosis

It is also necessary to check whether process control is being carried out properly, i.e., to check through the methods used as well as through the results obtained. Results can improve by chance or simply by people pulling their socks up and making a greater effort, but there is no guarantee that such results will be permanent. We should therefore check how everyone is thinking and acting, i.e., the process. This is one of the distinctive features of TQC.

(3) Checking process control in new-product development

New-product development is vital for companies, and if QC and process control at this stage appear to be going well, it means that TQC and process control have made considerable progress. It is therefore helpful to investigate whether new-product development is proceeding successfully, and to check the number of design changes made at each step, the situation during initial production of new products, their sales, and the number of complaints together with their details.

(4) Checking whether or not control charts are being used skillfully for process control

Checking on the use of control charts should be carried out at least once, preferably twice, a year.

(1) The items to check include:

(i) What kinds of work are the control charts being used to control?

(ii) Are the characteristic values appropriate?

(iii) Is there any confusion between eliminating the causes of abnormalities, adjustment, and inspection?

(iv) Are the control standards for the use of control charts appropriate?

(v) Are there any changes in the way in which the causes of abnormalities appear?

(vi) Are the standards for taking action suitable? Do they need to be improved? Is proper action always taken? Has the situation definitely improved as a result of recurrence-prevention action?

(vii) Are the present types of control chart, control limits, plotting and subgrouping methods, sampling intervals, and measurement methods satisfactory?

(viii) Is it still necessary to continue plotting each control chart?

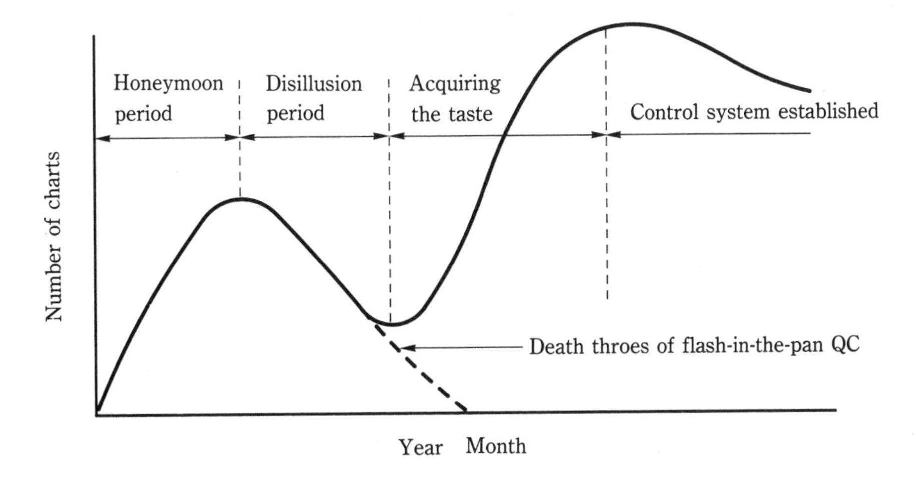

Figure 5.2 Change in Number of Control Charts

(ix) Have there been any changes in process capabilities?
(x) Are the work standards being appropriately revised?

(2) The number of different types of control chart used in each factory and in the company as a whole should be totalled annually; this provides valuable review data and is a good yardstick for assessing the progress of TQC. The number of control charts in use after the introduction of quality control generally varies over time according to the pattern shown in Figure 5.2.

In the beginning, a company generally experiences a "honeymoon" period. At this stage, people are under the mistaken impression that quality control means drawing control charts. They draw charts indiscriminately, forgetting about assignable causes, check points, eliminating the causes of abnormalities, adjustment, and work standards. This does have some motivational effect.

Next comes the disillusion period. At this stage, the novelty wears off, boredom sets in, and, as described later, people start jumping to the conclusion that control charts are useless. Although people continue to urge each other to draw usable control charts, their enthusiasm has in fact disappeared. If matters are left like this, QC will end as a flash in the pan and control charts will die a natural death.

After the disillusionment comes "acquiring the taste." People now begin to understand the true meaning of control characteristics and of control charts as

Table 5.2 Control Chart Diagnosis Report

Section or group	For control	For analysis	For adjustment	Graph	Lapse	$\bar{x}-R$	x	p, pn	c, u	Subtotal
	A A'B C	D E	F	G						
Subtotal Percentage										

a means of checking them. The control net spreads over the whole company, and the number of control charts in use increases rapidly.

Finally, the control system is established. At this stage, control is perfected or becomes more skillful, and the number of control charts in use gradually decreases to a certain extent.

The number of control charts in use and how they are being used should be audited companywide once every six months or one year. Reports like those shown in Table 5.2 should be submitted and the results plotted on graphs to show the long-term trends. The meanings of the grades in the table are as follows:

A (for control): Since the process has stabilized to the point where close control is unnecessary, the sampling interval may be lengthened, the number of samples reduced, or the control chart dispensed with in certain cases.

A' (for control): Since the process capability comfortably satisfies the specifications and targets and the process is well controlled, it is sufficient to continue controlling it as at present.

B (for control): The control chart is being used successfully for control, but the process capability is a little inadequate.

C (for control): The process has recently become unpredictable, and problems can be expected if nothing is done about it. Analysis and research are needed.

D (for analysis): The control chart is being used successfully for analysis and could also be used for control at the same time.

E (for analysis): Present analysis is insufficient; further analysis is needed.

F (for adjustment): This cannot be called a control chart. Eliminating the causes of abnormalities is being confused with adjustment. The chart should either be renamed an "adjustment chart," or the control items should be reviewed.

G (graph): This is not a control chart, and people are viewing it merely as a graph without taking any action.

Lapsed: The chart is recorded in the control chart ledger but is not being plotted at present.

(5) The long-term view of control charts: process control, process analysis and revision of standards

If one were to take from one to several years' worth of control charts for a particular characteristic and arrange them in time sequence, the control charts may simply be lined up as they are like this, or the control limits for each month can be calculated and plotted in order on a graph. Plotting the values for several different characteristics in parallel on the same graph gives an even better picture. This method of viewing control charts is effective for checking every kind of work, not just process control, to see whether it is proceeding satisfactorily over the long term. The method can also be used for QC diagnosis.

If the central lines of $\bar{x}$ charts are lined up, patterns like those shown in Figure 5.3 are usually obtained. *R* charts give the same kinds of pattern. The following conclusions can be drawn from these graphs:

Pattern a: Either control is being well exercised and a product of constant quality is being produced, or technology has stagnated and hardly any improvements are being made.

Pattern b: Control is not being exercised, nor is analysis being performed. The quality is being adjusted considerably in response to economic changes. At worst, this is a company that cannot be trusted.

Pattern c: The process is being analyzed thoroughly every month, work standards are being faithfully revised, and unceasing improvement and control are being effected. The concept of control is gradually permeating every part of the company. Saturation point is reached after approximately six to twenty months. When this happens, either equipment improvements must be introduced or the technology must be radically rethought.

Pattern d: Equipment investment and factory experiments are frequently carried out, but they are not fully utilized and control is absent. The concept of control has not sunk in.

Pattern e: Technical improvement is being actively pursued through improvement of equipment and properly designed experiments, and control is also good. This is a good example in the Euro-American style.

Pattern f: Continuous process improvement and technical innovation are being actively carried out through design of experiment methods and equipment investment, process analysis is being diligently carried out by analyzing each month's past data, and good control is also being exercised. This is *the best Japanese-style approach.*

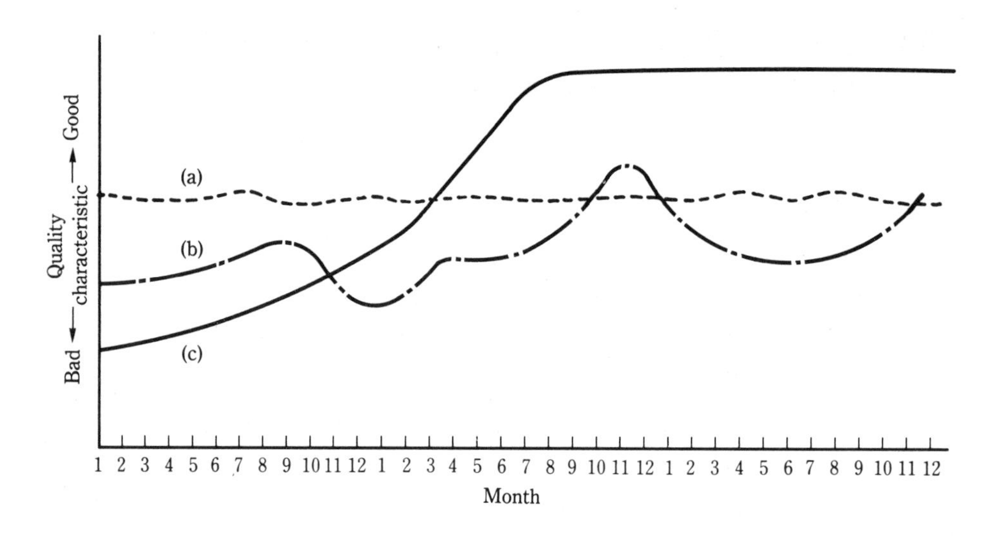

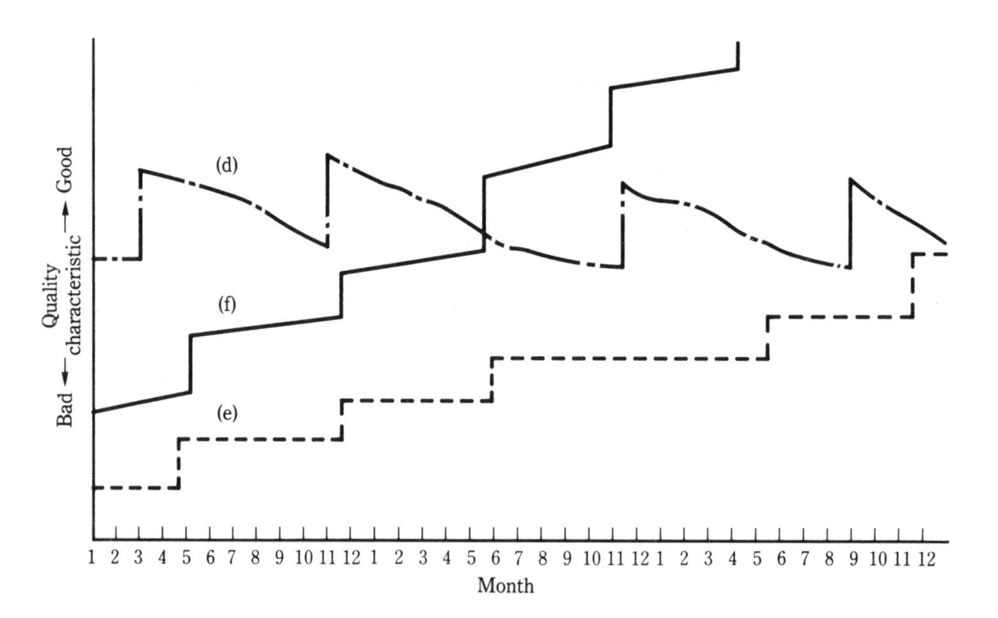

Figure 5.3 Long-Term Changes in Process (Long-Term View of Control Charts)

5.8 The Benefits of Control Charts and the Controlled State

If control charts are used competently as a tool for checking processes, they have the benefits described below compared with the masses of figures found in conventional daily reports:

(1) Expressing data graphically makes information easier to grasp. Temporal changes and abnormalities can be detected particularly quickly, and control is simplified.
(2) The work of a process over a long period can be appreciated at a glance.
(3) The concept of variation is more easily understood by workers and managers.
(4) Judgments become more objective, since they are made on the basis of control lines. Fruitless discussions and emotional disputes cease.
(5) It becomes easier to check whether action is being faithfully carried out with regard to particular causes. The results of such actions are also easier to check.
(6) Since abnormal variations can be tackled on a priority basis, it becomes easier to improve processes and bring them under control. Quality in its wide sense then improves.
(7) It becomes possible gradually to clarify authority and responsibility within the company.
(8) Work standards and other standards become more rational.
(9) It becomes possible to maintain processes in a state of control.
(10) It becomes possible to determine process capabilities scientifically.
(11) The work of engineers becomes easier to perform.
(12) It becomes easier for the workplace to accept support from the technical department.
(13) It becomes easier for workers, workplace supervisors, and QC circles to see whether work standards, jigs and tools, gages, measuring instruments, parts, etc., are suitable or not, which makes it easier for action to be taken.
(14) Workers, workplace supervisors, and QC circles become more interested in their work, and autonomous control becomes possible.
(15) Collecting statistics on the causes of abnormalities makes it possible to invest rationally in equipment and apparatus.
(16) Everybody becomes more careful with data, measurement, and sampling.
(17) Managers gain time to think about the future.

Stabilizing variations due to people, materials, machinery, and working methods, and bringing processes under control is one of the aims of management, and everything proceeds more smoothly if this is done. Some particular benefits of the controlled state are as follows:

(1) Processes are enabled to exercise their maximum capabilities.
(2) Work is standardized and becomes easier to perform.
(3) The number of measurements and tests performed can be reduced.
(4) Managers relax and become more able to delegate work.
(5) Future plans can be made with confidence, and contracts can also be signed with peace of mind.
(6) It becomes possible to effect technical improvements on a priority basis.
(7) Ample time and resources are freed for planning for the future.
(8) If the controlled state fully satisfies the certified grade, it becomes possible to guarantee quality without inspecting the product, thus cutting inspection costs.
(9) The company can confidently provide its customers with product warranties.
(10) Budget control and cost control are simplified.

5.9 Setting Adjustment Criteria

In Section 5.3 and elsewhere, I have frequently discussed the difference between eliminating the causes of abnormalities and adjustment. Here, I would like to talk about how adjustment criteria should be determined. 3-sigma control limits are suitable for deciding whether a process is out of control and eliminating the causes of abnormalities, but adjustment limits and criteria must be set statistically from a different standpoint, taking specification values and other factors into account.

To begin with, I would like to make the following points clear:

(1) Adjustment consists of monitoring a situation (e.g., length, thickness, humidity, etc.) and regulating something (e.g., turning a valve or mixing raw materials) according to certain criteria.
(2) A process has a certain variation that exists even when no adjustment is performed, together with a variation that changes when its assignable causes are adjusted.
(3) In general, adjustment means altering the variation and the mean of the distribution that exists in the absence of adjustment. In determining the

adjustment criteria, the following items must therefore be investigated:

(i) The standard deviation, variance, and other data relating to the variation in the absence of adjustment (moving range, rational subgrouping, histograms, sampling error, measurement error, etc., should also be found).

(ii) The relation between the amount of adjustment and its effect: this relation is not necessarily linear (estimation of difference in means, stratification and regression analysis may be required).

(iii) The time lag between an adjustment and the appearance of its effect (including the measurement time).

(iv) The time lag between taking data and making a judgment and performing an adjustment (the feedback time).

(v) The relevant costs, e.g., the losses arising from not noticing that a process has changed, the costs of adjustment, and the losses created by overadjusting.

It is also necessary for this purpose to study the philosophy and theory of automatic control and to obtain quantitative data using statistical tools.

Adjustment criteria are decided by combining the figures obtained from the above investigations and solving theoretical equations in such a way that the process is stabilized, profit is maximized and costs minimized. This will not be dealt with here because of the large number of possible situations. However, adjustment limits can often be simply determined even without using complex numerical expressions if the above values are known. It is therefore important to start by carrying out the investigations listed.

As adjustment criteria, the following items must be decided:

(1) What data will be monitored for making adjustments.

(2) How those data will be collected. In this case, since it is necessary to estimate the process average precisely, various devices in statistical analysis are needed when the variation, sampling error, and measurement error in the absence of adjustment are large. For example, the average may be found through the use of continuous sampling or systematic sampling, or by taking a large number of measurements, or the process variation in the absence of adjustment may be radically reduced.

(3) The adjustment limits. No adjustment is performed unless the values obtained in step 2 exceed certain adjustment limits. Usually, these limits are not clearly defined, and processes are often disturbed by the performance of unnecessary or harmful adjustments, i.e., by overcontrol. This happens when we allow our actions to be dictated by the process varia-

> tion in the absence of adjustment; in other words, it is the "hot-headed" type of error. If the adjustment limits are exceeded, an adjustment of the necessary amount is performed as calculated from the previously investigated relationship between the amount of adjustment and its effect. The adjustment limits and adjustment amounts are decided by analysis employing statistical methods.

The adjustment limits should be spaced on either side of a central line set at a certain target value; in my experience, the limits should be drawn not at the 3-sigma values of the distribution in the averages but at the 1.5-sigma to 2.5-sigma values.

Adjustments can be made continuously in some cases, while in others they must be made stepwise. When continuous adjustment is possible, an adjustment sufficient for restoration from the limit value to the target value should be performed. When only stepwise adjustment is possible, this should be taken into account when the adjustment limits are being calculated, and adjustment should be performed in one step or, if necessary, two. This approach is also useful when measuring instruments are being calibrated.

In performing the investigations, it is best to operate as normal, keep records of observations and adjustments, and collect data on the results over a relatively long time period. This is how adjustment criteria are determined, but once they have been set, the results should be checked and the criteria (which are a form of work standard) should be revised appropriately.

If the adjustments are complex and require various calculations each time they are performed, on-line computer control may be needed. In simple cases, when the same kind of adjustment is repeated, automatic control may be sufficient. However, adjustments are never anything more than controlling changes due to one assignable cause by altering a different assignable cause. Our ultimate aim must therefore be to select suitable control characteristics, prepare control charts, control the control system itself, and ultimately eliminate the true causes and institute a system of control that can control the process satisfactorily without adjustment. In other words, the ideal situation is one in which computer control and automatic control are absent.

Chapter 6
QUALITY ASSURANCE AND INSPECTION

6.1 WHAT IS QUALITY ASSURANCE?

Quality assurance (QA) is the heart and soul of quality control. Expressed simply, quality assurance consists of guaranteeing that a consumer can purchase a product or service with confidence and enjoy its satisfactory use for a long period. Quality assurance represents a type of promise or contract with the consumer regarding quality.

In practice, there are various problems with the way in which quality assurance is interpreted. Some common misconceptions are as follows:

(i) If strict inspection is being performed, quality assurance is being carried out.
(ii) Our company is practicing quality assurance, since we offer free replacement of faulty products with good ones.
(iii) Quality is assured if repairs are carried out free for a certain period.

Items (ii) and (iii) mean particularly taking responsibility for replacement or repair but not assuring quality. On the contrary, it is more like assuring customers that some products are bound to be defective or break down during use. Free repair and replacement are of course an important part of quality assurance, but they are far from being the whole of it.

Terminology:
Terms such as "security," "compensation," "indemnity," "assurance," "guarantee," and "warranty" abound in the field of quality assurance, and they mean slightly different things to people in different areas of business (the legal profession, general consumers, the construction industry, the quality control field, etc.). There are no clear rules about how they should be used.

Here, I use the word "warranty" in the sense of accepting responsibility and indemnifying the customer against something going wrong. The warranty period for a product thus means a fixed period from the time of purchase during which the product will be repaired free of charge under certain conditions. The warranty period for a car, for example, might be two years or 50,000 kilometers, whichever comes first.

In contrast to this, "assurance" means ensuring that customers obtain satisfactory use from a product over a long period. An assurance period should really be the number of years for which the product can be used. Since parts are subject to normal wear and tear, it means the number of years for which the company pledges to repair and service the product for a fee after the expiration of the warranty period. Of course, this will depend on how the customer uses the product and on how well checks and maintenance are performed. Automobiles, for example, are more durable these days, and I think their assurance period should be set at a minimum of 15 years. With some products, there is a legal obligation to make parts available for three years, five years, or seven years after the product is put on the market, but these are minimum responsibilities, and every company should decide on a policy with regard to assurance periods and create a system that enables it to supply parts for longer periods than these. Some companies pledge to provide a lifetime supply of parts, i.e., to make parts available for as long as any of their products continue to operate. Consumer durables come with warranty certificates showing a warranty period of one year, during which a unit will be repaired or replaced free of charge if it breaks down due to faulty manufacture, but a separate assurance period should also be set, since such products do not immediately break down or become unusable as soon as the one-year warranty period is over.

For products whose quality gradually deteriorates after they are shipped even when not used (e.g., photographic film, pharmaceuticals, foodstuffs, etc.), I believe an assurance period (which will depend on the storage method) should be set showing the number of years for which the product will remain usable if stored according to stated precautions.

6.2 The Principles of Quality Assurance

A company must practice quality assurance in order to guarantee customers and users that its products or services will perform satisfactorily before purchase, at the time of purchase, and for a certain period after purchase; in other words, its products or services must be sufficiently reliable to satisfy customers and win their trust. We can justifiably claim that quality assurance has been achieved if it becomes possible to sell trust in the quality of a company's products or services, or, even more broadly, trust in the quality of the company itself, so that consumers feel confident to purchase even new products or services from it.

To achieve this, the following principles must be followed:

(1) Adopt a 100% customer-first approach and obtain a firm grasp of consumers' requirements.

This involves clearly identifying what consumers are demanding and what types of guarantees they require. Since different countries have different situations and consumers themselves are polarizing and diversifying, high-variety, low-volume production will naturally be the norm from now on. Also, consumers are not generally marketing professionals, and their requirements are often nebulous or even subconscious. The sellers—i.e., the producers and marketers—are professionals, and they must question consumers to draw out and clarify their requirements. In contrast to this, there is no way in which the old producer-oriented, "product-out" approach of first making a product and then going out and selling it will satisfy consumers or be able to provide quality assurance.

(2) Hammer out a clear customer-first philosophy, and ensure that everybody from the company president down is concerned with quality. This means all employees including sales and service staff, as well as the company's suppliers and those in its distribution organizations.

Quality assurance is clearly impossible unless everyone is eager to practice it. Achieving this requires not just companywide quality control but groupwide quality control, in which everyone, including subcontractors and distributors, pulls together to tackle quality assurance as a team.

(3) Constantly rotate the quality PDCA cycle (see Section 1.4.1 and Figure 1.2).

However painstakingly we practice the customer-first philosophy, however comprehensive our market surveys, and however carefully we design quality and try to practice quality assurance, we can never be perfect; moreover, consumers' demands and expectations are constantly changing and rising. This means that we must ceaselessly rotate the quality PDCA cycle and never stop improving quality.

(4) Producers and marketers are responsible for quality assurance.

The responsibility for quality assurance lies with the producer, i.e., the company that makes the product. When a sales company, supermarket, or department store provides an outside company with specifications and commissions it to make a product, the responsibility for quality assurance rests with the producer, i.e., in this case, the sales company, supermarket, or department store. The manufacturer is responsible for quality assurance if it is part of the same group.

If a supplier practices TQC and has a properly functioning quality assurance system, a purchaser can confidently buy its products without inspecting them. Eighty to ninety percent of Japanese companies that are advanced in TQC purchase without inspection.

Within a company, the responsibility for quality assurance lies with the product planning department, the design department and the manufacturing department. In principle, the inspection department and quality assurance department are not responsible for quality assurance. The production department, for example, is responsible for practicing good autonomous control and inspection and for implementing quality assurance that will satisfy the consumer. If this is done successfully, the staff of the inspection and quality assurance departments can be greatly reduced. The pre-delivery inspection department and the quality assurance department check quality from the consumer's standpoint and investigate product liability problems (see Section 6.6). They actually have very little responsibility for quality assurance. This means that if complaints or expressions of dissatisfaction are received from consumers, top management should have the purchasing, planning, design, and manufacturing departments on the carpet, not the inspection or quality assurance departments. Before they can do this, however, they must have clearly advised those departments of their responsibility for carrying out quality assurance and must have ensured that a quality assurance system is put into place.

(5) The above quality assurance principles are set out from the consumer's viewpoint, but, within a company, "your next process is your customer," which means that everyone should strive to apply the principles with regard to the next process down the line.

6.3 Quality Assurance Methods and Systems

(1) Quality assurance methods

Since I have already described the progress of quality assurance in Section 1.3, I will only summarize it briefly here. Soon after quality assurance was introduced into Japan, we started to leave behind the Euro-American inspection-oriented type of quality assurance and move to the process-control-oriented type of quality assurance, in which tight process control is practiced in an attempt to achieve zero defects. We succeeded in improving quality and raising productivity, and were able to produce good products at reasonable prices. However, we then realized that this alone was not enough; if new-product planning and design or the selection of raw materials were poor, it would be impossible to assure quality no matter how hard we tried to implement process control. From the latter half of the 1950s, we therefore started to go further back in the production process, shifting the focus of our quality assurance efforts from the planning, design, prototype, and pilot production stages to the new-product development stage.

Even when a system for assuring quality during new-product development is established as described above, proper process control is needed as long as products are being produced, and 100% inspection is also needed as long as defective products come off the production line.

To create a true quality assurance system, we must go even further and implement a system of groupwide quality control in which subcontractors, suppliers, distribution organizations, after-sales service organizations, and all other related companies practice quality control as a single unit. Some companies take over ten years to establish this kind of quality assurance system covering all their activities worldwide.

(2) Quality assurance systems

How to implement a quality assurance system covering every stage from new-product development to marketing and after-sales service was discussed in Section 1.6.2. For a more detailed discussion of this, please consult works such as *A Quality Assurance Guidebook.**

I would like to concentrate here on new-product development, mentioning some particularly important items. Figure 6.1 is basically the same as Figure 1.16.

In practice, each quality assurance step should be standardized, checklists prepared, and specific control methods devised and implemented for each type of product. However, not all of these activities will be successful at the first attempt. Failures should be reviewed, methods of preventing their recurrence should be standardized, and technology and experience should be accumulated so that the same mistakes are not made with the next product. A quality assurance system will only emerge if this process is repeated in a controlled fashion many times.

The second step in Figure 6.1—after the actual "idea" has been formulated—is *new product planning*. In developing a new product, carefully thought-out new product plans should be drawn up and used as the basis for subsequent work. These plans should include the following items:

(1) The type of consumer at whom the product is aimed.
(2) The sales price and the targeted unit cost (cost planning).

---

* *Hinshitsu Hoshō Gaidobukku* (A Quality Assurance Guidebook), ed. Asaka and Ishikawa, JUSE Press, 1974.

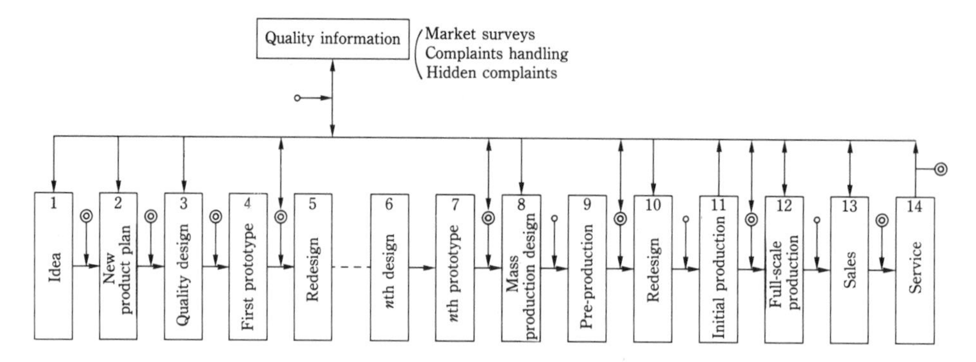

Figure 6.1 The Steps in Quality Assurance

(3) Monthly sales forecasts and sales periods.
(4) Quality plans (as far as possible, these should be expressed in consumers' own words, giving true quality characteristics).
(5) The desired launch date.

Quality design, the next step, means starting by carefully studying quality as discussed in Section 1.4, performing quality analysis, and performing the design and redesign steps described below (Steps 3–10 in Figure 6.1).

(1) Carefully investigate what characteristics (i.e., performance) should be guaranteed. Do not rely solely on market surveys, product research, and purchasers' specifications; listen carefully to consumers' true quality requirements. In particular, thoroughly investigate the degree of importance of each quality and the method of measuring it.
(2) Decide how long the product should be made to last.
(3) Check for problems concerning safety, possible misoperation, and product liability.
(4) Decide which parts should be replaceable, how long they should last, and how often they should be replaced. Set quality standards in accordance with this, select measurement and evaluation methods for checking whether or not the parts meet these standards, and decide on inspection and endurance testing methods and conditions.
(5) At this stage, prepare QC Process Chart I and lay down guidelines on how to manufacture the product and implement process control. Also set relevant raw material standards and parts tolerances. Check the company and supplier process capabilities needed for this.
(6) Fabricate prototypes and perform operating tests to check their perfor-

mance and lifetime. Have all departments cooperate to evaluate quality. If necessary, have the customer perform these operating tests.

(7) Prepare operating instructions and inspection and maintenance schedules.

(8) Analyze information and process capabilities from manufacturing, inspection, purchasing, sales, and service departments, and promptly redesign the quality.

In its narrow sense, a *design review* consists of reviewing a product's design drawings, but here I would like to approach the matter from a broader perspective, briefly mentioning the items that should be checked (shown by the symbols ◎ and ○ in Figure 6.1) form the new-product specification stage (Step 2) to full-scale production (Step 12). A design review, thus, should encompass the following:

(1) Assessment of performance, reliability, maintainability, serviceability, safety (particularly product liability prevention; see Section 6.6), environmental considerations, design, manufacturability, QC Process Charts, unit cost, life-cycle cost, laws and regulations, patents, etc.

(2) Determination of departments that should participate in the design review: sales, planning, design, quality assurance, inspection, prototype fabrication and testing, purchasing, production technology, manufacturing, packaging and shipping, cost accounting, legal, patent, etc.

(3) During the review, a check from the customer's standpoint into whether the product's quality and reliability selling points can be fully guaranteed, the new-product plans and planned quality targets are satisfied, the product is easily manufactured and can be produced at the targeted cost, and the items listed in (1) above are satisfactory. The purpose of this review is not to criticize the design and prototype fabrication departments but to work together to create a good product.

(4) Setting out clearly the available data on performance and reliability tests, subcontractors, manufacturability, and process capabilities, with particular emphasis on incomplete data relating to targeted quality and unit costs and on outstanding design and prototype fabrication problems.

(5) Having as many relevant people as possible take part in the first design and first prototype fabrication in order to pick up as many problems as possible and ensure that the number of design changes follows curve A in Figure 6.2 and reaches zero by the pilot production stage, or at least by the initial production stage. If the number of design changes follows curve B in this figure, the company is incompetent at new-product development and will achieve a reputation for unreliability in new products.

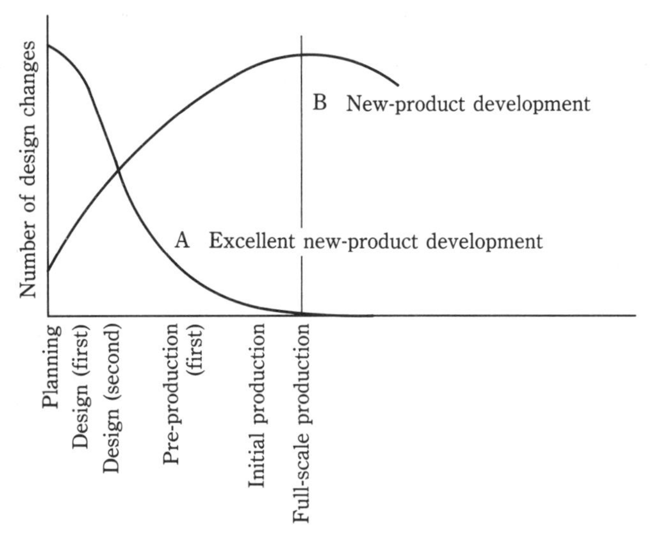

Figure 6.2 Trend in Number of Design Changes

(6) The number of test items and conditions for assuring quality and reliability during new product development should be several hundred for a simple product such as a ballpoint pen or two to three thousand for a more complex product such as an automobile. These test items and conditions constitute a body of knowledge for the company, and companies that gradually perfect them and build up their expertise in this area will succeed in developing new products with guaranteed quality and reliability. These kinds of data should also be carefully filed and preserved for purposes of product liability prevention.

(7) Companies manufacturing materials, parts, and products for industrial use should find helpful users with whom to carry out joint research, development, and evaluation in tandem with in-house product research.

The following are key steps in *process design and control*:

(1) Start preparing QC Process Chart II before preproduction and complete it before initial production.

(2) Purchase raw materials and parts that can satisfy quality standards and the process.

(3) Implement thorough control and build in stable quality via the process. Product lifetimes and reliability in particular are often difficult to guarantee by inspection, and reliability can really only be achieved by securing

a state of control with small dispersion. We could even go so far as to say that achieving reliability depends totally on obtaining the controlled state.

(4) Eliminate defectives at as early a stage as possible through in-process inspection.

(5) Decide in advance what kind of endurance tests are to be performed during the process, what they are to test, and when, at what stage, and by whom they are to be performed.

The steps in *inspection* are:

(1) Standardize where, when, and what kind of inspections are to be performed.

(2) Whenever possible, 100% inspection and inspection of substitute characteristics should be carried out by the production department.

(3) The inspection department should carry out performance tests, inspections from the consumers' standpoint, and inspections to check the inspections carried out by the production department.

(4) An order of priority should be assigned to selected characteristics from among quality specifications or drawing dimensions, and one should determine which will be designated as inspection items, who will make this decision, and how the inspections will be regulated.

(5) Inspection standards should be revised in accordance with information provided by consumers and all departments within the company. It is particularly important with sensory inspections to match inspection levels to consumer levels.

(6) If necessary, lifetime and performance tests should be done. One should also investigate the state of control of reliability and consider substitute inspection characteristics.

(7) 100% inspection should be automated.

The following are the steps to be taken in *service*:

(1) Investigate service methods and establish a service network.

(2) Train and develop service technicians.

(3) Prepare for the long-term supply of replacement parts and consumable items.

(4) Feed back information on parts replacement, repairs, servicing, and complaints in an easily analyzed form to all relevant departments.

If necessary, one should set up a quality assurance group in a company department capable of taking an objective standpoint (for example, the head office quality control department), make information on quality freely available to it from every area and have it carry out quality audits covering everything from design to how customers use the company's products. This group should also be given authority to stop product shipments. It should be made a center for rotating the quality assurance cycle companywide.

A few points worth noting here are first, costs tend to rise during the development stage, and it is necessary to control them and use various stratagems to ensure that they stay on target by proceeding in exactly the same way as above and practicing cost control at every step.

Second, it is difficult to keep new-product development on a set schedule; it tends to fall behind, resulting in skimped quality and reliability assurance. Solid progress control and deadline control are essential. Since many new-product development projects will resemble previous ones, it is best to decide on a standard schedule. For example, pilot production could be scheduled for the eighth month after the new product plan has been issued, full-scale production for the tenth month, and product launch for the eleventh month.

6.4 Why are Defectives Produced? Some Suitable Countermeasures

We can say that a company is practicing quality assurance if it produces no defective products as far as the customer is concerned, not as far as the manufacturer is concerned. I would now like to discuss briefly the kinds of situation in that defectives occur and some countermeasures that can be taken. Defective products manifest themselves in the following situations:

(i) During the process or at the pre-delivery inspection stage.
(ii) When products of a quality different from that required by the consumer are made.
(iii) When a product becomes defective before it reaches the customer.
(iv) During the customer's incoming inspection.
(v) When a customer finds that the product is defective on attempting to use it.
(vi) When a product becomes defective during use by the customer.
(vii) When a product becomes defective as a result of incorrect use by the customer.

As this list suggests, there are various problems in defining what a defective product is. In order to assure quality, we must be sure to do the following:

(i) Identify the quality characteristics and levels required by the consumer.
(ii) Ensure that no defective items are shipped.
(iii) Decide on the definition and period of guaranteed product lifetime (reliability).
(iv) Devise countermeasures to be taken if defective items are produced.

Let us consider the causes of defectives and the countermeasures to be taken.

(1) Why are defective items shipped?

(a) Because defective items are manufactured.
Solution: Control the production process so that this does not happen.
(b) Because of hazy quality and inspection specifications, inappropriate selection of characteristics (true and substitute) and their values, and inadequate market surveys and product research, i.e., unclear definitions of what constitutes defective and non-defective products.
Solution: Perform quality analysis and quality function deployment and check that the results of this analysis are correct. Even when quality is analyzed by means of cause-and-effect diagrams and quality function deployment tables, it is dangerous not to check that the results correspond with the facts.
(c) Because consumers' true requirements are not known, or pre-delivery performance tests are impracticable.
Solution: Carry out more product research and further investigation of operating tests and inspection methods.
(d) Because 100% inspection is not feasible.
Solution: Place priority on control or identify good substitute characteristics and automate inspection. If automated inspection is introduced, be careful about sensor errors and the reliability of automatic inspection equipment.
(e) Because of mistakes in design, prototype fabrication, production, purchasing, inspection, packaging, measuring instruments, sampling, measurement, data handling and computation, etc.
Solution: Eliminate errors through tight control of such work. Mistakes such as delivering the wrong product are particularly glaring evidence of lack of control.
(f) Because of errors due to sampling inspection.

Solution: If it is company policy to set certain inspection levels and carry out sampling inspection, some defectives are inevitable, since there will be statistically or probabilistically a certain proportion passing the inspection (see Section 6.9). The sampling plan must be statistically designed and written into the customer contract. Sampling inspection is not used very widely in Japan today, since the minimum fraction defective levels that can be guaranteed by sampling inspection are unacceptably high.

(g) Because defective products are shipped deliberately. Such actions are, of course, completely beyond the pale.

(2) The question of product lifetime

(a) First of all, what is product lifetime, and what happens when a product comes to the end of its useful life? It is often said that a product's life is over when it breaks down and can no longer be used, but this is wrong. A product reaches the end of its useful life when it can no longer give its full performance; in the case of a machine or measuring instrument, for example, this means when it can no longer exercise its full process capability or operate to the guaranteed precision. This confusion of definitions means that we must start by clarifying the definition of lifetime for each of our products, being sure to involve our customers in the discussion.

(b) Products must have a guaranteed life. In the past, there was a tendency to take the irresponsible view that it was good enough if a product was functioning properly when shipped or if the customer noticed nothing wrong at the time of purchase. However, this is a surefire way to lose the trust of customers—a trust that takes ten years to build but can be lost overnight. This problem is dealt with extensively in the guise of reliability as an integral part of quality control. It is particularly important with high-priced articles and consumer durables.

The number of years for which each product is guaranteed to exercise its various performance characteristics should be clarified at the quality design stage as a matter of company policy.

(c) Storage methods, etc., must be specified. In defining a product's life, it is necessary to specify storage methods, inspection and maintenance methods, methods of use, parts supply methods, after-sales service methods, etc. These must be spelled out simply and comprehensibly in the product's operating manual. It is the responsibility of the manufacturer, particularly with machines and measuring instruments, to supply inspection and maintenance details (i.e., preventive maintenance methods) equivalent to equipment control standards. When the product's users are amateurs, a service organization with service stations and service engineers must be created so as to allow periodic inspection and maintenance to be performed easily. Items whose quality can easily deteriorate during distribution, i.e., during transportation, in warehouses, or at wholesalers or retailers, require quality control of packaging and the provision of storage methods and facilities. Furthermore, quality assurance is impossible without a long-term supply of replacement parts.

(d) The variation in lifetime must be kept small. It is not enough for a product simply to have a long life; in fact, it is more important to minimize the variation in lifetime, especially with rapidly advancing products such as televisions. For example, people would of course make a fuss if their television tubes only lasted for one month, but there is little sense in giving them a lifetime of 100 years. It is more in the customer's interest to be able to buy a reasonably priced television with a lifetime of from five to ten years than to have to pay through the nose for a set that will last a century.

Lifetime and performance are also sometimes inversely related. For example, under certain conditions, making a light bulb brighter may reduce its life, while making it dimmer may extend it. The balance between lifetime and performance must therefore also be decided as a matter of policy in consultation with the consumer.

There may also be a considerable variation in the lifetime of individual products. For example, one bearing may wear out after three months, while another of the same type may last for a full year. When there is such a large variation, the user does not know when to replace the product and is uneasy and distrustful because it is impossible to guess when it is likely to fail. Things would be far easier for the user if it were known that all the bearings would fail after about 200 ± 10 days, since it would then be sufficient to replace them all on the 190th day. In other words, every product must have an appropriate lifetime that should not vary widely.

(e) A balance must be struck among the lifetimes of the different components. In products consisting of assemblies of various parts, the question of balance and variation among the lifetimes of the different parts and performance characteristics arises. For example, it may constitute superfluous quality for certain parts to have lifetimes longer than that of a particular critical part. A refrigerator will not act as a refrigerator if its door hinges break or its insulation crumbles, even if its refrigerating mechanism still works. Of course, replacements must be made available for short-lived parts as discussed above.

(3) Action to take when a defective is delivered to a customer

As discussed above, defective products appear at various stages of the manufacturing and sales process, but here I would like to talk about the action to be taken from the quality assurance standpoint when a customer discovers a defective product. This is the same thing as complaints processing (see Sections 1.4.1 and 6.14). As illustrated by Figure 4.2, we must of course uncover latent dissatisfactions and complaints as well as deal with overt ones.

(a) Speed and sincerity
Whatever the rights and wrongs of the situation, the customer feels dissatisfied, and the most important thing is to adopt a conscientious attitude and analyze the complaint without delay.

(b) Prompt replacement
It is important to replace the defective product promptly with a good one, but it is wrong to think that quality assurance ends there. The cause of the defect must be identified and recurrence-prevention countermeasures of the type described in item (i) below must be carried out to ensure that the same kind of defective product is not delivered to customers again.

(c) Payment of compensation if specified in the contract
In Japan, contracts rarely contain written penalty clauses specifying what should happen if the contract is breached. Efforts should be made to make contracts clearer and more rational.

(d) Free repair period
Since the costs of free repair are naturally included in the product price, it is necessary to review whether to use that money to make the product less likely to break down in the first place, or to have a product of slightly lower quality that is easily repaired.

Depending on the product, it may be advantageous from the standpoints of cost, the acquisition of repair technology, and customer service to replace complete subassemblies or products when they fail and carry out repairs en masse, rather than repairing products one at a time. This has become particularly important with the recent widespread adoption of electronic parts. Another possibility might be to eliminate the free repair period and charge for all after-sales service, but sell the product at a lower price.

(e) Provision of service stations

With many products, especially today's long-lived consumer durables, a network of service stations and a staff of properly trained service technicians must be provided to service products whose performance has deteriorated or that have broken down after the end of the free-repair period, to carry out preventive maintenance checks, and to supply replacement parts.

(f) Public relations and the provision of operating manuals setting use standards

A common cause of products breaking down or failing to give their full performance is poor use. Preventing this requires that companies provide operating manuals that detail methods of use that will be read and followed; it also requires quality design that takes possible misoperation fully into account. This also has a close bearing on product liability, discussed in Section 6.6.

(g) Preparation of periodic inspection standards for preventive maintenance

Products are often not supplied with these, and, sometimes, even when they are provided, the inspection interval is too short, either because there is a lack of confidence in the product, maintenance is too complicated and too difficult to carry out, or special service skills are needed but too few service stations or well-trained service personnel are available. A particular problem in Japan is that even when such standards exist, people have the bad habit of not following them, and tend to wait until a product breaks down before doing anything about it. Although there may be problems with maintenance standards being too detailed and maintenance costs being too high, public relations efforts are needed to advertise the fact that periodic maintenance benefits the consumer in the long run. This type of public relations effort has been slow to get off the ground in the past.

The serviceability of products should be evaluated during the design and prototype fabrication stages and appropriate countermeasures put into effect if problems are discovered. When users cannot be persuaded to undertake the necessary maintenance in spite of all these efforts, either products must be designed so that inspection and maintenance are simplified or rendered unnecessary, or the manufacturer must take positive steps to send its own technicians out to service its products.

(h) Long-term supply of replacement parts

The life of a product varies considerably depending on how it is used and whether it is well maintained, but ensuring that replacement parts are available for a relatively long period is an important part of quality assurance. A company that fails to provide replacements for the worn-out or broken parts of models launched the previous year is a typical example of an unreliable outfit that does not practice quality assurance. As mentioned in Section 6.1, a "lifetime supply" parts policy must be adopted. However, if care is not taken, the amount of capital and interest tied up in parts stocks may swell to enormous proportions, and administrative expenses may also balloon. When launching new products, serious consideration must therefore be given to parts standardization, costs, and servicing.

(i) Measures to prevent the recurrence of complaints
These are exactly the same as the recurrence-prevention measures discussed in Section 1.5.3. I would like to re-emphasize here that the old approach of simply replacing a defective product with a good one, repairing it free of charge or checking and fixing it at a service station does nothing to improve quality assurance. The crux of the matter is whether or not the information obtained from these actions is fed back to the places that need it. However excellent a company's complaints handling regulations may be, they will come to nothing unless this type of information is properly organized and reported to the required place at the required time, and recurrence-prevention action is taken. Information gathered by service stations and repair shops, e.g., data on parts lifetimes and failure modes, constitutes valuable information for improving reliability. This information must of course be subjected to Pareto analysis, and recurrence-prevention action must be carried through to a successful conclusion.

6.5 RELIABILITY

Japanese Industrial Standards define two terms meaning "reliability": "shinraisei" and "shinraido." The former is: "The ability of an item to perform a required function under stated conditions for a stated period of time," while the latter is: "The probability that an item will perform a required function under stated conditions for a stated period of time." However, from the common-sense viewpoint, reliability is a question of whether a product can be bought with confidence and used for a long time with confidence.

Reliability is thus a quality characteristic, and achieving it is a quality assurance activity. In this section, I would like to discuss reliability as an aspect of quality assurance.

The difference between reliability as a quality characteristic and the normal quality characteristics is that it emphasizes different conditions of use and the time factor and takes a lot of time and money to measure. It is also often difficult or impossible to test.

This means that well-planned reliability tests must be carried out at every step of new-product development from the design stage onward, for parts, subassemblies and the complete product. Also, while one is controlling processes satisfactorily and bringing them into the controlled state, market information should be skillfully fed back and recurrence-prevention action taken against the proximate and root causes of unreliability. When Shewhart first started practicing statistical quality control, he said that predictability and reliability are determined by control and that the statistically controlled state is basic to reliability.

Historically, reliability became a problem for the following four reasons:

(1) Products (for example, telephone cables) began to require longer lifetimes.
(2) Conditions of use became harsher.
(3) New-product development times became shorter.
(4) The number of products using very large numbers of parts increased. Products with over a million parts, such as space rockets, mainframe computers, and complex chemical plant with intricate automatic control systems, will not operate properly with a parts defect rate as low as even 1 per million. Defect rates must be kept below even this extremely low level.

If a rocket has 1,300,000 parts and the parts defect rate is 1 per million, the chance of it taking off successfully if the failure of a single part will cause it to crash is:

$$\left(1-\frac{1}{1{,}000{,}000}\right)^{1{,}300{,}000}\approx 0.27$$

The chance of success is therefore 27%; i.e., on average, only 27 out of 100 such rockets will take off successfully. A simpler example might be a product with 100 parts and a parts defect rate of 1 per 100. In this case, the chance of success is

$$\left(1-\frac{1}{100}\right)^{100}\approx 0.37,$$

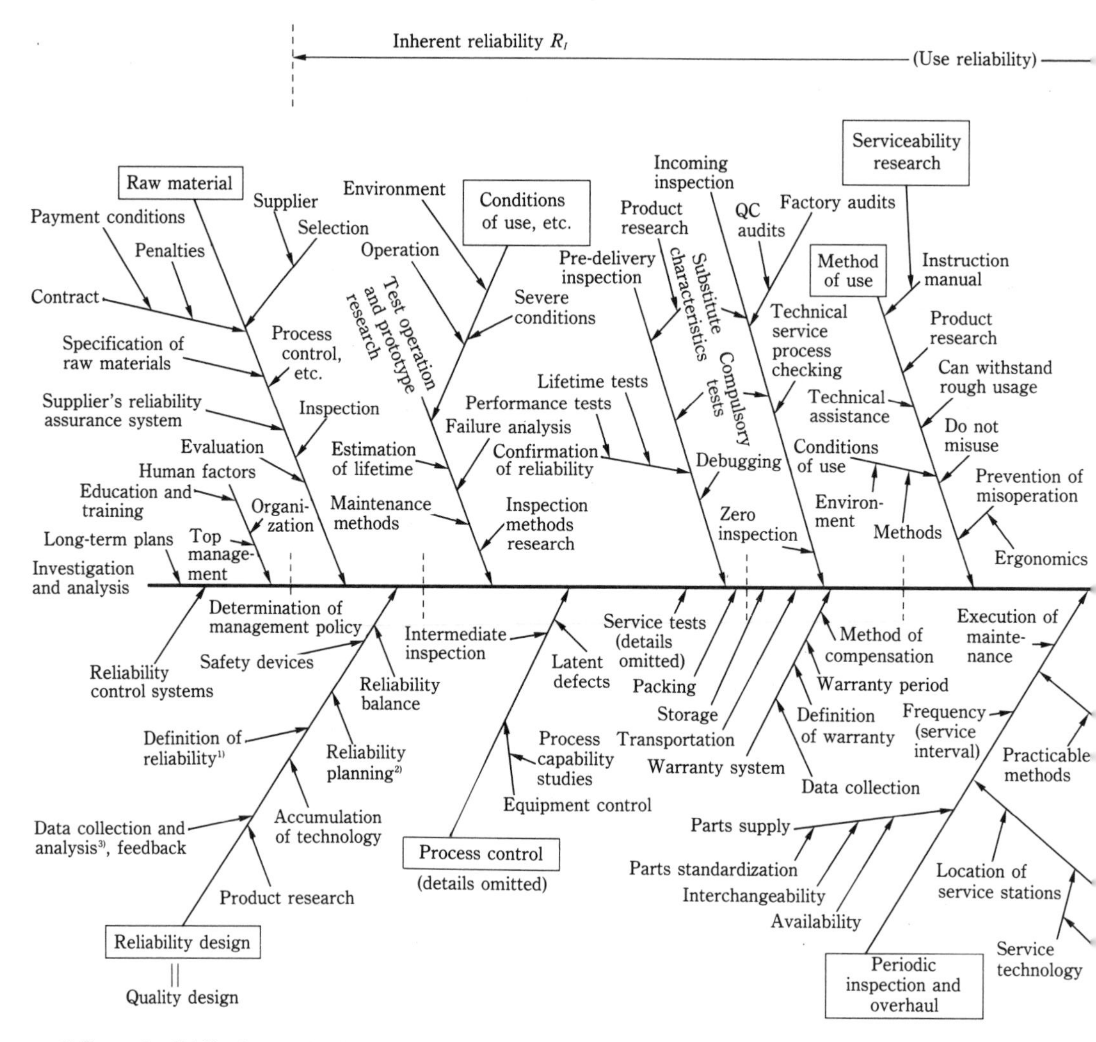

1) For each reliability item at the right-hand ends of the arrows
2) Individual plans for reliability assurance relating to 1)
3) Collection and analysis of data relating to consumers and 1) and 2). This also applies to all main arrows.

Figure 6.3 Cause-and-Effect Diagram for Reliability

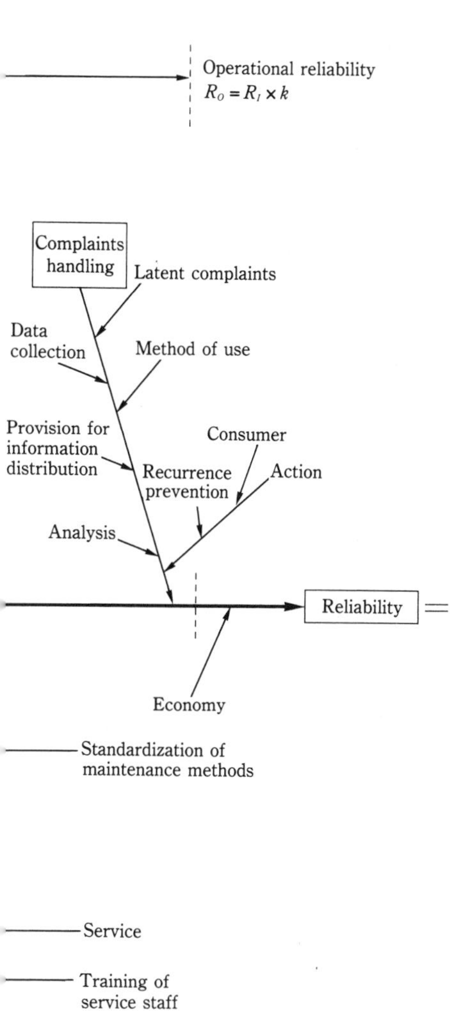

High reliability, commonly expected reliability.
Reliability of parts and systems.
Reliability of products, pricing, equipment, the company itself, the company's country of origin, etc.
Reliability before purchase, at time of purchase, and after purchase.
Product quality is consistent; products can be bought and used with confidence.
Feeling of trustworthiness.
Absence of variation in products.
Buying product is not a gamble.
Reliability is predictable.
Design is simple.
Product life: 1) Appropriate life; no change in performance or process capability, no change in quality. Definition of product life.
2) Small variation in product life.
3) Balanced lifetimes.
Extremely low fraction defective.
Inspection: 1) 100% inspection, automated inspection.
2) Testing under severe conditions.
3) Accelerated tests.
Failure: 1) Zero failures.
2) Failures easy to detect and correct.
3) Repair costs low.
4) Clear judgment criteria; education and training.
Zero abnormalities and abnormal articles.
No interaction, can be used anywhere.
Redundancy: 1) Stand-by circuits and devices.
2) Spare parts.
Safety factors, safety coefficients and allowances.
Protective functions and safety devices.
Method of use:
1) Withstands even rough use.
2) Foolproof design.
3) Instruction manual, packing away, do not use unreasonably.
4) Sales engineers and instructors.
Warranty system.
Periodic inspection and overhaul, after-sales service:
1) Maintenance-free.
2) Standardization, education and training in maintenance methods (periodic inspection and overhaul).
3) Long maintenance interval, simple maintenance.
4) Maintenance performed according to standards.
5) Low inspection and maintenance costs.
Parts supply: 1) Parts readily available.
2) Parts available for an unlimited period (even when product becomes obsolete)
3) Parts standardization.
Packing, storage, transportation.
Technical progress (balance between reliability and obsolescence).
Few model changes.
Price stability.
Credibility of company:
1) Company remains solvent.
2) No exaggerated advertising.
3) Verbal pronouncements of sales staff.

i.e., 37%. With only ten parts, the success rate will be

$$\left(1-\frac{1}{100}\right)^{10} \approx 0.904,$$

or approximately 90%. As these examples show, it is impossible to make complex mechanisms highly reliable with conventional parts or designs, which means that we have to think of reducing parts defect rates to extremely low levels or of introducing redundancy into the design (i.e., including backup circuits, standby parts, etc.). This has resulted in competition to increase reliability among different countries and companies. The problems of advanced reliability should be studied in specialist works. Here I would like to consider reliability from the common-sense, basic standpoint.

From the viewpoint of the consumer or user, we can start by dividing reliability into the following three types:

(1) Reliability before purchase: this is the reliability of a particular company, i.e., whether customers regard its products as always good and believe they can buy them with confidence.
(2) Reliability at time of purchase: whether a product is good at the time it is purchased, and whether its characteristics are initially satisfactory.
(3) Reliability after purchase: whether a product can be used for a long time with peace of mind.

The various factors creating reliability are illustrated in the cause-and-effect diagram of Figure 6.3.

Statistical data analysis is of course important, but it is self-evident that high reliability cannot be obtained without closely combining specific technology with quality control, improving individual materials and parts, and reducing their defect rates to near zero, and ensuring that processes are in the controlled state.

Reliability terminology

Here I would like to briefly explain some of the terms used in reliability (see JIS Z 8115-1981).

Intrinsic reliability, R_I, is reliability built into an item through design, manufacture, testing, and other processes. Quantitatively, it is the targeted or predicted value of the reliability set at the design stage, or the reliability characteristic value obtained from the results of reliability tests (see Figure 6.3).

Operational reliability, $R_o = R_I \times k$, is the reliability of an item under operation or in use. k is a coefficient that depends on the conditions of use and maintenance. Normally, $k<1$ (see Figure 6.3).

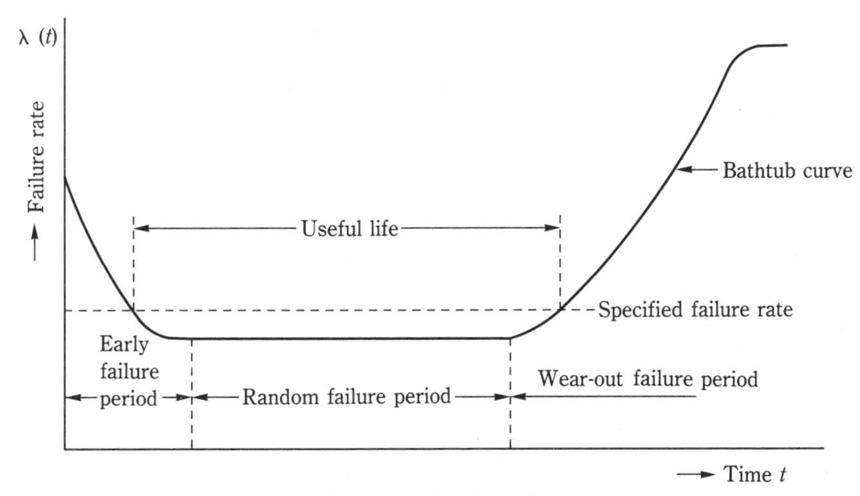

Figure 6.4 Failure and Its Terminology

Use reliability, a term not included in Japanese Industrial Standards, is that part of intrinsic reliability built into an item through testing, etc. It is part of intrinsic reliability in a broad sense.

Mean time between failures, MTBF, is the average operating time between successive failures.

Initial or early failure is failure occurring at a relatively early stage after first use as a result of design or manufacturing faults or unsuitability for the environment in which the product is used.

Random or chance failure is sporadic breakdown that occurs between the initial failure stage and the wear-out failure stage.

Wear-out failure is breakdown that increases with time as a result of fatigue, wear, or deterioration.

Useful life is the period for which a product can be usefully used before the failure rate rises to an unacceptable level and economic operation is no longer possible.

Gradual failure is failure in which characteristics deteriorate gradually with time and can be predicted by inspection or monitoring.

Sudden failure is breakdown that occurs unexpectedly and cannot be predicted by inspection or observation.

Maintainability is the probability of the maintenance of an item being completed under specified conditions during a certain period of time.

Failure is the loss of ability of an item to perform a specified function.

Failure rate is the frequency of failure of an item during a continuous period where the item has functioned normally up to a certain point in time.

Redundancy is the provision of extra structural elements or means of achieving a specified function to ensure that the overall system does not fail even if one of the components fails.

Parallel redundancy is redundancy in which all structural elements are functionally connected in parallel.

Stand-by redundancy is redundancy in which redundant constituent elements are held on stand-by until switched on when the primary constituent element fails.

Figure 6.4 shows the relation between initial failure, random failure, and wear-out failure. Because of its shape, this curve is called a bathtub curve. It also resembles the stages of human life: early failure represents infant mortality as a result of a congenital condition or lack of resistance; random failure represents death during youth or early adulthood as a result of a traffic accident, infectious disease, etc.; while wear-out failure represents death due to old age.

6.6 Quality Assurance and Social Responsibility (Product Liability and Environmental Disruption by Product)

In connection with quality assurance, it is necessary to consider consumer safety (the absence of risk of a product causing injury, sickness, death, fire, explosion, etc.) and environmental disruption (exhaust gases, noise, vibration, electromagnetic radiation, waste materials, etc.) caused by products. It is the social responsibility of companies to implement a reliable quality assurance program, but they should also prepare contingency plans in case of possible litigation. These questions must be dealt with extremely carefully since they concern not only individual companies and people but also whole industries and the public at large. This means returning to respect for humanity as our starting-point and assuring quality from the viewpoint of its effects on society.

Since many of the problems of environmental damage and disruption by products are common-sense matters, they will not be discussed here.

(1) What is product liability?

I would like to discuss product liability (PL) briefly here, since Japan is a society where problems have traditionally been solved by mutual discussion, and the concept of product liability is an unfamiliar one. Product liability involves legal considerations, and Japanese companies, particularly those that export to the U.S. and other countries, should station PL experts in their quality assurance or legal departments in order to study the matter in depth and devise appropriate strategies. I refer readers to the many specialist works published on this subject.*

Product liability concerns the responsibility of the seller for making reparation for bodily injury or damage to property incurred by the final consumer of a defective product sold. As I have mentioned, Japan is said to be a society in which problems are solved more by discussion than by litigation, and lawsuits are relatively rare. However, there have been incidents such as the thalidomide affair, the Kanemi affair and the "hiso-miruku jiken," [Translator's note: the Kanemi affair was a 1968 incident in which 126 people died after ingesting PCB-contaminated cooking oil; the latter was another poisoning incident, in 1955, in which 126 babies died and over 1,000 others were affected after drinking arsenic-contaminated powdered milk.] as well as problems with defective automobiles, food poisoning, etc.

America is said to be a litigious, contract-conscious society in which people are very ready to take their grievances to the courts. Product liability lawsuits have increased dramatically, particularly since the latter half of 1961, and the total has now reached over a million. The size of settlements has also increased, some going as high as seven million dollars. One reason for this is said to be that America has a large number of lawyers who began to actively seek more PL work around 1960 when the need for their services in automobile accident cases was curtailed. Another reason is that PL work is good business for them, as they can expect to receive from 30%–50% of the award in a successful suit. Japanese manufacturers and distributors must therefore take a fresh look at their product liability prevention strategies, not only when exporting to America and other countries, but also as a matter of social responsibility. To achieve this, they must take the following three main steps:

(1) Do not ship products (including published materials), or issue advertisements, or articulate performance promises liable to result in PL claims.
(2) To guard against possible claims, collect data proving that products are not defective.
(3) Take out product liability insurance in case the worst should happen. Recently, however, the premiums for this type of insurance have gone

---

* I recommend the following three books as an introduction to this subject: *Purodakuto Raiabiritii — Seihin Sekinin Mondai o Saguru* (Tackling the Product Liability Problem), ed. Kaoru Ishikawa, pub. JUSE Press, 1973; *Seihin Sekinin Jidai no Hinshitsu Hyōji* (*Good Quality Labeling System in the PL Age*), ed. Shigeru Mizuno, JUSE Press, 1974; *Hinshitsu Hoshō Gaidobukku* (A Quality Assurance Guidebook), ed. Tetsuichi Asaka and Kaoru Ishikawa, Ch. 18, "*Hinshitsu to Shakaiteki Sekinin*" (Quality and Social Responsibility), pub. JUSE Press, 1974.

> through the roof in the U.S., and some companies, doctors, and others in particularly high-risk lines of work have been blacklisted and refused insurance. Some doctors have even given up attending childbirths or performing operations because of the enormous sums being awarded in damages.

Without waxing too technical, I would like to start by describing some brief examples of product liability claims that have occurred in the United States.

Case 1: American do-it-yourself stores sell many ladders for housepainting. One manufacturer thought that it would make the painting job easier if its ladders were fitted with a platform on which a paint pot could be rested. However, a householder stepped onto this platform, which bent under his weight and threw him to the ground, injuring him severely. He filed a suit against the company, claiming that the ladder carried no warning that it was dangerous to stand on the platform. The courts found for the plaintiff, and damages were awarded against the ladder manufacturer because it had not provided a suitable warning notice.

Case 2: A customer started a fire and burnt his house down by filling an oil stove with gasoline instead of kerosene. He filed a suit against the manufacturer, claiming that the stove carried no instructions warning people not to use gasoline. The manufacturer lost the case because, although there was a notice on the cap of the stove's fuel tank instructing users to fill the tank with kerosene, the notice did not actually say that it was wrong to use gasoline.

Case 3: Some time ago, the glass used in the front windshield of a certain make of automobile was changed from toughened glass to laminated safety glass. A salesman sold one of these cars to a customer with the claim that the new front windshield was "absolutely safe." However, the car's owner was blinded when a shovel fell off a truck in front and smashed through the car's windshield. The owner filed proceedings against the manufacturer because the salesman had claimed that the windshield was absolutely safe. The manufacturer lost the case. Salespersons should never use expressions such as "absolutely safe," and companies should compile lists of such taboo expressions.

Case 4: A motorbike manufacturer put out an advertising poster showing a beautiful girl in flared pants riding one of its bikes. It turned out that flared pants like these were dangerous to ride in; the firm had to scramble to withdraw all the posters. Using such a poster would constitute proof that the company was claiming it was safe to ride in the clothing shown. Great care must be taken with the contents of advertising materials.

Case 5: A student driving to campus had a minor accident involving another vehicle. His car was only slightly damaged, and nobody was injured. On arriving at college, he immediately went to a professor specializing in product liability and asked whether it would be possible for him to claim compensation. The professor told him he could only claim for the cost of repairs, but the student apparently suggested that he might see his doctor, obtain a note to say that he was suffering from whiplash, and claim compensation for mental suffering and the loss of ten days' part-time wages. Spurious PL claims can sometimes be cooked up in this way.

Case 6: A worker was injured by a machine made by a certain manufacturer. This machine had no safety device to prevent that particular type of accident from happening, while machines made by other manufacturers did. A claim was entered against the manufacturer on the ground that no safety device had been fitted, and the verdict went against the defendant. The sophistication of safety devices that should be fitted to machines is decided by the level of technology and social awareness prevailing at the time. This is of course a question of degree, since trying to make a car perfectly safe, for example, would result in a product costing millions of dollars. The American Professor J. M. Duran once claimed angrily that fitting cars with crash bags and arrays of warning devices as a result of consumer campaigns by Ralph Nader and others would simply mean consumers having to buy a more expensive product, and that it would be far more effective to make the wearing of seat belts compulsory and outlaw drinking and driving.

These six cases probably convey a general sense of the problems involved in PL. Even in the United States, moves are now being made to reappraise PL litigation.

(2) Legal terminology relating to PL

Legal terminology is always somewhat troublesome, but here I would like to give brief explanations of some of the terms used in the PL field.

Negligence is failure to perform the duty of care normally considered necessary in a particular situation; the plaintiff in a negligence suit used to have to prove that the defendant had been negligent. However, the principle of no-fault liability was subsequently adopted in PL litigation, and the burden of proof fell on the manufacturer to show that no negligence had occurred and that the product had not been faulty. This was an extremely significant development for product liability prevention.

Warranty may be explicit or implied. An explicit or actual warranty arises when the seller of a product makes some representations about the product. An explicit warranty exists if such claims naturally lead the purchaser to buy the product. Examples include guaranteeing that a product is "absolutely safe," "completely effective," "safe," etc. An implied warranty is a guarantee that a particular brand-labeled product is fit for its customary purpose when sold on the open market. This concept of warranty became current in the U.S. in the 1930s.

Strict liability is the term used to express the concept that manufacturers have a liability verging on no-fault liability, irrespective of the presence or absence of any contract. In other words, the plaintiff in a PL suit merely has to prove (1) that the injury or damage resulted from the product; (2) that significant danger is associated with the product; and (3) that the defect already existed when the product left the control of the manufacturer. This judicial precedent was established in 1944, and the principle of strict liability was then established in 1966, on the ground that the costs of injuries resulting from defective products should be borne by the manufacturers that put such products on the market rather than by the injured persons who are powerless to protect themselves. This resulted in a tremendous increase in the number of PL lawsuits.

Japan has various PL-related legal definitions, such as tort liability (The Civil Law Act, Clause 709) and warranty liability (The Civil Law Act, Clause 570).

Product liability prevention expresses the fact that, in view of the above, defensive measures must be taken. Just as with TQC, everybody must be educated in the importance of product liability prevention (PLP) from the planning and design department through development, purchasing, manufacturing, quality assurance, the PLP department and the sales and service department on to subcontractors and dealers. Everyone should participate in devising and implementing defensive measures, and the following steps should be taken:

(1) To ensure that PL problems do not arise in the first place, the manufacturer can perform the following actions: Thoroughly investigate the safety and possible misuse of products in order to eliminate design faults. Check that all products are at least as safe as any comparable products on the market. Put important parts that are highly likely to give rise to PL claims (e.g., safety parts on automobiles) into a special category, perform reliability and failure analysis, investigate the safety of these parts during the life of the product, and perform proper tests (PL problems usually occur some years after a product is first launched). Manufacture products using processes with ample capability and tight process control, and devise sound foolproofing measures. Subject bought-out parts to strict quality and reliability assurance. Provide products with easily understood warnings about likely dangers or misuse. Carefully check operating manuals, catalogs, service manuals, maintenance procedures, advertising posters and other written materials, photographs, and illustrations from the PLP standpoint. Clearly indicate how to use the product safely and the first-aid measures to be taken in the event of injury. Devise methods of ensuring that the user cannot lose operating instructions or warnings when using the product. Warn sales staff about making exag-

gerated claims, and prepare lists of forbidden words and phrases. When written materials concerning the product are provided in foreign languages, especially English, they should be checked by a lawyer who is a native-speaker.

(2) Take out PLP insurance.

(3) Prepare for litigation: Prepare evidence to show that products are defect-free. Investigate methods of recalling defective products rapidly. Retain data on safety and reliability tests performed during new-product development, including methods and results. Ensure that drawings correspond to the actual product. Store process-control data by lot, and prepare proof that the work is being done according to work standards. Preserve inspection data (from incoming inspections, intermediate inspections, pre-delivery inspections, and dealer inspections) for each product lot. This sometimes means performing inspection for product liability prevention purposes even when it is not required for assuring quality. Give careful thought to collecting other types of quality assurance data.

We have consulted with Japanese lawyers concerning the problem of who should pay compensation for claims arising from bought-out parts or through inadequate pre-delivery inspection by dealers, and the conclusion was that the manufacturer should probably take full responsibility for paying compensation if proper contracts have been drawn up between subcontractors, the manufacturer, and its dealers. In such situations, the manufacturer should of course take out PL insurance, and the subcontractors and dealers should bear part of the cost of the premiums.

6.7 WHAT IS INSPECTION?

The term "inspection" is commonly used rather loosely with a variety of meanings, but in quality control, the inspection function is defined in Japanese Industrial Standards as follows:

Inspection consists of judging whether an individual article is defective or non-defective by comparing the result of a test carried out by some means or other with a quality criterion, or judging whether a particular lot is acceptable or rejectable by comparing a test result with an acceptability criterion.

The terms "measurement," "test," and "inspection" are often confused. In quality control, testing and measuring merely mean measuring something and obtaining data, and are strictly distinguished from inspection. Outside Japan, the word "inspection" is used with various meanings, which has introduced confusion into the practice of QC.

As should be clear by now, the purpose of inspection is to assure quality. However, as explained in Section 6.3, although inspection is one step in quality assurance, it is only a small part of the quality assurance function. The function of inspection referred to here is not the exclusive province of the inspection department, and neither should the inspection department concern itself solely with inspection. In other words, as discussed further in Section 6.12, the questions of what inspection is and what the inspection department should do should be considered separately.

It is clear from the basic principle of quality control (that quality is built into the product in design and the process, not through inspection), that good, cheap products cannot be made by inspection, no matter how strict it may be. In particular, inspection is unable to produce reliable products. Defective products are not immediately converted into good ones by inspection, and they must either be reworked or scrapped, simply increasing costs. This is not only unprofitable for the producer, but is also expensive for the purchaser, who ultimately has to bear these costs.

Again, even if a process is shown to be in a state of control by its control charts, this certainly does not constitute an assurance of quality. Pre-delivery inspection is needed even when the process is in the controlled state if the process capability, i.e., the product, does not satisfy the specifications.

As explained above, process control and inspection are different functions. We must never forget that action taken on a process should be based on work standards and control limits, while action taken on products or product lots should depend on inspection criteria.

6.8 Inspection Type

Inspections may be carried out by various methods and classified in various ways. When carrying out quality assurance, it is necessary to review inspection plans occasionally and investigate what kind of inspections should be carried out at each stage of the manufacturing process.

(1) Classification according to number of items inspected

(a) 100% inspection (screening): In this type of inspection, every unit of product is checked individually to sort the good pieces from the bad. Since inspector errors are generally high in this type of inspection, the inspection process must be analyzed and controlled using the process-control approach, stratifying the product according to priority, and double-checking the inspection work by sampling. 100% inspection is often sensory inspection, which requires constant control of inspection criteria.

(b) Sampling inspection: The sampling inspection referred to here is that based on statistical theory; it does not simply mean checking samples taken haphazardly, as was often done in the past. It is the type of inspection in which a sample of a product is examined in order to make a decision about the action to take on a complete product lot. There are various types of sampling inspection.

(c) Check inspection: This type of inspection is for checking large changes in quality levels with extremely small samples. In most cases, it is not performed for taking action on a product but is used for control purposes in conjunction with process control or for checking normal inspection work.

(d) Zero inspection: No inspection is needed if a process is in the controlled state and all the product satisfies the quality standards.

(2) Classification according to stage in product flow

(a) Incoming inspection: This type of inspection is carried out to ensure that materials conforming to specifications are purchased and to prevent nonconforming materials from entering the process. However, it is difficult to purchase conforming materials economically simply by using this type of inspection, and purchasing contracts should be rationalized and suppliers carefully selected, or inspection methods should be chosen that will encourage suppliers to implement quality assurance and quality control. The most effective approach is to emphasize quality control at the supplier.

(b) Intermediate inspection: This is the type of inspection carried out between processes to decide whether a product or lot can be passed from the previous process to the subsequent process. It is also known as "process inspection." Performing measurements in order to supply information to a process is also sometimes known as process inspection, but such activities do not have the true function of inspection and would be better termed "process testing" or "process measurement," although they may be one of the inspection department's responsibilities.

(c) Product inspection: This type of inspection is for deciding whether a completed product should be accepted or rejected. It is often identical to pre-delivery inspection, and may also be called "final inspection." It is combined with pre-delivery inspection when a completed product is shipped without further modification.

(d) Pre-delivery inspection: This type of inspection is used for deciding whether, at the time of shipping, a product meets the certified grade, whether it will satisfy the customer, and whether or not it should be shipped. It is generally difficult to achieve rational quality assurance with pre-delivery inspection alone; good process control is also needed. When performed separately from product inspection, pre-delivery inspection focuses on critical defects, serious defects, and characteristics liable to change during storage. It is best to arrange matters so that sampling inspection will be sufficient for this.

(e) Hand-over inspection, witnessed inspection: These are inspections carried out at the time a product is handed over to a customer.

(f) Stock inspection: This is inspection carried out on stock that has been warehoused for long periods. The characteristics to be inspected will depend on the storage time.

(g) Audit inspection: This is inspection to check and diagnose whether quality assurance and normal inspection are proceeding smoothly. It is generally carried out by the quality assurance department.

(h) Inspection by third parties: Examples of these include export inspections and other government inspections, and inspections by private inspection companies and associations or consumer groups. This type of inspection is carried out on general consumer goods either for consumer protection or as an impartial refereeing procedure to prevent exaggerated advertising and unfair competition. Japan still lags behind other countries in introducing this kind of system.

(3) Classification according to inspection details

(a) Authorization/formal inspection: This is inspection to decide whether a prototype or a new product delivered for the first time has the required capability. This type of inspection is used mainly for inspecting quality of design and process capability.

(b) Performance inspection: This is inspection to check whether an item can perform as required.
(c) Endurance inspection: This is inspection to check whether something can perform as required over a long period. It could also be described as reliability inspection. Inspections of type (b) and (c) are often simply tests.
(d) Severe inspection: This is inspection under severe conditions, and is mainly used for inspecting reliability.
(e) Inspection through substitute characteristics.
(f) Analytical inspection (precise inspection): In inspections to determine whether a product should be accepted or rejected, inspection is terminated as soon as one of the many quality characteristics to be inspected is found to be rejectable; the remaining characteristics are not examined. The data from this kind of inspection cannot be used for real process analysis and improvement. Again, sampling inspection is sometimes terminated as soon as enough failed items have been found to make a lot rejectable. In order for inspection data to be useful for analysis and control, data from all the characteristics of all the items in a sample must be obtained. This type of exhaustive inspection is called "analytical inspection." Because of the above, much conventional inspection data suffer from the problem that they cannot be used satisfactorily for analysis and control. Analytical inspection is also needed for adjusting inspection tightness.

(4) Classification according to method of judgment used

(a) Inspection by variables is inspection in which judgment is made on the basis of variables.
(b) Inspection by attributes is inspection in which product items are compared with gages, standard samples, specifications, etc., and individual products are judged good or bad or assigned different grades.

(5) Classification according to whether inspected items are usable or not

(a) Destructive inspection is inspection in which measurement or testing destroys the product. 100% inspection is of course impossible with this type of inspection.
(b) Non-destructive inspection is inspection in which the product is not destroyed by measurement or testing.

(6) Classification according to inspection location

(a) Centralized inspection is inspection in which products are collected for inspection at a fixed point.
(b) Roving inspection is inspection in which inspectors move round and inspect at different locations. As QC advances, this type of inspection is gradually being replaced by autonomous inspection or process checking.

(7) Classification according to whether or not the supplier can be freely selected

(a) When there is no choice of supplier: With inspection between processes in a factory, pre-delivery inspection, or incoming inspection of material from a designated subcontractor, there is no choice of supplier. This also applies to inspection of made-to-order parts, machinery, and equipment. In such cases, acceptable lots are shipped or accepted as they are, while rejectable lots must be subjected to 100% screening, reworked, used for different purposes, downgraded, or scrapped. In such cases, rather than rely exclusively on inspection, it is far more effective to ensure that the supplier practices good process control.
(b) When there is a choice of supplier: With incoming inspections carried out by ordinary companies or government agencies not purchasing from specified suppliers, there is generally a wide choice of suppliers and specialist manufacturers. In this case, it is necessary to carry out thorough preliminary investigations of suppliers' reliability and quality control practices, besides adopting inspection plans that will permit the results to be examined and those companies able to supply acceptable lots in the controlled state to be selected. Inspection tightness should then be adjusted as appropriate.

It is generally best when purchasing to arrange matters so that method (b) can be adopted, provided that suppliers can be surveyed and selected rationally. In many cases, method (a) is adopted because of a special relationship between supplier and purchaser, even when purchasing by method (b) would be more beneficial.

6.9 WHAT IS SAMPLING INSPECTION?

In this section, I will discuss the general principles of sampling inspection based on statistical theory. However, it is generally difficult with sampling inspection to guarantee lot percent defectives of less than 1%, and is particularly difficult to guarantee those of less than 0.1% or of ppm order. For this reason, apart from a small amount of destructive inspection, sampling inspection is not widely used in Japan today. I will, however, discuss it briefly here, since it is part of the basic knowledge of statistical quality control.

6.9.1 Sampling Errors

Let us consider a lot consisting of 1,000 pieces, of which 100 are defective, i.e., a lot with a percent defective of 10%. If a sample of 10 items is taken at random from this lot,* what sort of result will we obtain?** Common sense tells us that the sample may contain from 0 to 10 defective items. The probabilities of the sample containing specific numbers of defectives are shown in Table 6.1.

Assuming we wish to reject lots that have a percent defective of 10% or more (this value is called the lot tolerance percent defective, or LTPD), let us adopt a sampling plan in which we take a sample of 10 items from each lot and accept the lot if the sample contains no defectives (in sampling inspection, this is expressed by $n = 10$, a (or c) $= 0$***) but reject it if the sample contains one or more defectives (i.e., if $r = 1$). Even if we do this, approximately 35% of lots with a percent defective of 10% will be accepted. Sampling errors of this kind are unavoidable with sampling inspection.

Table 6.1 Probability of Defectives Appearing in Sample ($N = 1{,}000$, $P = 10\%$, $n = 10$)

Number of defectives in sample	0	1	2	3	4	5	6	7	8	9	10	Total
Probability	0.35	0.39	0.19	0.06	0.01	–	–	–	–	–	–	1.00

– Dashes represent extremely small probabilities.

* The following remarks all concern samples taken at random from whole lots. Sampling inspection tables are usually calculated on the assumption of perfectly random sampling from complete lots.

** In the following discussion, we will assume properly controlled inspection with no inspection errors.

*** In sampling inspection, a is the "acceptance number," while $a + 1$ or r is the "rejection number." c is sometimes used in place of a.

Table 6.2 Probability of Defectives Appearing in Sample ($N = 1,000$, $P = 5\%$, $n = 10$)

Number of defectives in sample	0	1	2	3	4	5	6	7	8	9	10	Total
Probability	0.60	0.32	0.07	0.01	0.001	–	–	–	–	–	–	1.00

In this example, there is a 35% probability that lots with a percent defective equal to the lot tolerance percent defective (10%) will be accepted against the customer's wishes; this kind of error is called "consumer's risk,"* The consumer's risk, expressed by β, is the proportion of "bad" lots accepted, i.e., the probability (35%) of their being accepted. The probability of accepting lots with a percent defective higher than the LTPD will be less than β.

Let us now consider the situation in which a 1,000-piece lot with a percent defective of 5% arrives for inspection. If the LTPD is 10%, the producer is allowed to pass this lot, but what will happen if we are still using the same sampling plan, with $n = 10$ and $a = 0$? As can be seen from Table 6.2, this lot will be rejected with a probability of $100\% - 60\% = 40\%$. In other words, the lot may be rejected even though it should be accepted. This type of error** is called "producer's risk,"*** expressed by α, and in this case, $\alpha = 40\%$.

As the above examples show, these two types of error are unavoidable when we take a sample from a lot and use it to decide whether to accept or reject the lot. In this case, we can reduce a in order to reduce β or increase a in order to reduce α. To reduce both α and β, we can increase n. However, this will increase the inspection costs. We must therefore identify the optimum sampling plan from the technical and economic aspects, taking into account the quality we wish to receive, the quality we wish to assure, and the various probabilities and management policies involved.

As is clear from the above discussion, sampling inspection is performed in order to decide whether to accept or reject a lot on the basis of a sample taken from it.

The above discussion concerns sampling by attributes, but the same considerations apply to sampling by variables when the distribution is taken into account.

---

* This is equivalent to a type I error, the "wool-gathering" type of error.

** This is equivalent to a type II error, the "hot-headed" type of error.

*** When an acceptable quality level (AQL) has been fixed, this is the probability of rejecting a lot of AQL quality. See Section 6.9.5 for a discussion of AQL.

6.9.2 OC Curves

However well-controlled a process may be, the percent defective in product lots shipped or accepted is subject to dispersion, and it is not unusual for it to vary, for example, from 3% to over 10%. Figure 6.5 shows the probability of a lot being accepted or rejected for different values of the lot percent defective p and different values of n and a. These curves are called "operating characteristic curves," or "OC curves."

The figure shows that when a sampling plan with $n = 10$ and $a = 0$ is used, as in the above example, the probability of accepting a lot with a percent defective of 5% is 60% (the distance from the operating curve to the horizontal axis or the reading on the left-hand vertical axis), while the probability of rejecting this lot is 40% (the distance from the curve to the upper horizontal axis or the reading on the right-hand vertical axis). These curves are therefore useful for showing the general situation occurring when a certain sampling plan is used.

Sampling with a small n—e.g., taking a sample consisting of one item and passing the lot if this sample is good (i.e., using a sampling plan with $n = 1$, $a = 0$)—is clearly at the opposite end of the spectrum from 100% screening of the whole lot.* As n increases, the operating curve gradually becomes steeper and the sampling plan's power of detection improves. It may also be seen that the inspection characteristics differ considerably with OC curves having the same ratio of acceptance number to sample size (e.g., $a/n = 1/10$) but different sample sizes (e.g., $n = 10$, $a = 1$, and $n = 50$, $a = 5$). For the same value of n, the OC curve shifts to the right as a increases.

The OC curve is more or less decided if the number of pieces to be inspected, n, and the acceptance number, a, or the rejection number, r, are fixed. If economics are not considered, the steeper the OC curve, the more effective the inspection, since this reduces the incidence of both types of error. In the ideal case—e.g., when we want to accept lots with a percent defective of 3% or less and reject lots with a percent defective of more than 3%—the OC curve should be a perpendicular line such as line A in Figure. 6.5. This can only be achieved by carrying out completely error-free 100% inspection.

A distinction should be noted between the terms, "acceptance inspection" and "receiving inspection." The term "acceptance inspection" is often used to describe a sampling inspection whose results are used to accept or reject lots. The term can thus be used for any type of inspection in which lots are accepted or rejected, e.g., receiving inspection, between-processes intermediate inspection,

---

* Sampling inspection with $n = 1$ is still meaningful when the lot percent defective swings wildly, e.g., from 0% to 100%.

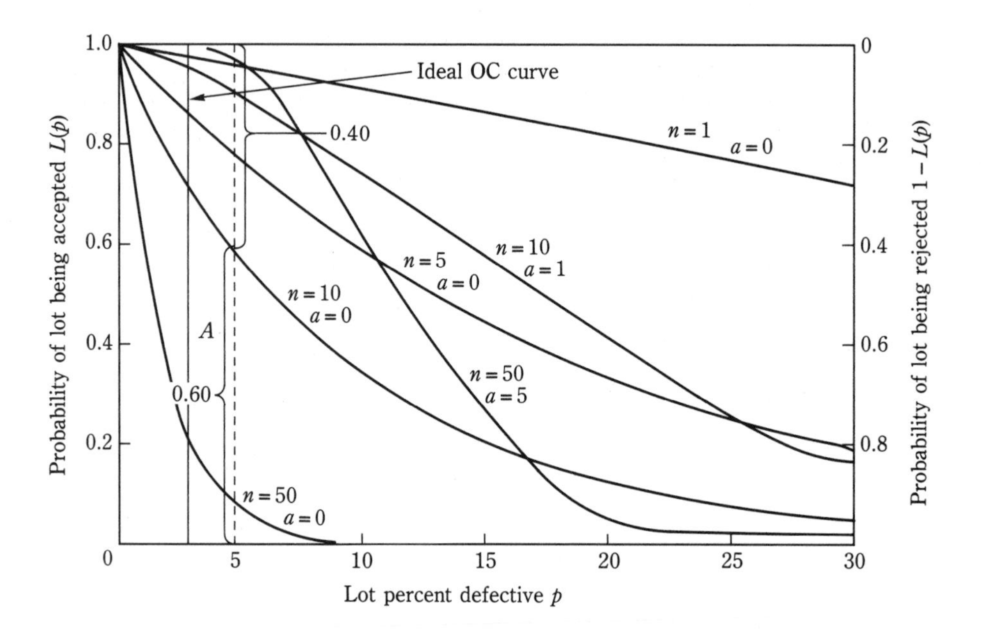

Figure 6.5 Examples of OC Curves

or shipping out inspection. Acceptance inspection and receiving inspection can easily be confused, but the tems are used in their strict sense here, although the use of the former has been avoided as much as possible.

6.9.3 Average Outgoing Quality (AOQ)

When inspection is performed according to a proper sampling plan, the OC curves discussed in the previous section show that the probability of rejecting lots with a low percent defective is small, while the probability of rejecting lots with a high percent defective is large. This means that the average quality of accepted lots shipped over a long period of time will be higher than the average quality of lots prior to inspection.

It should be remembered that if a succession of lots with a percent defective of 3% is presented for sampling inspection, the average percent defective in the accepted lots will not change, even if 20% of the lots are rejected and are scrapped. In other words, if the lot percent defective does not change greatly, sampling inspection alone will not greatly reduce the percent defective and is therefore not very meaningful.

The difference in the average quality before and after inspection increases

Table 6.3 Method of Calculating AOQ (Sampling Inspection with Screening; $N=100$, $n=5$, $a=0$)

Percent defective before inspection (%)	Proportion of lots accepted (%)	Proportion of lots rejected (%)	Method of calculation	AOQ (%)
5	77	23	5% × 0.77 + 0% × 0.23	3.9
10	59	41	10% × 0.59 + 0% × 0.41	5.9
15	44	56	15% × 0.44 + 0% × 0.56	6.6
20	33	67	20% × 0.33 + 0% × 0.67	6.6
25	24	76	25% × 0.24 + 0% × 0.76	6.0
30	17	83	30% × 0.17 + 0% × 0.83	5.1
40	7.8	92.2	40% × 0.078 + 0% × 0.922	3.1
50	3.1	96.9	50% × 0.031 + 0% × 0.969	1.6
60	1.0	99.0	60% × 0.010 + 0% × 0.990	0.6
70	0.2	99.8	70% × 0.002 + 0% × 0.998	0.2

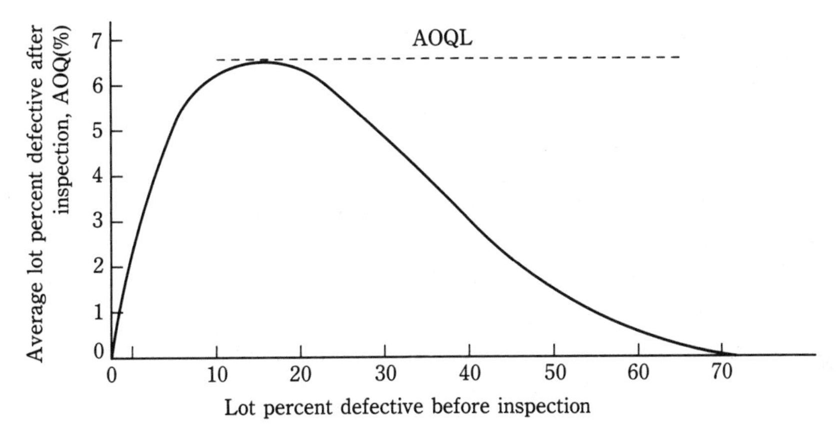

Figure 6.6 Lot Fraction Defective before Inspection and AOQ
(Sampling Inspection with Screening; $n=5$, $a=0$)

as the number of pieces inspected, n, increases; in other words, the OC curve steepens. The power of a sampling plan to distinguish whether a lot is acceptable or not depends mainly on n and bears little relation to the lot size, N. For example, when a sample of 50 pieces is inspected, there will be little difference in the probability of accepting the lot from which the sample is taken whether the lot size is 1,000 or 2,000, provided that the lot percent defectives are the same. It used to be widely believed that the sample size should be proportional to the lot size, e.g., the ratio of sample size to lot size should be 1/20. However, this is often unsatisfactory, as it would mean n changing with the lot size, and, as Figure. 6.5 shows, the ability of the plan to detect unacceptable lots would then vary greatly.

The average percent defective in product lots after inspection over a long period, mentioned above, is called the "average outgoing quality," or AOQ. As noted before, the AOQ will improve (i.e., the average percent defective will decrease), even with normal sampling inspection, when the percent defective of the incoming lots varies widely, and it will improve even further if rejected lots are subjected to 100% screening and only good products from these are passed.

Example: If an $n=5$, $a=0$ sampling plan is used on lots of size $N=100$, the AOQ can be calculated as shown in Table 6.3 if rejected lots are subjected to 100% screening and defective products are replaced with good ones. If the AOQ is plotted against the pre-inspection lot percent defective, p, a curve like that shown in Figure 6.6 is obtained. This AOQ curve illustrates one of the characteristics of sampling inspection with screening, discussed later, that the AOQ has a maximum value. This upper limit (i.e., the maximum possible value of the average percent defective) is called the average outgoing quality limit (AOQL; see Section 6.9.5).

6.9.4 Types of Sampling Inspection

The research of H. F. Dodge and H. G. Romig of Bell Telephone Laboratories in the early 1930s dragged sampling inspection out of the bad old days of hit-and-miss sample sizes and decisions into the new age of sampling inspection based on statistics. Many different types of sampling inspection plans were subsequently published to meet different applications, but here I would like to classify only the more important of these, shown in Table 6.4. For details of these plans, please refer to relevant specialized works.

(1) Classification according to decision method

(a) In *sampling inspection by variables*, decisions are based on variables.
(b) In *sampling inspection by attributes*, decisions are made on the basis of numbers of defectives or defects.

(2) Classification according to method of performing inspection

(a) *Sampling inspection based on operating characteristics*, is inspection carried out according to a particular sampling plan, in which product lots are simply accepted or rejected depending on whether or not they meet the plan's criteria. It illustrates the basic philosophy of sampling inspec-

Table 6.4 Sampling Plans

Test method	Type	Typical sampling plan	Type	Possible method	Year
By attributes	Sampling inspection based on operating characteristics	JIS Z 9002	Single	$p_0, \alpha ; p_1, \beta$	1956
		JIS Z 9009	Sequential	$p_0, \alpha ; p_1, \beta$	1962
		Paul Peach's tables	Single, double or sequential	$p_0, \alpha ; p_1, \beta$	1947
	Sampling inspection with screening	JIS Z 9006	Single	LTPD or AOQL	1956
		Dodge-Romig tables	Single or double	LTPD or AOQL	1944
	Sampling inspection with adjustment	JIS Z 9011	Single	P_b (critical fraction defective)	1963
		MIL-STD 105D	Single, double or multiple	AQL, inspection level (3 or4)	1963
		(ISO 2859)			1974
		JIS Z 9015			1980
		MIL-STD 1235A			1974
	Sampling inspection for continuous production	JIS Z 9008	—	AOQL (with individual products as unit certified by inspection)	1957
		Dodge CPS-1	—	〃	
		〃 〃 -2	—	〃	1951
		〃 〃 -3	—	〃	
		〃 SKSP-1	—	〃	1955
By attributes	Sampling inspection based on operating characteristics	JIS Z 9003 (known σ)	Single	$p_0, \alpha ; p_1, \beta$	1979
		JIS Z 9004 (unknown σ)	Single	or	1955
		JIS Z 9010	Sequential	$m_1, \alpha ; m_2, \beta$	1979
	Sampling inspection with adjustment	MIL-STD 414	Single	AQL, inspection levels (5)	1957
		(ISO 3951)	Single	〃	1980
		NTT's specifications for sampling inspection by attributes	Single	〃 (3)	1952

tion and is used in designing various more complex inspection plans, but is rarely used without modification.

(b) *Sampling inspection with screening* is inspection conducted according to a particular sampling plan, in which product lots satisfying the acceptance criteria are accepted as they are, while lots failing to meet these criteria are subjected to 100% screening. This method is used for predelivery inspections, between-processes intermediate inspections, and receiving inspections where the supplier cannot be selected. It is also possible to restrict the 100% screening, that is, to apply it only to some lots.

(c) In *sampling inspection with adjustment*, inspection is first carried out by the normal method (i.e., normal sampling inspection based on operating characteristics is performed). The inspection data are then analyzed, and the inspection procedure is either tightened for suppliers obtaining poor results or relaxed (e.g., by reducing the sample size) for suppliers posting good results. The strictness of the inspection is generally classified into three grades: normal, tightened, and reduced. However, the inspection tightness could also be adjusted to anywhere in the range from zero inspection, skip-lot inspection, and check inspection, up to 100% inspection. This method is used for receiving inspection in which suppliers can be selected and is a useful way of motivating them to practice quality control.

The above three methods are most effective when used as follows: Sampling inspection based on operating characteristics can occasionally be used for inspecting purchased lots. However, methods (b) and (c) are generally used more often. Sampling inspection with screening is used mainly for pre-delivery inspection, between-process inspection, and receiving inspection when the supplier cannot be selected. Sampling inspection with adjustment is used for receiving inspection when the supplier can be selected.

(d) *Sampling inspection for continuous production* is used for inspecting continuously produced product as it moves along a conveyor belt. Its aim is to hold the average percent defective in the product passing through the inspection station below a certain value (the AOQL). At first, the product is subjected to continuous 100% inspection; this is discontinued and fixed-interval sampling inspection introduced when the number of defectives drops below a certain specified value. If a specified number of defectives are again found, 100% inspection is reinstituted. The skip-lot sampling plan designated SKSP-1 applies the above inspection scheme to whole lots rather than just to individual products.

(3) Classification according to number of inspections performed
Inspection can also be classified according to the number of sampling stages taken when inspecting a lot.

(a) *Single sampling inspection* is inspection in which the disposition of a lot is decided on the basis of a single sample of n items, e.g., 10 pieces.
(b) *Double sampling inspection* is inspection in which the decision to accept or reject a lot may be deferred until a second sample has been taken.

A sample of n_1 items (e.g., 5 items) is inspected first, and the lot is accepted if the number of defectives in the sample is equal to or less than a_1 (e.g., 1) and rejected if the number of defectives is equal to or greater than r_1 (e,g., 3). If the number of defectives is greater than a_1 but less than r_1(i.e., 2 in this example), a second sample of n_2 items (e.g., 10 items) is inspected, and the lot is accepted if the total number of defectives in the first and second samples combined (i.e., in a sample of $n_1 + n_2 = 15$ items in this example) is equal to or less than $a_2 = r_1 - 1$ (2 in this example) and rejected if it is equal to or greater than r_1 (3 in this example).

(c) *Multiple sampling inspection* is similar to the double sampling method, but up to k successive samples may be inspected, where $k>2$. If $k=4$, for example, the inspection is carried out as shown in Table below:

Sample number	Number of items inspected (n_i)	Cumulative total of items inspected	Acceptance number (a_i)	Rejection number (r_i)
1	50	50	1	6
2	50	100	3	8
3	50	150	7	11
4	50	200	13	14

(d) *Sequential sampling inspection* is similar to multiple sampling, in that products are tested individually in succession or in sequential groups of n items, and the cumulative results are compared at each step with the acceptance number a_i and the rejection number r_i. A lot is accepted if the cumulative total of the number of defectives is equal to or less than the acceptance number and rejected if it is equal to or greater than the rejection number. If it lies between the acceptance and rejection numbers, a further sample is taken and the test is continued.

When the above four schemes are compared, the average sample number (ASN) is generally found to be smallest for sequential sampling and to increase as we move to multiple sampling, to double sampling, to single sampling. However, since the multiple and sequential methods are more complicated, single or double sampling may be more suitable for the particular purpose of the inspection and the test method used.

6.9.5 Quality Level and Disposition of Lot after Inspection

Sampling inspection is performed with the aim of taking action on product lots. An accepted lot is passed as is, but a rejected lot may be dealt with in one of the following ways:

(1) The complete lot may be returned to the supplier.
(2) The lot may be subjected to 100% screening, the defective items removed, repaired, or replaced with good ones, and then the lot may be passed.
(3) The lot may be returned to the supplier, who will have to carry out 100% screening, rework or replace defective products, and resubmit the lot for reinspection. When this is done, the screening and rework must be controlled and monitored with particular care, since often a rejected lot that has been submitted for sampling inspection a sufficient number of times will eventually be accepted, even if no rework is carried out.
(4) The lot may be scrapped.
(5) The lot may be downgraded and its price reduced.

The method of disposition as described above must be specified in standards or made clear at the time of signing contracts, and then must be faithfully followed. If no action is taken on products, lots, or prices, inspection becomes a pointless exercise.

If the above is done, what will happen to the percent defective of the product passing inspection? Keeping the percent defective below a certain value is one of the aims of sampling inspection. The average outgoing percent defective (the AOQ) was discussed in Section 6.9.3, but various other factors must also be considered, e.g., the purchaser's quality requirements, the shipper's policy concerning the quality level it wants to guarantee, how to dispose of rejected lots, etc. In many cases, the process average $\bar{p}$ is first considered, and one or two of the following parameters are then set: AQL; LTPD; AOQL; p_0, α or p_1, β; or $p_{0.50}$. These are discussed briefly below.

Acceptable quality level (AQL) is the lowest quality (i.e., the maximum percent defective) that can be accepted. It is a quality level often used in receiving inspection and is the maximum level of defectives the purchaser is willing to accept on average in view of the purpose for which the raw material is to be used. It is sometimes determined according to the supplier's capabilities. For example, setting an AQL of 2% and devising a suitable sampling plan is satisfactory when continuing to purchase from suppliers whose process average percent defective is approximately 1.5%. However, there will be a higher probability of rejecting lots from suppliers with a process average percent defective of 2% or more, and this sampling plan will consequently weed out such firms. As Table 6.4 shows, in sampling inspection with adjustment, the AQL is fixed and the sampling plan is adjusted according to the average percent defective in the observed samples, $\bar{p}$, (MIL–STD–105D).

Lot tolerance percent defective (LTPD) is the lower limit of the range of values of lot percent defective that should preferably be rejected. The consumer's risk, β, is the probability of accepting a lot with a percent defective equal to or greater than the LTPD. When lots passed by inspection are to be used as they are by the next process or by the consumer, a sampling plan can be chosen on the basis of specified values of LTPD and β. In such cases, sampling inspection with screening is often used (see Figure 6.7).

Average outgoing quality limit (AOQL) is the worst possible value of the average outgoing quality (AOQ) of lots that have passed sampling inspection; in other words, it is the upper limit of the percent defective in the outgoing product.

As Figure 6.6 shows, the AOQ of passed lots has a maximum value beyond which it cannot become any worse; this worst-case limit is termed the AOQL. When a sampling plan is being based on the AOQL, factors such as the process average fraction defective and inspection costs must also be taken into account. This limit is often used for sampling inspection with screening and sampling inspection for continuous production, i.e., for between-processes inspection or for pre-delivery inspection where product lots are broken down into individual units (see Figure 6.8). The Dodge-Romig sampling inspection tables are well-known sampling plans indexed on the LTPD and AOQL. They are often used for between-processes or pre-delivery inspection when the supplier cannot be selected. These plans are based on the premise that all defectives detected are either reworked or replaced by non-defective products (see Figure 6.8).

In the p_0, α; p_1, β method, the sampling plan is chosen by setting the percent defective p_1 of lots that it is desired to reject, the probability $L_{p1} = \beta$ (the consumer's risk) of accepting such lots, the percent defective p_1 of lots that it is desired to accept, and the probability $1 - L_{p0} = \alpha$ (the producer's risk) of rejecting such lots. These parameters are often set when using sampling inspection based on operating characteristics (see Figure 6.9).

The $p_{0\cdot 50}$ is a method in which the sampling plan is based on the "indifference quality," $p_{0\cdot 50}$, the quality that has a probability of acceptance of exactly 50%.

Which of the above quality levels to choose depends on factors such as the purpose of the inspection, the conditions governing the lots being inspected, and company policy. This question must be investigated and discussed thoroughly, since it is the key point as far as inspection is concerned.

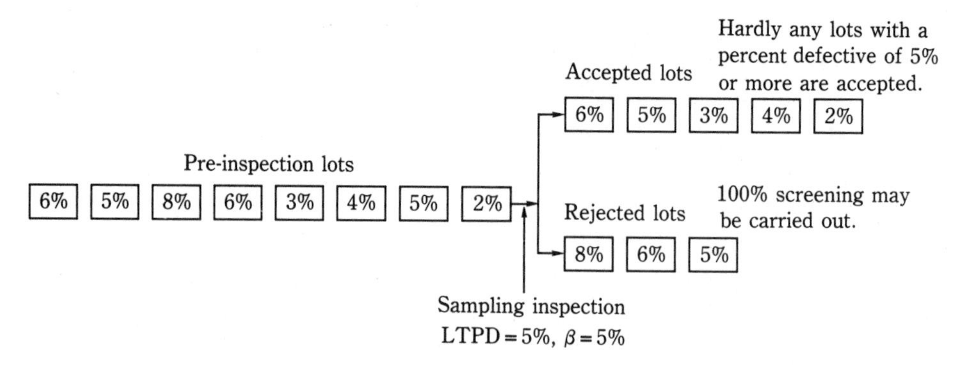

Figure 6.7 Example of Sampling Inspection Indexed on LTPD

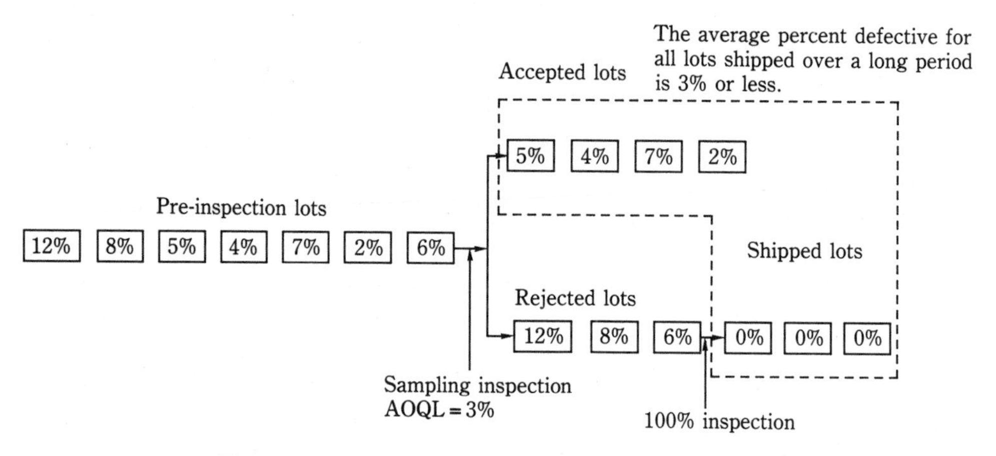

Figure 6.8 Sampling Inspection Indexed on AOQL

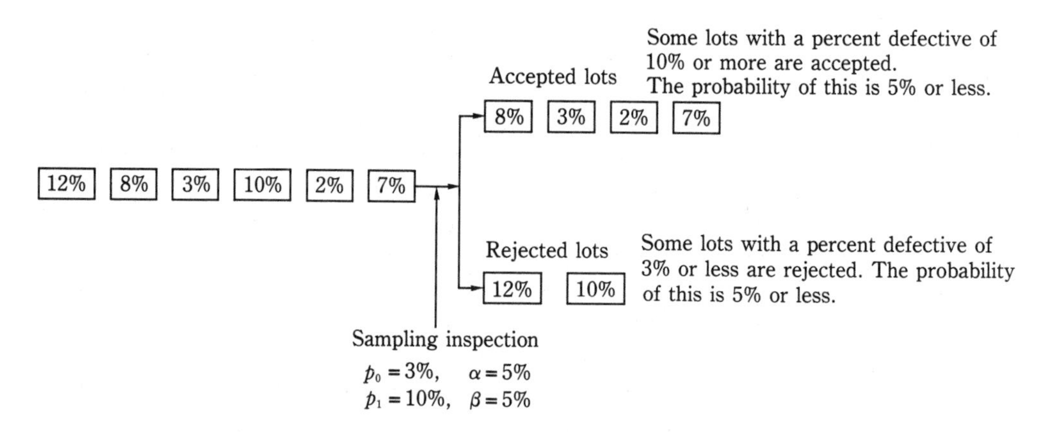

Figure 6.9 Example of Sampling Inspection Indexed on $p_0, \alpha; p_1, \beta$

6.10 100% Inspection or Sampling Inspection?

To guarantee a perfect product, 100% inspection is needed unless the process is in the controlled state with a process capability index C_p of at least 1.67. However, 100% inspection by human inspectors is very error-prone, and about 10%–30% of products are mistakenly identified as non-defective when they are defective or vice-versa. 100% inspection is carried out for form's sake in many Japanese industries, but in most cases, lots checked after passing inspection will be found to contain quite large numbers of defectives, while non-defective items will be seen in the defective tray. To guarantee a percent defective of zero with 100% inspection by human inspectors, the inspection process must be carefully repeated seven or eight times; this makes it extremely expensive. Since machines can perform inspection more accurately than people, the inspection process should be automated. But even automatic inspection equipment has reliability problems and is subject to error, and to be absolutely safe, it is necessary to check the results of such inspection or pass the product through twice. Furthermore, 100% inspection is impossible with products that can only be tested destructively.

Depending on the type of product or the particular characteristic being inspected, the passing of a small percentage of defectives may cause no great harm during the product's use. The question when inspecting is therefore which is best: 100% inspection, sampling inspection, or even zero inspection? This question is a problem of economics, and its solution requires a comprehensive view of the need for quality assurance. Along with the automation of inspection, the following should be considered:

(1) When is 100% inspection required?

(a) 100% inspection is needed when passing even a single defective will cause trouble, e.g., when a defect is critical and may endanger human life, or when shipping even one defective product will have a great impact either economically or from the viewpoint of credibility. One should note, however, that there is little point in carrying out 100% inspection designed to eliminate defectives completely if the product is liable to become defective during transport.

(b) When the product is extremely valuable, e.g., complete aircraft, ships, cranes, automobiles, etc.

(c) When the product's performance cannot be tested until it is assembled: in this case, 100% inspection should be carried out after assembly.

(d) When lots prepared for delivery include defective items, pre-delivery inspection should in principle be 100% inspection.

(2) When is 100% inspection advantageous?

(a) When 100% inspection can be carried out easily and reliably (e.g., checking whether light bulbs work) or when reliable automatic screening equipment can be used.
(b) When the lot size is small and there is no economic advantage in carrying out sampling inspection.
(c) When the lot average percent defective is far greater than the percent defective required.

(3) When is sampling inspection advantageous?

(a) When 100% inspection is imperfect.
(b) When there are many characteristics to be inspected.
(c) When inspection costs are high.
(d) When it is desired to guarantee the quality of product lots at a specified percent defective.
(e) When it is desired to motivate a producer to raise quality and practice quality control.
(f) For receiving inspections.

Purchased products should in principle either be subjected to sampling inspection or accepted without any inspection, and 100% inspection should be carried out by the supplier.

(4) When is sampling inspection required?

(a) For destructive inspection.
(b) For chemical analyses.
(c) When 100% inspection is impossible because of the sheer volume of products passing through.
(d) When the purchaser cannot rely on the supplier's quality assurance.

(5) When is zero inspection acceptable?

Zero inspection is only acceptable when the process is stable and in the controlled state, and:

(a) When it can be concluded that every characteristic of the product amply satisfies the specifications and that all products are good, e.g., when the process is in the controlled state at a process capability index C_p of 1.67 or above.
(b) When the lot percent defective is clearly better than the percent defective required. However, even when this is so, it is better to perform occasional inspections as a check.

6.11 Process Control or Inspection?

The question as to whether priority should be given to process control or inspection can be answered simply, as discussed in Sections 1.3 and 6.3, by saying that process control should always be carried out, while inspection should only be carried out as needed.

(1) When is inspection unnecessary?
Inspection need not be carried out when a process is in the controlled state and its product comfortably meets the certified grade (generally when $C_p = 1.33$, 1.67, or higher).

(2) When is process control unnecessary?
Process control is never unnecessary. It is of course possible to spread out or terminate certain quality measurements or the plotting of control charts when good work standards have been prepared, thorough education has been conducted, the process is well controlled, and the controlled state has continued for a long period, but process control itself must still be continued.

(3) When is it disadvantageous to practice quality assurance through inspection?
The answer is surprisingly often. Some examples are:

(a) When it is company policy never to deliver even a single defective product. Virtually the same thing applies when it is desired to keep the guaranteed percent defective at an extremely low level.
(b) When reliability, particularly durability and other factors affecting product life, is important.
(c) When inspection costs are high.
(d) When there is a large variation in the product, or when measurement or sampling errors are large.

(e) When inspection is destructive.

In these situations, assuring quality through inspection is often uneconomical and it may be better to adopt one of the following approaches:

(i) Give priority to process control.
(ii) Concentrate on improving the process technically.
(iii) Use more reliable parts and designs.
(iv) Lower the certified grade if existing technology makes this inevitable.

6.12 The Inspection Department

6.12.1 The Duties of the Inspection Department

As mentioned in Section 6.7, the inspection function and the duties of the inspection department (i.e., the work the inspection department should do) should be considered separately, since confusion often arises when they are discussed without first being clearly distinguished. Let us therefore first look briefly at the various types of work that inspection departments perform.

(1) The inspection department performs measurements on raw material lots, semi-finished product lots, and finished product lots, compares the measurements with set criteria to judge whether to accept or reject each lot, and decides on the disposition of each lot according to standardized procedures. This is the true function of inspection, and it may be necessary to make it a duty of the inspection department. Great care must be taken to ensure that inspectors and the inspection department do not introduce subjectivity. If a process improves, if the quality of the product becomes amply able to satisfy the specifications, and if the process continues in the controlled state for a long period, the severity of inspection may be relaxed, separate inspection processes may be combined, or inspection may be replaced by simple checking or abolished entirely, thus reducing the inspection department's workload. The inspection department will become busier if a process goes out of control and becomes unable to satisfy the specifications, since inspection will then have to be tightened up. Because of this, the inspection department must also keep an eye on control charts in the workplace and keep track of process data.
(2) The inspection department performs measurements on individual articles (raw materials, semi-finished products, finished products, etc.), com-

pares the results with set criteria, decides whether to accept or reject each article, and screens out the defective ones. This duty consists of performing 100% inspection and eliminating individual defectives. However, as TQC spreads through a company, the company should institute a system of autonomous control and inspection and make the production department responsible for inspecting semi-finished and finished products. If the inspection department does this job, those on the shop floor know that it will screen out any defectives for them, no matter how many they produce; they will grow apathetic about quality and become fixated on quantity. Information feedback will also take longer, making control more difficult. 100% inspection should therefore be carried out on the shop floor, where the responsibility for quality assurance lies, and the inspection data obtained should then be used for process control. Check inspection must also be performed, since 100% inspection is extremely error-prone.

(3) The inspection department performs measurements on raw materials, semi-finished products, and finished products, and provides the data obtained to the departments that need it. As a manufacturing process becomes more industrialized, there is increasing division of labor within factories, and inspectors are sometimes put in charge of taking and recording measurements, while workers no longer take them, or else they take them but do not record the results. Taking and recording measurements, however, is really the responsibility of the workplace, and it is better to assign inspectors to other duties and move toward autonomous control under the supervision of process checkers responsible for monitoring. If the inspection department is responsible for taking measurements, the workplace cannot obtain information on product quality directly; the inspection department must supply this information to the workplace before it can be used for process control. With such a setup, the inspection department is responsible for the service of providing information for process control. Some points to be noted when this is done are:

(a) The inspection department is an information provider and in principle has no authority to take action on a process.

(b) The data supplied to the workplace are for the purposes of process control and analysis, not for product inspection. Since these three purposes are different, it is necessary to review whether or not the inspection data being taken are in fact suitable for process control or analysis. The data are to be used in the workplace for controlling processes and eliminating the causes of defects, and care must be taken that the emphasis does not drift toward using such data for lot disposition or the reworking of defectives, thereby confusing control with inspection.

(c) When providing data for process control, the inspection department has a duty to ensure that the data are well stratified and supplied promptly.

(d) Data can be provided in one of two ways; it can be supplied to the shop floor raw and plotted on control charts by those responsible for control in the workplace, or inspection personnel can plot it on control charts displayed in the workplace and inform those responsible when an abnormality is indicated. The former method is usually better. If the latter method is adopted, the inspection personnel should be attached to the workplace as process checkers.

(4) The job of packing may be assigned to the inspection department in order to reduce costs, but care must be taken when doing this, since attention will be diverted from inspection to packing, and inspection may become a hit-and-miss procedure. This work should normally be assigned to the production department.

(5) When inspectors are responsible for plotting points on control charts, they should immediately contact the production department when a process shows an abnormality or when the number of defectives suddenly increases, but the inspection department should not generally take action on a process, advise about such action, or direct the search for causes. However, the inspection department may be given authority to stop a process when certain criteria are met as a result of inspection performed according to the inspection department's work standards, provided that these standards have been perfected and this has been agreed with the production department.

(6) From the QC viewpoint, handling complaints is an extremely far-reaching and important task, and it should, if possible, be assigned to the quality assurance department or the quality control department.

(7) Auditing products from the consumer's standpoint is the duty of the quality assurance department, QA group, or quality control department, and it is normally better not to assign it to the inspection department.

(8) Since the true function of the inspection department is to perform inspection according to inspection standards, it, like the shop floor, needs few engineers. The duties of engineers in the inspection department consist of liaising with other departments and researching, preparing, improving, and controlling measurement methods, inspection standards, and inspection criteria.

(9) The department responsible for preparing and finalizing various plans and standards related to inspection differs depending on the organization of the company, but is probably one of the following:
   (i) The technical department.
   (ii) The quality control department.
   (iii) The inspection department.

As quality control gradually develops and the organization becomes rationalized, the inspection department may formulate inspection standards but should not finalize them; this should be done by the technical department, the quality control department, or the quality assurance department.

(10) Inspection control is always the job of the department performing the inspection. Inspection control consists of controlling inspection methods, inspection standards, and the inspection task itself, as well as measuring instruments, jigs and tools.

Controlling inspection methods means altering the severity of inspections (to reduced, normal, or tightened) whenever necessary. Once enough inspection data have been accumulated, it is best to arrange for this adjustment to be performed automatically using computers or other means.

Controlling inspection standards is done in exactly the same way as controlling work standards.

Controlling the inspection task obviously consists of controlling the inspection process, and it is one of the most important duties of managers and supervisors in departments performing inspection.

6.12.2 Errors to which Inspection and the Inspection Department are Susceptible

Since I have discussed the functions of the inspection department, I would now like to voice some words of caution about common mistakes made in connection with these. The spread of TQC has recently improved matters in Japan, but other countries still fall into the traps mentioned below.

(1) It is easy for the inspection department to forget its principal duty, i.e., performing the inspection task according to inspection standards. Although inspection managers should manage the work of inspection (including sampling, measurement, and testing), they sometimes fail to create or operate proper systems for controlling the inspection process. We could go so far as to say that if inspection work standards are properly

prepared and inspection is carried out in accordance with them, any defectives that happen to be shipped are not the responsibility of the inspectors but of the departments that finalized and approved the standards.

(2) Sometimes the effect of a particular inspection plan on the next process or the consumer has not been sufficiently investigated. For example, abolishing receiving inspection is a bonus for the inspection department, but the process may still produce many defectives and customer complaints may remain at the same level even if pre-delivery inspection is carried out. In such cases, information on defectives and complaints is not being fed back and used to rationalize the inspection standards.

(3) There may be no clear definition of what constitutes a lot, and mixed lots are formed from materials that should be kept separate.

(4) Although action should be taken on individual products or lots as a result of inspection, the action laid down in standards is sometimes not taken, or lots are specially accepted at the whim of individual inspectors. In the worst case, there may be no standards governing the disposition of lots. It is also out of order for production department managers or section managers to make such decisions on their own initiative. Sometimes, no stop is put on shipments, and defective lots are sent out.

(5) Although the inspection department is responsible for supplying data to be used for process control and analysis, it sometimes forgets this and only tells the production department about rejected lots or products. There is not enough investigation of what kinds of data would facilitate process control and analysis. In the worst case, the inspection department is not interested in providing such data and does not perform analytical inspections.

(6) Inspection engineers may forget their true inspection task and concern themselves exclusively with statistical sampling inspection.

(7) Inspection plans that promote process control (e.g., MIL–STD–105D) should be selected; not to do this is a mistake.

(8) Inspection should be adjusted flexibly for each process by using a variety of inspection plans to suit the prevailing conditions, taking economic considerations into account. However, people often feel bound to use conventional published inspection plans, or tackle things the wrong way round by selecting the inspection plan on the basis of the number of inspectors available. They may also neglect to relax or tighten inspection as needed.

(9) Inspection is sometimes tightened when complaints are received and then relaxed again after some time has elapsed, and inspection criteria may

be changed according to the passing fancies of inspection staff. They may also be changed arbitrarily at the request of the sales department. Quality standards are vitally important and should be treated as a matter of company policy. It is extremely odd to allow the inspection or sales department or individuals such as inspectors to alter these standards at their own discretion. Standardization and control are just as important in inspection as they are in process control.

6.13 INSPECTION STANDARDS AND HOW TO SET THEM

Since inspection is a form of work, it requires work standards. There are various types of inspection work standards, e.g., experimental standards, measurement standards, instrument control standards, jig and tool control standards, sampling standards, inspection standards, inspection implementation regulations, inspection standards control regulations, etc. These standards are classified in various ways by different industries and companies; here, I prefer to focus on some important inspection-related items that must be included in the standards.

(1) Items that must be specified in inspection standards
At the very least, inspection standards should do the following:

(a) Clearly state the rules for lots and how they should be composed.
(b) Clearly specify sampling units, units certified by inspection, and units of product to be measured or tested. The differences or relationships among these should also be specified if necessary.
(c) When specifying sampling methods, state not only sample sizes but also specific methods for collecting samples.
(d) State the items to be inspected, their degree of importance (major, minor, or slight), the order in which they are to be inspected, and the measurement, and testing methods to be used. Specify whether or not analytical inspection is to be carried out.
(e) Set forth clear inspection criteria for individual products and product lots. Ensure that there are no inconsistencies among design values, certified grades, and inspection criteria. Ensure that these are rational and, particularly in sensory testing, ensure clarity by providing limit samples as inspection criteria for each test sample, and prescribe periodic checks.
(f) Clearly define the methods by which individual defectives and rejected lots should be disposed of, and state who is responsible for doing this.

Rejected lots and defective products are often disposed of or specially accepted at the discretion of department managers or factory managers, but it is better to standardize this as far as possible.

(g) Standardize items relating to the submission and circulation of inspection reports, particularly the speed with which this should be done, and, if necessary, standardize methods of utilizing inspection records, e.g., for adjusting inspection plans or for process control or analysis.

(h) If necessary, specify methods of using information from complaints, the quality assurance department, the next process, etc., in revising inspection standards.

(i) Specify methods of checking and controlling inspection work.

(2) Selecting inspection plans

As there is not enough space in this book to describe all the various methods of choosing inspection plans, I will restrict myself to discussing the basic procedure and philosophy. The order given below may be reversed depending on the particular situation.

(a) Decide on the purpose of inspection. Purpose should be decided in line with company policy after investigation of the way in which the product is used by the next process or by the consumer, the economic losses arising from the production of defective products or lots, and the possible damage caused by loss of trust. Consider the relationship between the product's true characteristics (its performance) and substitute characteristics and the division of inspection work (inspection characteristics and methods) between the production department and the inspection department, and ensure that inspection is not duplicated unnecessarily.

(b) Perform statistical analyses of past data. If necessary, collect and analyze fresh data. Control charts are extremely useful when doing this.

(c) Decide on the assurance unit. This decision is particularly important in the production of bulk materials.

(d) Prepare inspection and quality assurance process charts. Considering the flow of materials, decide on the location of inspection stations and create an inspection plan covering receiving inspection, between-processes intermediate inspection, final inspection, pre-delivery inspection, etc., that specifies the points at which the flow of various types of defectives should be stopped. It is usually best to perform inspection in the previous process whenever possible. In addition, decide which inspections are to be carried out by the production department and which by the inspection department, and whether patrol inspection or centralized inspection is needed.

(e) Set the sampling and measurement units.

(f) Decide on the quality characteristics to be inspected and the order in which they are to be inspected. Select the measurement and testing methods. When the drawings show large numbers of dimensions, decide which of these are to be inspected and who is to do this.

(g) Set the inspection criteria for each characteristic of each individual product.

(h) Decide on the quality levels required. In sampling inspection, for example, decide whether to judge lots on the basis of AQL, LTPD, AOQL, (p_0, p_1, α, β) or $p_{0\cdot50}$. Since choosing one of these and deciding the level at which to set it is extremely important, it is necessary thoroughly to study and investigate the purpose of inspection (see Section 1 above), together with the economic aspects of sampling inspection, process control, and the disposition of defective items and rejected lots.

(i) Decide on how lots are to be composed. It is best to stratify the product in order to make individual lots as uniform as possible.

(j) Decide whether to adopt 100% inspection, sampling inspection, zero inspection, inspection by variables, inspection by attributes, etc.

(k) If sampling inspection is chosen, decide the following:
  (i) The sampling plan (n_i, a_i, r_i, etc.).
  (ii) The criteria for judging whether to accept or reject lots.
  (iii) The method of taking samples (be as specific as possible).

(l) Decide on the method of disposing of accepted and rejected lots and items. However carefully inspection criteria are set, inspection is virtually meaningless unless these methods of disposition are clear. Also decide how accepted and rejected lots and items are to be indicated.

(m) Set up a system for controlling the inspection work.

(n) Decide how the inspection will be adjusted. If necessary, set criteria for shifting among zero inspection, check inspection, reduced inspection, normal inspection, tightened inspection, and 100% inspection for each inspection characteristic.

(o) Lay down methods of revising inspection standards. For example, decide how inspection data or information from the previous process, the next process, complaints, market surveys, etc., will be fed back, and by whom, when, and by what procedure inspection standards will be revised.

(p) Decide on the style, circulation route, and method of use of inspection report forms, when they should be completed, and who should do this. Remember that upper-level managers should not have to see overly detailed data.

6.14 Complaints Handling and Special Acceptance

6.14.1 What are Complaints?

Complaints handling has already been discussed in Sections 1.4.1 and 6.4, but since it is such an important aspect of quality assurance, it is summarized here.

(1) Types of complaint:

- Complaints about quality and service
- Complaints about quantity
- Complaints about delivery dates
- Complaints about price
- Complaints involving money
- Complaints not involving money
- Actual complaints
- Hidden complaints
- Wrong choice of product
- Wrong product delivered
- Misoperation, misrepair, misjudgment
- Defective product
- Complaints about individual products
- Complaints about product lots
- Complaints about design quality
- Complaints resulting from poor inspection
- Complaints resulting from poor production process
- Complaints resulting from poor bought-out parts
- Complaints resulting from poor after-sales service
- Complaints resulting from poor sales methods

(2) Problems with handling complaints: Some cautionary notes
I would now like to discuss some problems relating to complaints and their handling, and to give some general words of advice, from the TQC standpoint.

(a) Sales departments and distribution organizations often pride themselves on clearing up complaints as painlessly as possible, believing it better to keep the lid on anything likely to cause trouble. This simply keeps complaints under the surface and leads to no action being taken.

(b) In QC, positive efforts are made to elicit customers' grievances and dissatisfactions, to disclose hidden complaints and eliminate their causes. This usually means an increase in the number of complaints when a companywide QC campaign is initiated (see Figure 4.2).
(c) Some companies draw up complaints handling regulations and go through the formalities when complaints are received, but still suffer from the following problems:
  (i) Only information on complaints involving money is received.
  (ii) No effort is made to bring hidden complaints to the surface.
  (iii) The opinions of people who want to complain do not reach those in the company who really need to hear them.
(d) Complaints are often not dealt with carefully and considerately enough, particularly at the end of the chain where the company meets the customer. If complaints are not dealt with skillfully and satisfactorily, they will turn into product liability problems.
(e) Some companies still take the attitude that if a complaint is received, it is sufficient to replace the offending article with a new one.
(f) Dealing with the monetary side of complaints often takes too long, and feedback in the form of complaint information and quality information can be extremely slow.
(g) Too much time and effort may be spent on trifling complaints that somehow happen to have reached the ears of top management.
(h) Not all complaints are justified. It is important to check the circumstances of complaints carefully. For example, the number of complaints rises tenfold in recessionary periods. Of complaints made about parts, the parts are only really to blame in about 1/10 to 1/100 of cases.
(i) The number, details, and degree of seriousness of all complaints should be examined. If these fall within certain limits, quality is in a healthy state.
(j) The fact that if no complaints are received does not necessarily indicate that quality is satisfactory. Actually, it is a danger signal. It means that no information on complaints is being received and that customers have resigned themselves to poor quality.
(k) The higher the price of a product, the more complaints are likely to be received about it. People do not complain so much about cheap items; they simply keep their complaints to themselves and stop buying the merchandise.
(l) Sales and distribution organizations sometimes have the following problems:
  (i) They have no desire to disclose hidden complaints and take no positive steps to do so.

(ii) They lack the product knowledge and technical knowledge required to understand complaints and collect information that will be useful for taking action.

(iii) They employ few sales engineers or none at all. Sales engineers are an absolute necessity, particularly for primary products.

(m) There may be no system for having sales-related departments actively collect quality information.

(n) Some companies are too hasty, mistakenly ordering their design departments to take countermeasures whenever a complaint is received.

(o) In QC, the next process is your customer. This means that quite a large number of complaints should arise within the company itself, between different departments. Internal complaints must also be controlled by formulating handling regulations in the same way that outside complaints are handled.

6.14.2 Complaints Handling

Complaints handling should be investigated and faithfully executed from the following two aspects:

- *The external aspect* – keeping the customer satisfied: this requires speed, sincerity, prevention of recurrence, and sound investigation.
- *The internal aspect,* which encompasses:
  (1) recurrence prevention:
    (a) eliminate the symptom
    (b) eliminate the immediate cause
    (c) eliminate the root causes
  (2) handling the accounting side: institute a system of accountability.
  (3) disposition of products about which complaints have been made.

Dealing with the external aspect of complaints was discussed in Section 6.4, and dealing with the internal aspect, particularly recurrence prevention, was discussed in Section 1.5.3. To reiterate, however, in reliable complaints handling, the following action should be taken:

(1) Formulate complaints handling regulations and draw up a complaints handling flow chart.

(2) Design standard report forms and list the items to be investigated for

each product. These should include the product name, the circumstances surrounding the complaint, the conditions of use, the place of use, the name of the person investigating the complaint, the name of the user, the date of occurrence of the fault, the date of completion of the report, the history of the product (date of manufacture, lot number, etc.), the date the product was received, the method of transport, the quantity, the price, the technical details of the complaint, the inspection method used by the recipient of the product, the inspection criteria, the apparent causes, emergency measures taken, recurrence-prevention action, whether or not the defective article was returned, etc.

(3) Set up a special complaints section.

(4) Decide who is to screen complaints and notify the departments responsible for dealing with them, and how this is to be done. It is best to put the quality assurance department in charge of this.

(5) Designate a department in charge of preparing complaints statistics, reporting to top management, and making recommendations to other departments. Again, it is best to make the quality assurance department responsible for this.

(6) The quality assurance department should ensure that the complaints handling control cycle is properly rotated, from the receipt and recording of a complaint through each step of dealing with it (particularly recurrence prevention) to its final resolution.

6.14.3 Special Acceptance

Special, or "as-is," acceptance means specially passing items that have failed raw-material, intermediate, or pre-delivery inspection. Some possible reasons for allowing special acceptance are:

(i) Because the specifications are too strict.
(ii) Because the designed quality or drawings are unreasonable.
(iii) Because the process capability is insufficient.
(iv) Because process control is inadequate.
(v) Because of failure to rank quality characteristics appropriately in their order of importance (major, minor, or slight).

The most common reason is that the specifications are too strict and there will be no effect on quality or cost if failed items are accepted as is. If this happens, it normally indicates irrationalities in the specifications and other standards, or, as mentioned below, is allowed because a product does not suddenly become unusable when a certain specified value is not met. In fact, as-is acceptance is probably more common than not, and the problem is how to control it.

It is generally sufficient to observe the following precautions:

(1) Stratify as-is acceptance into various levels. For example, there may be no objection to specially accepting items with slight defects, but it is in principle unacceptable to accept as is items with critical or major defects (if the specifications have been drawn up on a rational basis). Certain conditions may also be attached to as-is acceptance. Some examples might be to use an article after drying it, or to restrict its use to a particular purpose. Such conditions should be set by the department responsible for ordering as-is acceptance.
(2) Standard rules should be laid down for as-is acceptance, making it unnecessary to seek the department manager's or factory manager's permission every time.
(3) Since as-is acceptance is a type of experiment, the details should be properly recorded and arrangements made for stratified data to be taken to enable follow-up on the results. Analyzing the results may make it possible to relax the specifications.
(4) Measures should be devised to prevent recurrence of the need for as-is acceptance.
(5) The possibility of as-is acceptance arises because specifications have definite values. For example, an item with a maximum specified length of 10.00 mm may still be usable even if its length is 10.01 mm. It may, however, be more difficult to use, and the introduction of penalty clauses in conjunction with as-is acceptance should be considered.
(6) Great care is needed with as-is acceptance, since its thoughtless use will adversely affect employees' quality consciousness.
(7) If necessary, an as-is acceptance committee should be set up.

6.15 Conclusions

Since quality assurance is the heart and soul of TQC, as well as its *raison d'être*, we must keep on practicing it as long as we make and sell products. Mere inspection is not quality assurance, but it is needed as long as defectives continue to be produced. Unfortunately, inspection is still needed in most cases.

We must also remember that the work of inspection must be controlled, just like the work of the shop floor. The following points should be noted, particularly with 100% inspection and sensory inspection:

(1) Match inspection levels to consumer requirements.
(2) Provide limit samples rather than standard samples.
(3) Rationalize inspection standards.
(4) Control inspection carefully, treating it as one kind of process. For example, check whether screening is being carried out properly by taking samples from the good and bad product groups and preparing control charts.
(5) Compare inspection information with QC data and complaints.

As long as we are practicing quality control, we must assure quality through many kinds of activities, as well as inspection. Quality control that does not actually guarantee quality is quality control in name only.

Chapter 7
THE SYSTEMATIC IMPLEMENTATION OF TOTAL QUALITY CONTROL

7.1 Total Quality Control

As I have repeatedly emphasized, total quality control must be implemented systematically across the whole company in conjunction with personnel management, cost control, profit control, and control of production volume and delivery schedules. To achieve this, all employees must have a good understanding of the topics discussed in the first six chapters of this book, while top management must consider their company's history and present circumstances, set out clear policy guidelines, and exercise their wisdom and ingenuity to put TQC into effect.

This book is concerned chiefly with the details of implementing TQC, i.e., its technical aspects; for a discussion of the management aspects of systematic TQC implementation, I refer readers to my book, *TQC towa Nanika? — Nipponteki Hinshitsu Kanri* (What is Total Quality Control — The Japanese Way; revised and enlarged edition, pub. JUSE Press, 1984; English translation by David J. Lu, pub. Prentice-Hall, Inc., 1985). I restrict myself in this final chapter to summarizing some of the guiding principles.

7.2 The TQC Organization

Like real technical progress and industrial rationalization, TQC is impossible without organizational rationalization. Since TQC is an all-embracing activity requiring the participation of every department and every employee of a company, the following steps are essential:

(1) Owners and top management must state policy clearly.
(2) The organization must be rationalized and authority and responsibility clarified.
(3) The scope of delegation of authority and methods of controlling this must be researched with particular thoroughness.
(4) A system of cooperation can be established by reminding ourselves that an organization does not exist for the benefit of individuals but for the smooth running of the company.
(5) The organizational setup should be clarified for cross-functional management as well as, of course, for intradivisional management.
(6) Clear distinctions must be drawn between line personnel and staff personnel. Unreasonable behavior by line and staff personnel, a particular problem in Japan, must be restrained. The staff's next process (i.e., its customer) is the line.
(7) Management must try to adopt a more scientific approach, freely appointing technical personnel to administrative departments, and vice versa, as needed.

Contact between the producer and the customer lies at the root of quality control. This means that personnel rotation among sales, materials, and cost control departments, together with the appointment of talented engineers to these departments, will be indispensable for companies in the future.

To implement quality control properly, it is ultimately necessary to establish a TQC promotion office, quality control department, or quality assurance department that can devote itself solely to QC activities. However, it may be better to start by assigning the responsibility for QC to a staff department close to top management (e.g., the planning, technical, inspection, survey, or works department) in addition to its normal work. As QC steadily disseminates through the company and a conducive atmosphere is established, a TQC promotion office or other organization should be set up and its authority gradually increased. Setting up such an organization from the start can produce the opposite of the intended result, since it tends to convey the mistaken impression that QC is this organization's exclusive preserve, thereby creating sectionalism and hampering the development of QC. Any department can be made responsible for promoting QC at first, but it should be made known to everybody in the company. Of course, when the top management group is of one mind and the company president is prepared to give a strong lead, the TQC promotion office should be set up as a presidential staff department from the outset.

(1) The duties of TQC and the quality control department

The department at the center of TQC promotion has various duties including the following:

(A) TQC promotion and administration
- (i) policy management staff work
- (ii) planning and execution of QC education and training
- (iii) planning, execution, and promotion work as top management QC audit secretariat
- (iv) promotion of QC circle activities
- (v) surveying and assisting with intra departmental TQC promotion
- (vi) promoting groupwide TQC (subcontractors, distribution organizations, and affiliated companies)
- (vii) liaison and cooperation with outside QC activities

(B) Quality assurance
- (i) acting as a quality assurance center
- (ii) acting as a complaints processing center
- (iii) acting as a center for quality assurance in new-product development, possibly including new-product development cost control and progress control
- (iv) performing quality audits
- (v) performing incoming, intermediate, pre-delivery, and other inspections
- (vi) occasionally exercising authority to stop shipments of existing and new products

(C) QC staff work
- (i) acting as general staff, giving assistance and advice on quality problems to the company president, head-office department managers, factory managers, branch managers, etc.
- (ii) providing a quality assurance service to all departments
- (iii) assisting with QC promotion for organizations such as subcontractors, distribution organizations, and affiliated companies, and promoting groupwide quality assurance.

All of the above duties may be assigned to a single TQC promotion office or, in very large companies, the work may be shared out among three different organizations, e.g., a TQC promotion office, a quality assurance department, and an inspection department, or two organizations if the inspection department is part of the quality assurance department. If this is done, the duties listed in (A) should be carried out mainly by the TQC promotion office, while those listed in

(B) and (C) (except for inspection) should be done by the quality assurance department. Inspection should be a line job for the inspection department.

The work of the TQC promotion office and quality assurance department consists mainly of general staff and service staff duties.

(2) Who should be put in charge of TQC promotion?

People who know a lot about statistical methods are usually put in charge of quality control, but this is a mistake. From now on, all employees will have to learn about statistical methods as a basic qualification for remaining in employment, and it is wrong, and may even be harmful, to put people in charge of TQC or QC promotion simply because they know about these methods or have started investigating them a little sooner than anyone else. Learning to use statistical tools is no more than acquiring a basic item of knowledge that happens to have been missing in the past.

The people in charge of promoting TQC are those responsible for implementing it in their particular post, e.g., workplace foremen and supervisors, section managers, department managers, factory managers, branch managers, directors-in-charge, the company president, etc.

(3) Staffing the TQC promotion department

In companies on the point of introducing TQC, the TQC promotion staff are responsible for disseminating the concepts and methods of TQC and QC and educating people in these, as well as for promoting their implementation. Since they will be at the heart of TQC promotion and implementation, it is important to select people with as many of the following qualifications as possible:

(a) *Experience*: Staff should have at least three years' work experience in a line department such as the factory floor or a branch office and should have extensive technical work knowledge. If possible, they should have experience of the construction and management of factories, offices, etc. A knowledge of statistical techniques is helpful but not essential, since these can be learned later.

(b) *Personality*: Staff should be well accepted within the company and should be cooperative and diplomatic in the best sense. They should not be easily carried away by idealistic speculation and groundless hypotheses, but should base their conclusions on an undistorted appreciation of the facts. They must be tireless and hardworking and neither overemotional nor intolerant.

(c) Physical and emotional makeup: They should be both physically and mentally tough.

The company may have nobody who meets these conditions perfectly, but top management should select those who come closest to staff the TQC promotion office. People with a range of experience should be appointed, and the TQC promotion staff should work together to complement each other's strengths and weaknesses and satisfy the above requirements as a whole.

The selection of the person who will head the TQC promotion department, play the leading role in the TQC promotion office, chair the TQC study group, or hold some other position of authority during the introduction of TQC is particularly important. If the wrong person is selected, or he or she goes about things in the wrong way, the full implementation of TQC will be greatly delayed. This has often happened even in Japan. Of course, this is true in every situation; only the very best people should be put in charge of important new departments.

The TQC promotion office staff should also include one or two people who have a superior knowledge of statistical techniques. In Japan at present, factories and larger departments require at least one or two statistical experts, who may also be trained up after their appointment. Old-style inspection staff are often unsuitable as the staff of TQC promotion offices or quality assurance departments.

(4) TQC or QC committees

The TQC or QC committee plays the central role in implementing and promoting quality control in the company and its factories when QC is first introduced, and great care should be taken over its organization and management. Whether to form a TQC committee or a QC committee is a question that will decide itself naturally when the company or its top management decides on the kind of standpoint from which they intend to promote TQC. The following types of organization are possible:

I. Company committee:

a) Chair: president or vice-president (the company's No. 1 or No. 2).
b) Members: directors or department managers in charge of sales, production, technology, purchasing, accounting, personnel, and quality control, together with factory managers.
c) Secretary: quality control department or section manager.

II. Factory or branch office committee:

a) Chair: factory manager or branch manager.
b) Members: department managers, or section managers where there is only one department (technical and administrative staff should be included).
c) Secretary: quality control department or section manager.

In factories or branch offices with very large departments and sections, each department should establish a departmental committee and subcommittees along the lines of the above examples. These committees should start by regarding themselves as educational groups, and should meet regularly (at least once a month) to discuss the items listed below:

(1) Finalization of TQC or QC promotion programs, including education and standardization programs.
(2) Matters relating to policy management.
(3) New-product quality targets, quality levels, prototype fabrication, etc.
(4) Important quality problems, quality standards, and targets.
(5) Quality items for analysis as a matter of priority.
(6) Trouble in individual areas and handling of complaints and claims.
(7) Reports on elimination of abnormalities from processes.
(8) Other important matters relating to quality control, e.g., setting up cross-functional management committees or suspending product shipments.
(9) QC circle activities and QC team activities.

In addition, quality control promotion groups and secretariats, cross-functional management committees, QC teams, the manager-in-charge system, etc., may be adopted as necessary.

7.3 TQC Promotion Programs

(1) Long-term TQC programs

In promoting TQC, a long-term program (for example a five-year plan) must be set up as a matter of management policy. It is also important to make the program an integral part of the company's long-term business plan. Japanese companies have for a long time had long-term profit, sales, and production plans but have notably failed to include long-term quality programs in these. If quality programs are not unified with other business plans, TQC and management will be regarded as separate entities and people will easily fall under the delusion that TQC is something apart from their normal daily work. TQC must be made an inseparable part of everybody's routine duties. The following items should be included in a long-term TQC program:

(i) Policy management (Note: policy management is included when TQC is being implemented in its wide sense).
(ii) Plans for new-product development and the discontinuance of obsolete products.
(iii) Quality improvement programs.
(iv) Quality assurance programs (in the wide sense).
(v) QC education and training programs, organization and personnel plans.
(vi) Standardization promotion plans (materials and regulations).
(vii) Subcontracting, purchasing, and raw materials plans.
(viii) Sales, distribution, service, and consumer plans.
(ix) QC circle activity promotion plans.

(2) Education and training programs (see Section 1.6.7)
The education and enlightenment of all employees is a vital part of promoting quality control. Without it, quality control would simply become a pastime for a select group of people. The education of the workforce should have the following aims:

(a) To make quality everyone's concern.
(b) To ensure that everyone understands the new philosophy of quality control (quality *and* control; see, for example, Sections 1.4 and 1.6).
(c) To have everyone understand the statistical approach (see Section 2.2).
(d) To enable everyone to grasp the philosophy and methods of QC circle activities (see Section 1.10).

However, since people at different levels of the organization need to know different things, education should be tailored to the different grades, as shown below (the word "philosophy" below refers to the four philosophies of quality, control, statistical methods, and QC circle activities):

(i) Top management needs mainly to understand philosophy (see Chapters 1 and 2).
(ii) Middle management must understand philosophy, the use of control charts, and some statistical techniques (the whole of this book).
(iii) General engineers should understand philosophy, plus introductory statistical techniques, including control charts.
(iv) High-level engineers, like general engineers, need to understand philosophy and statistical techniques, but their knowledge of the latter should be at a slightly higher level.

(v) Administrators should understand at least as much as middle managers, with some people going as far as the level of high-level engineers.
(vi) Workplace supervisors should understand philosophy and the seven QC tools, if possible as much as general engineers.
(vii) Workers should initially understand philosophy and some of the seven QC tools; eventually they should understand all seven.
(viii) Statisticians should be well versed in advanced statistical techniques, design of experiments, operations research, marketing research, etc.

Education may be more effective if outside experts are invited to conduct training programs or people are sent to outside seminars, but in the end, much depends on the efforts of the people in charge of quality control within the company. The way in which it is introduced is particularly important, as is adequate follow-up.

Three-year or five-year education programs for all of the above levels should be prepared and scheduled in advance. If possible, all employees should have completed their education by the final year. Also, since the workforce is constantly changing, these education programs must be continued for as long as a company exists. Education and personnel appointing programs should be linked, and each person's educational history should be included in his or her personnel record. This educational history should be taken into account when considering the organization and its staffing. QC circle activities are extremely effective for doing this.

(3) Standardization programs

The following concerns must be addressed in the organization of standardization programs:

(i) How should the different standards be classified and what standardization system, should be established?
(ii) What is the deadline for completing each standard?
(iii) Control regulations for standards must be prepared and forms and filing methods specified.

See Sections 1.5.2 and 5.4 for the kinds of items to be included in standards and regulations.

Names for standards may be freely selected according to company custom. Some examples might be "regulations," "specifications," "procedures," or "orders." The basic standards of a company are its articles of incorporation; other standards are formulated around these. Standards concerned particularly closely with quality include the types described below. There are almost an infinite number of different types of standards and ways of classifying them, but in principle it is best to start by preparing the minimum number of effective, practicable standards and to add to these as necessary.

(1) Product quality standards, i.e., standards regulating the quality of individual processes: standards for final and intermediate products. Standards for sampling, measurement, and testing may either be included in these or kept separate.
(2) Raw material quality standards, i.e., standards regulating the quality of all types of purchased materials, supplementary materials, parts etc. These may include regulations for ordering, delivery dates, and handling of materials.
(3) Test method standards, measurement method standards, measurement control standards, sampling method standards, inspection standards, standard inspection plans. Inspection standards such as inspection implementation regulations, test methods, measurement methods, and sampling methods may be written separately (in which case, the individual product standards will indicate which inspection standard is to be used), or they may be included in the individual product standards. Inspection implementation regulations are a kind of work standard for inspection and should specify the combination of inspection methods to be used. They should include inspection criteria, handling of defectives, disposition of non-conforming lots, and a description of responsibility and authority for matters such as as-is acceptance.
(4) Technical standards (including standard operating ratio, standard amount for a product unit, standard yield, etc.), design standards, design technology standards, new-product development regulations.
(5) Work standards, work instructions, work guidelines, control standards. In the broad sense, work standards specify what every employee should do. As well as dealing with the routine work performed on the shop floor, they should therefore also include: inspection work standards (inspection implementation regulations); sampling, measurement, and test analysis work standards; preparation of contract documents; measurement control; complaints handling; sales management; stock control; market surveys; quality information; standards for controlling processes using control charts; equipment, plant and machine control; control of jigs and tools; factory experiments; safety and hygiene management; education, training and skills management; transportation and production volume control; manpower control; budget control; cost control; personnel

management; administration; standard forms for all kinds of reports, vouchers, etc. and work standards for organizing and filing these.

(6) Organizational standards, committee regulations (for quality control committees and new product committees). Organizational standards specify the duties and standard work of upper-level employees (e.g., from staff level up to director) and administrative personnel. These standards are sometimes called "job descriptions" or "management guidelines." Their particular purpose is to delegate authority and clarify the relationships among different levels of the organizational hierarchy.

(7) Standards for policy management and information transmission, control item standards, reporting system standards.

(8) Standards control regulations.

To make effective use of the above types of standards, it is necessary to specify the method of controlling each standard. Standards that do this are called standards control regulations.

They should specify the following:

- Who should formulate the standards, by when, and in what way, and whose approval must be obtained; who is to prepare draft education programs, when this is to be done, and whose approval must be sought for revisions.
- How the standards are to be filed, organized, disseminated, revised and checked.

Standards should be prepared in accordance with management policy with the aim of achieving specific objectives. Preparing them is the duty of engineers and administrative specialists. In principle, standardization only really starts to make progress after management policy and objectives have been decided. Superfluous standardization with unclear objectives can easily turn into standardization for its own sake.

(4) Organizational rationalization programs and cross-functional management programs

As we proceed with standardization and quality control, we eventually come up against the problem of organizational rationalization. It is therefore best to decide in advance when to start tackling this. Given the present state of many companies, it may be difficult to create the ideal organization in a short timespan, particularly in the areas of production, technology, inspection, and control. It is probably better to prepare a program which proceeds gradually.

Although individual departments may be quite well organized, Japanese companies traditionally exhibit a strong sectionalistic tendency with extremely weak cross-functional links. It is therefore a good idea, simultaneously with the introduction of QC, to establish cross-functional management committees (e.g., for quality assurance, new-product development, profit, cost, production volume, sales, personnel, subcontracting, and affiliated companies), to devise and expedite a program for building a cross-functional management system.

Since sectionalism tends to appear in all areas of human society, the following are some hints that may help with breaking it down:

(a) The responsibility for crushing sectionalism lies with top management.
(b) Fundamentally, establishing horizontal contacts is the duty of middle managers, section heads, and department managers.
(c) All employees should understand and act in accordance with the idea that "the next process is your customer."
(d) Cross-functional committees should be set up to clarify the duties and responsibilities of each function and to establish cross-functional links.
(e) Authority should be extensively delegated to QC teams and full use should be made of QC team activities.
(f) Joint QC circles should be formed and one must ensure that they remain active.
(g) One may form small management units such as divisions system.

7.4 Design Control

Since design control was touched on in the discussion of quality assurance systems in Sections 1.6.2 and 6.3, and some pertinent maxims were also given in Section 1.1.2(5)., here I will discuss only the main points.

Planning, design, prototype fabrication, and evaluation are not the exclusive province of the design department. They should be carried out in groups or teams that include other related personnel. Design means bringing to fruition product plans decided on a companywide basis.

Design work is high-variety, small-volume production of the products of drawings, and it must therefore start from the customer's standpoint. Drawings should only be prepared after conditions of use have been carefully checked, product research has been performed and production methods and process capabilities have been investigated.

The preparation of drawings is inevitably accompanied by mistakes and an increase in the variety of parts required. To obviate this, we should promote design standardization and parts standardization as well as eliminating drawing errors, tightening up the system for checking drawings, eradicating drawing and design changes, and preparing drawings that enable products to be produced without adjustment.

The use of QC tools such as Pareto analysis and check sheets will help with this, and we should think of prototype fabrication as an experiment and make use of design of experiment methods. Statistical methods are also useful for determining tolerances and safety factors.

7.5 Raw Material Control, Subcontractor Control, and TQC for Small and Medium-Sized Enterprises

(1) Subcontractor control

An average of 70% (from 50% to 85%) of the costs of Japanese industrial production is spent on raw materials, parts, and processing. There is no way to produce good products if materials, parts, or processing are poor, and this is why large companies in Japan joined forces with subcontracting factories and other small and medium-sized enterprises to promote TQC from the late 1960s onward. This has enabled industry to produce high-quality products at reasonable prices, reduce parts stocks, and beat world competition. Cooperative associations and other groups have also been formed, and buyers and sellers have become allies.

In contrast, the proportion of production costs accounted for by raw materials, parts, and processing in some countries is slightly over 50%, less than that in Japan. This is because suppliers are regarded as enemies who cannot be trusted, and purchased parts have high defect ratios. As a result, parts stocks are high, with correspondingly high interest charges.

Example: Company A has the following basic policy for fostering and developing subcontractors:

1. We do not purchase from companies that do not also supply other manufacturers. In future, limit the proportion of products supplied to us to a maximum of 50% of your total production.
2. We do not purchase from companies that do not offer us their opinions and suggestions.
3. Since we are adopting a guaranteed purchasing system (a zero-inspection purchasing system), subject your supplies to proper quality assurance.

Total quality control means thinking of subcontractor quality, quantity, delivery date and cost control on a long-term basis. It takes time to develop good subcontractors.

(2) Ten quality control principles for buyers and sellers

To rationalize the relationship between buyers and sellers and to improve quality assurance, the following ten quality control principles (entitled "Ten Principles for Vendee-Vendor Relations from the Standpoint of Quality Control") were formulated in 1960; some were revised in 1966:

Preface: Both purchaser and supplier should sincerely practice the following ten principles, while fostering a spirit of mutual trust, cooperation and tolerance and a sense of social responsibility:

(i) Purchaser and supplier are responsible for understanding each other's quality control systems and working together to implement quality control.

(ii) Purchaser and supplier should preserve their individual autonomy while respecting each other's independence.

(iii) The purchaser is responsible for presenting his requirements to the supplier in such a way that the supplier clearly understands what he should manufacture.

(iv) Before starting any business transactions, purchaser and supplier should draw up and sign proper contracts covering matters such as quality, quantity, price, delivery deadlines, and payment conditions.

(v) The supplier is responsible for guaranteeing that the materials supplied are of a quality that will satisfy the purchaser during use. He is also responsible for providing the objective data needed to confirm this, as required.

(vi) When the purchasing contract is drawn up, purchaser and supplier should decide on evaluation methods satisfactory to both parties.

(vii) When preparing the contract, purchaser and supplier should decide on methods and procedures for the amicable settlement of disputes.

(viii) Purchaser and supplier should give due consideration to each other's standpoints and exchange the information needed by each party for quality control.

(ix) To ensure a trouble-free relationship, both purchaser and supplier should at all times properly manage business activities such as ordering, production, inventory planning, administration, and organization.

(x) In their business transactions, both supplier and purchaser should always take full account of the final consumer's interest.

Table 7.1 summarizes the relationship between supplier and purchaser in quality assurance. The QC relationship can be regarded as more advanced as it proceeds from Step 1 to Step 8 in the table. Do not become obsessed with the idea of zero-inspection purchasing and fail to carry out inspection when it is in fact needed.

(3) Ten items for your VA checklist

Value analysis (VA) is useful for controlling raw materials. The following is a checklist used by General Electric: (Publisher's Note: Since this is a back-translation from Japanese, its wording may not correspond exactly with the original)

(i) Does the use of the raw material add value?
(ii) Is the raw material worth its cost for that particular application?
(iii) Is there any waste inherent in the form of the material?
(iv) Is anything more appropriate available?
(v) Is there any way of doing it cheaper?
(vi) Can a standard item be used?
(vii) Can the material be used with a setup appropriate to the production volume?
(viii) Is the cost appropriate when compared with the total of materials costs, labor costs, indirect costs, and profit?
(ix) Are there any more reliable or cheaper suppliers?
(x) Is anyone else buying the same material cheaper?

Table 7.1 Supplier-Purchaser Quality Assurance Relationship

Supplier		Purchaser	
Production	Inspection	Inspection	Production
1. –	–	–	100% inspection
2. –	–	100% inspection	
3. –	100% inspection	100% inspection	
4. –	100% inspection	Sampling or check inspection	
5. 100% inspection (autonomous inspection)	Sampling inspection	Sampling or check inspection	
6. Control (autonomous control)	Sampling inspection	Check inspection or zero inspection	
7. Control	Check inspection	Check inspection or zero inspection	
8. Control	Zero inspection	Zero inspection	

(4) Criteria for selecting suppliers from the QC standpoint*
The following items relate to outsourcing and purchasing:

(a) Has a basic policy been established? Do you aim to function as a specialist manufacturer or as an industrial group? Are existing suppliers to be selected or are new ones to be developed?
(b) See the "Ten Principles for Vendee-Vendor Relations" descibed in Section 7.5(2) area.
(c) The following criteria are considered in ranking suppliers: organization and degree of development of quality control and quality assurance; abilities and personal qualities of top management; level of business management; independence; financial status; technical level; state of equipment; number of years of previous dealings; degree of dependability; use of subcontractors; labor-management relations; degree of cooperation (in meeting delivery deadlines); price.
(d) Is purchasing to be done through QC audit or through inspection?
(e) Categorize parts (e.g., A, B, C, D) and vary the ordering system appropriately.
(f) Some Japanese companies have had an international purchasing strategy since the late 1950s, and most companies have now started to purchase and subcontract outside Japan. Does the company under consideration have the capacity to do this, and have they developed people with the requisite ability?
(g) The selection of suppliers should be reviewed regularly. The following should be considered in the review:
- Supplier education: group education (on own initiative); committees; cooperative groups; joint QC teams; QC circles; QC study groups and mutual visits; individual advice; suggestion systems; inspection adjustment; bonus and penalty systems; rationalization of contracts; termination of contracts with unsuitable suppliers; planned price reductions.
- Ordering and delivery systems: fixed-time delivery; fixed-day delivery; bulk delivery; fixed-quantity ordering; fixed-period ordering; planned ordering; spot buying.
- Stock control systems.
- Purchasing systems: estimate system; open-tender system; private-tender system; individual negotiation.

---

* Kaoru Ishikawa, *Hinshitsu Kanri* (Quality Control), Vol. 15 (1964), no. 8, p. 567 (in Japanese).

- Is joint testing needed?
- Contract rationalization.
- Who decides whether to make parts in-house or to purchase from outside, and when and how is this done?
- Outsourcing of finished products: whether to manufacture in-house, order from an outside supplier, act as OEM, manufacture abroad, etc.

(5) The Ten Commandments for small and medium-sized enterprises

(i) A business that does not contribute to society will be cold-shouldered by the public.
(ii) Develop and foster successors, rationalize personnel selection, and remove ineffective people from top management.
(iii) Establish constructive and cooperative labor-management relations and take responsibility for employees and their families.
(iv) Develop top management's awareness of quality control and improvement, concentrate on new-product development, and become a specialist manufacturer.
(v) Master the statistical approach, base policies and plans on statistical data, and make use of market surveys. Know your own company's process and manufacturing capabilities.
(vi) Do not depend exclusively on orders from a single company. Retain your independence and accept no more than 50% of your orders from one company; if possible, reduce this figure to 20% or less.
(vii) Do not hold too many fixed assets or overinvest in equipment; this can lead to lack of assets or insufficient liquidity when needed.
(viii) For the same reasons, carefully control inventories and credit sales.
(ix) Do not rely on cheap labor.
(x) Avoid irrational business customs such as lack of enthusiasm or leadership on the part of top management, ignorance, indecisive handling of problems, lack of experience, helping oneself to too much of the profits, insufficient investment in education, lack of personnel development, and failure to select and appoint competent individuals.

(6) Delivery control

The following are recommendations to be considered in addition to raw material and subcontractor control:

(a) Clearly define what is meant by delivery, set specific deadlines, and ensure that everyone enters the spirit of meeting those deadlines.

(b) Clearly define what is meant by a missed deadline or an incorrect delivery. Unreasonable deadlines and delays in issuing drawings or raw materials are frequent causes of missed delivery dates. Trouble in supplier-purchaser relationships and too many failed lots, defectives, or incorrect deliveries are 60% to 70% the responsibility of the purchaser or contractor and only 30% to 40% the responsibility of the supplier or subcontractor.
(c) Use Pareto analysis.
(d) Provide quality information feedback to suppliers.
(e) Check changes in quality of goods supplied, process capabilities, failed lots, and defective products.
(f) Examine the percent defective after incoming inspection and decide whether the incoming inspection method or the supplier should be changed.

7.6 Equipment Control, Jig and Tool Control, and Measurement Control

The philosophy behind all three of these types of control is basically the same (see Sections 1.6.4 and 1.6.6).

(1) Historically, equipment control methods started out with a plan to repair equipment because it had broken down; this developed into carrying out preventive maintenance to prevent equipment from breaking down. The next development involved ensuring that equipment maintains and improves its process capabilities; the most recent development has been a focus on improving equipment reliability through TPM.
(2) To perform process capability studies, one must decide who is responsible for investigation, maintenance, and improvement. Preventive maintenance will come to nothing without an awareness of dynamic precision, static precision, and the statistical approach.
(3) Considerations involving inspection and maintenance standards for equipment and measuring instruments are: Who is to prepare these, the manufacturer? And is the required inspection technology available?
(4) If there are frequent breakdowns after inspection and maintenance, one must question whether the test operation standards are adequate.
(5) Which department is responsible for equipment control? Does the work-

place use equipment unreasonably and fail to carry out regular servicing? Grass-roots improvement is impossible if people are simply forced to follow orders. What about stock control of spare parts?

(6) Priority control must be carried out.

(7) What are the equipment renovation standards like? The process capabilities of old equipment can be considerably improved. People tend to want to invest more than is needed in new equipment and to install measuring instruments unnecessarily. Before purchasing equipment, question whether it is really necessary. It is no good trying to write off costs through the tax rules; technical progress will decrease the value of equipment much faster.

(8) Are there proper work standards for using equipment, jigs and tools, and measuring instruments?

(9) It is totally wrong to assume that control is being implemented just because periodic servicing and calibration are being carried out. Adjusting and repairing equipment that has gone wrong is crisis management; it is no more than shutting the stable door after the horse has bolted.

(10) What about reliability control?

(11) Is equipment investment really necessary from the viewpoints of cost control, process capability and manufacturing capability? Through proper control, manufacturing capability can be increased by 50% to 100% and process variation can easily be reduced to one-half to one-third of its original value.

(12) Are the results of equipment investment being checked properly? Ensure that people are not escaping their responsibilities by simply spending their investment budgets or using the equipment but failing to check the results.

(13) Is error control firmly established?

(14) Before automating or introducing robots, ensure that proper process analysis and process control are carried out and effective quality control process charts are prepared.

7.7 TQC in Marketing, Sales, and Service

The following is a list of some common defects of conventional marketing departments from the TQC viewpoint:

(1) They fail to realize that marketing is the entrance and exit of TQC.
(2) They think that marketing and TQC are unrelated, and are therefore ignorant of TQC and QC.
(3) They lack data on the factors underlying why some products sell well and others do not.
(4) The customer is king, but many kings are blind. It is the job of the sales staff to educate customers correctly about products, but sales staff themselves often lack sufficient product knowledge and work on the basis of "intuition, experience, and audacity" alone.
(5) Their staff are no different from the sales staff of distributors or wholesalers.
(6) Their policies and plans are vague and not properly disseminated. No effort is made to sell in accordance with sales plans.
(7) They take only easy orders and avoid difficult ones.
(8) They have no feeling for quality assurance and lack a sense of responsibility.
(9) They know nothing of product line-up management or product line-up research.
(10) They lack the sense to sell better-quality products at higher prices.
(11) They do not appreciate that marketing is redundant when one sells at give-away prices. We should sell through quality.
(12) They lack the feeling for securing a profit.
(13) They believe that everything is rosy as long as sales are increasing.
(14) They apply pressure-selling tactics without considering possible payment problems, concentrating on the sales figures and forgetting about interest rates.
(15) They have few or no sales engineers, and their sales staff lack product and technical education.
(16) They are too keen to sell special models, and neglect sales of standard versions.
(17) They are not interested in selling items that are loss-makers under formal cost-accounting procedures.
(18) They fail to understand the details of orders sufficiently.
(19) They are ignorant of process capabilities, manufacturing capabilities, and factory conditions.
(20) They fail to consider the company as a whole.
(21) They ignore costs and cash flows.
(22) They have no idea of acting as the company's feelers and collecting market quality information, and they lack the ability to do this.
(23) They are unreliable (on quality, price, and delivery).

(24) Their advertising is often exaggerated and susceptible to PL problems. They do not study product liability.
(25) Their before-sales service is inadequate and they lack the concept that they are selling service.
(26) After-sales service is inadequate; a product should not be sold if no service is available for it.
(27) They do not know how to control product inventories, and they fail to perform adequate Pareto analysis on the basis of quality, quantity, and price.
(28) They have little idea of market surveys and do not know how to perform them.
(29) They fail to investigate the kinds of distribution organizations through which it is best to sell, and do not provide distribution organizations with sufficient TQC education and development.
(30) They fail to study purchasers and consumers carefully enough.
(31) They fail to provide adequate tie-up with advertising and publicity.
(32) Their advertising and publicity are not in the spirit of TQC.
(33) They lack the courage to discontinue obsolete products.
(34) They provide insufficient service and quality assurance for discontinued products.
(35) They claim that sales are impossible unless new products are launched, while they forget their own responsibilities for new-product planning.
(36) They accept orders without considering whether the goods can actually be delivered.
(37) They are ignorant of TQC-style complaints handling.
(38) Their sales data often cannot be used for analysis because it is lumped together and not properly stratified.
(39) Their operating manuals (including methods of use and maintenance), catalogs, parts lists, etc., are not based on the QC approach.

Some hints for practicing TQC in the marketing department can be arrived at by practicing the opposite of many of the above faults. Briefly summarized, these are:

(1) Become able to act as the eyes and ears of your organization in collecting market information and consumer information.
(2) Prepare sales plans for each product, considering quantities and profits as well as sales values. Sell products and collect payment in accordance with these plans.
(3) Have technical information about products and their methods of use, and

provide technical services or carry out joint experiments to enable consumers to make appropriate choices and to use products correctly.

(4) Take orders properly using the QC approach. Clarify the grade that it is important to guarantee for each product, as well as its method of use, conditions of use, guarantee period, and warranty period, i.e., clarify consumers' requirements and prepare proper contracts.
(5) Encourage consumers to buy standard versions rather than just special models.
(6) Provide advice and feedback on the kinds of new products and quality improvements needed.
(7) Inculcate devotion to selling quality, and sell high-quality goods at higher prices.
(8) Analyze large amounts of data statistically.
(9) Set up a self-supporting accounting system for marketing.

7.8 TQC and Distribution Organization

Even if a company produces good-quality products, the quality of these products cannot be guaranteed and their sales and manufacture will not proceed smoothly if quality control in the company's distribution organizations is poor. The hard-won quality of primary products—e.g., textiles, plastics, metal products, etc.—will be lost if their processing (this is included in distribution in its widest sense) is poor. The effect is particularly damaging for small and medium-sized enterprises. Even the makers of primary products are being obliged to think about guaranteeing their quality until after they have been turned into final products. With general merchandise too, unskillful stock control will lead to product deterioration, increased levels of defective stocks, and more frequent returns. The appearance of defective products may also cause valuable sales opportunities to be lost. Furthermore, if product knowledge is lacking, customers will be sold unsuitable products and this will lead to complaints and claims. Customers will not be satisfied if product testing at the time of sale, pre-delivery inspection, installation, and other services are poor, and it will prove impossible to sell products in the future if after-sales service is unsatisfactory.

The following recommendations are important in connection with the above:

(1) Educate distribution organizations (e.g., trading companies, agents, wholesalers, and retailers) in QC.
(2) Ensure that distributors are well-versed in the philosophy of stocking

up with good-quality products, selling good-quality products, and guaranteeing quality (including after-sales service) even after purchase.

(3) Select distribution organizations carefully.
(4) Unsatisfactory transportation, packaging, storage, and stock control by distributors will not only make quality assurance impossible but will also jeopardize your company's business. Ensure that all these aspects are closely controlled.
(5) Educate distributors thoroughly in complaints handling. Many distributors do not know the QC approach to complaints handling and fail to take recurrence-prevention action when a complaint is received; their usual procedure is usually no more than an apology and an offer of a replacement for the faulty item.
(6) Ensure that distributors carry out proper pre-delivery checks when handing the goods over to the customer, in addition to performing careful incoming inspection.

7.9 Research and Development Control

We are now in an era of intense competition in new-product development, but good new-product development and good products are impossible without effective research. The story of Columbus and the egg seems appropriate in connection with research and development. Columbus had a dream, and he succeeded in realizing his dream. People often make lists of ideas and plans for new products and then proceed to evaluate these ideas, but such evaluations are scarcely trustworthy. Ideas that have passed through this kind of evaluation process will produce run-of-the-mill results. When an idea occurs, it should be tried out unhesitatingly, since the really good ideas only emerge through a process of trial and error. It is important to create a corporate culture in which there is no fear of failure and in which ideas can be freely implemented. A company with such an atmosphere will create an abundance of good new products and new technology. The following are important guidelines in controlling research and development:

(1) Split up research into different categories (basic research, applied research, developmental research, service research, product research, short-term research, medium-term research, long-term research, and "rush" research), and devise systems for controlling these. Basic research should have a free range of topics and an open-ended budget. (Hardly any Japanese companies practice basic research at present.) The other types of research should have properly set and controlled topics, goals, targets, organizations, schedules, and budgets.

(2) Decide whether to carry out research in-house, or to contract it out, purchase patents, head-hunt people from other companies, or merge with other companies. Determine who is to decide this.
(3) Set up a system that minimizes the need for unplanned, last-minute, "rush" research.
(4) Create a corporate culture in which there is no fear of failure and in which ideas can be put into practice freely.
(5) Decide how research topics, aims, and targets will be decided.
(6) Carry out activities in teams. Arrange for people to be gathered into project groups as required.
(7) Actively select and exchange research personnel.
(8) Encourage the development of creative "idea-people."
(9) Develop people's scientific and statistical analysis capabilities and the ability to write research reports that are easily understood by top management. People should happily accept occasional evaluations during the course of research.
(10) Be ready to decide when to suspend a certain line of research and have the determination to do so.
(11) Remember that research is a long-term investment.
(12) Strengthen research laboratories' internal service departments (administration, control, library resources, investigation, testing, equipment, analysis, and measurement).
(13) Use statistical tools, the PERT technique, etc.
(14) Developmental research should start on a broad front and gradually narrow down (I call this "cone-shaped research") rather than simply starting at a single point and proceeding straight down ("well-shaped research"). Experiments should be designed to demonstrate the effect of variation in various factors.
(15) Do not forget product research.
(16) Standardize methods of evaluating research results and of allocating resources.

No truly new products or technology will emerge from the kind of organization in which top management heaps criticism on failure and gives scant praise to success; only copycat products and technology will be produced. Fewer than 5% of ideas are successful in their original form, but an idea will ultimately succeed if we are brave enough to go through a long process of trial and error, modifying the idea after each failure. Because of this, I think it is better to appoint as head of a research and development laboratory a sensible person who has come up through the sales department of the company, rather than someone who started out as a scientist or engineer.

7.10 Quality Audits

A quality audit is a kind of diagnostic procedure in which quality itself is checked by sampling products and services inside and outside the company and performing various experiments on them in order to find out whether or not their quality is good and consumers are satisfied. Some notes on performing quality audits are:

(1) When implementing quality control and quality assurance, it is essential to form a quality audit or quality assurance department that reports directly to top management. This department should be given full freedom of action and adequate authority to carry out quality audits. For example, it should be able to visit and obtain data from anywhere in the company as well as being authorized to suspend product shipments when necessary.
(2) The quality assurance department should be free of any responsibility for design, production, costs, or scheduling.
(3) It is advisable to institute a program for developing quality specialists ("Mr. Quality"), and foster people with experience in developmental research, sales and marketing, service, design, production, QC, and inspection.
(4) It is no good simply to put up a new signboard on the inspection department; the department itself must undergo a complete change of attitude.
(5) Complaints and quality information from inside and outside the company should be easily collected and communicated directly to the quality assurance department. Quality information should be purchased if necessary.
(6) The quality assurance department should participate in the evaluation of new-product planning, design, prototype fabrication, production, storage, and market quality.
(7) All other departments must act on the advice of the quality assurance department.
(8) The quality assurance department should be authorized to suspend prototype fabrication, production, shipments, or product sales as necessary.
(9) The quality assurance department should be authorized to set up testing panels and to perform sample testing.
(10) Quality audits of primary products should be performed through quality evaluations and quality assurance of the secondary manufactures and the products made from them.
(11) Quality audits should be carried out from the standpoint of the consumer.
(12) Methods of performing quality (including reliability) audits should be gradually perfected and standardized.

(13) Quality audits should be carried out regularly on a company's own products and on those of its competitors, and reports should be prepared with recommendations for action.
(14) The necessary equipment for performing evaluations must be provided.
(15) A written report must be prepared whenever an evaluation is carried out.

7.11 QUALITY CONTROL AUDITS AND TQC AUDITS

A quality control audit consists of examining the processes and methods by which quality control is being carried out, pointing out any weaknesses, advising on methods of curing these weaknesses, and taking appropriate action. When the scope of a QC audit is made even wider, covering the whole company, it is called a TQC audit. Company presidents carry out TQC audits to look at TQC in its wide sense, i.e., to look at the company's business management as a whole.

QC audits can be classified as follows:

(1) QC audits by outsiders

(i) Audits of suppliers by purchasers (e.g., the U.S. Armed Forces, the Japanese Self-Defense Force, Nippon Telegraph and Telephone Corporation, Japanese Railways, and other public and private corporations)
(ii) Audits for purposes of certification, e.g., the JIS mark, ASME certification, etc.
(iii) TQC examination for the Deming Application Prize and the Japan Quality Control Medal
(iv) QC or TQC audits by consultants

Of the above four types of audit, only the third is unique to Japan. The others are carried out in many other countries. With the first two, particularly when the people performing the audit are not experts and lack experience in implementing QC, and when the organization being audited adopts the attitude that everything is fine as long as the customer buys the products or it obtains the qualification it wants, the whole attempt at TQC can easily end up as a paper exercise, with the QC department kept busy producing reams of documents but no actual QC being done. When a company receives an external audit, it is much better to use this as an opportunity to promote QC and TQC across the board, to carry out an effective and meaningful QC review, and thereby to obtain useful results.

(2) Internal QC audits

(i) President's QC or TQC audits
(ii) QC or TQC audits by heads of department
(iii) QC audits by QC staff
(iv) Mutual QC audits (e.g., by one process and the next)

Internal QC audits are rare outside Japan. The president's audit in particular is almost unknown elsewhere, because the presidents of foreign companies know little about QC. Japanese companies performing TQC properly hold presidential audits at regular intervals and achieve excellent results through this practice. If the presidential audit and other QC audits carried out by company staff are done skillfully, the following significant benefits can be obtained:

(a) The people receiving the audit are stimulated by it, and their QC activities and quality assurance activities receive a boost. TQC activities tend to go in cycles, sometimes riding high and sometimes becoming stereotyped, and it is a good thing to review the situation from time to time to ensure that TQC continues steadily without fizzling out.
(b) Relations within the company improve. Top management and upper-level managers usually have little opportunity to meet section managers, supervisors, and workers at the bottom of the chain of command face-to-face and listen to their opinions. QC audits are therefore a good chance for them to hear what these people have to say and to get to know the true situation.
(c) QC audits are an excellent opportunity for top management to find out what is actually going on in the company. Top management usually knows surprisingly little about the actual situation. The person who learns most from a presidential audit is the president himself, and audits often show presidents exactly what poor shape their companies are in and turn top management into zealous QC leaders.
(d) If department and section managers and staff are made to attend these audits, they will find out about other departments in the company and will be able to see things from a broader perspective. This should allow them to grow personally and is useful in fostering the next generation of managers.
(e) If top management is to perform QC audits, it must understand QC. The presidential audit is therefore a good chance for top management to study QC and experience what it is all about.

(3) Notes on carrying out a presidential QC audit

(a) The president must always lead the audit group. However, his deputy may take his place if absolutely necessary.
(b) The audit group should include not only the director in charge of the department being audited but also directors in charge of other departments, the director responsible for QC, and other QC staff. Department and section heads should attend as necessary. It is also a good idea to invite an outside QC consultant to attend at first.
(c) The aims of the audit should be made clear.
(d) A QC audit should cover quality control in its wide sense, but should focus as closely as possible on quality. The scope of a TQC audit is somewhat broader.
(e) QC audits should be carried out from the companywide, long-term standpoint.
(f) All departments and all places of business should be covered. In addition, audits should be extended outside the company to purchasing, sales and marketing, and other external activities.
(g) The presidential audit should be included in the company's quality control program as an annual event. The schedule, audit team, and scope of the audit should be announced in advance (as early as possible, but in any case at least two months ahead). Proper preparation for an audit is extremely useful; however, when there is a tendency to make things look good just for the audit, or when TQC has permeated the organization to a high degree, it is better to examine things in their normal state without much preparation.
(h) In some cases, it is a good idea to specify priority items to be audited.
(i) Although the term "QC audit" is in general use, the word "audit" does have the slight flavor of inspection, and it might be better to adopt a different title, such as "top-management QC clinic." It is important to create a relaxed atmosphere in which top management is open to all opinions and everybody puts their heads together to think about the best ways to improve the company's overall quality control.
(j) Every member of the audit team should write a report giving his or her opinions and advice. The team members should be informed of the need for this report beforehand and should be given standardized report forms, checklists, etc. The team members themselves will benefit from doing this.
(k) The individual reports should be collated and an audit report form clearly showing the items requiring action should be prepared. Copies should

Table 7.2 Quality Control Audit Checklist (for Deming Application Prize)

Item	Checkpoint
1. Policy	(1) Management, quality, and quality control policies. (2) Policy-setting methods. (3) Suitability and consistency of policies. (4) Use of statistical tools. (5) Dissemination and permeation of policies. (6) Checking of policies and their degree of implementation. (7) Relationship with long-range and short-range plans.
2. The organization and its management	(1) Clear-cut authority and responsibility. (2) Appropriate delegation of authority. (3) Cooperation among different departments. (4) Committee activity. (5) Use of staff. (6) Use of QC circle (small-group) activities. (7) Quality control audits.
3. Education and dissemination	(1) Education programs and results. (2) Quality-consciousness, control-consciousness, degree of understanding of quality control. (3) Education and degree of permeation of the statistical approach and methods. (4) Identification of results. (5) Education of subcontractors and other outside organizations. (6) QC circle (small-group) activities. (7) Improvement suggestions.
4. Collection, communication, and utilization of information	(1) Collection of information from outside the company. (2) Transmission of information between departments. (3) Speed of information transmission (use of computers). (4) Data organization, statistical analysis and utilization.
5. Analysis	(1) Selection of important problems and topics. (2) Suitability of analytical methods. (3) Use of statistical tools. (4) Linkage with specific technology. (5) Quality analysis and process analysis. (6) Use of analytical results. (7) Positiveness of improvement suggestions.

be sent to the central quality control committee, the department being audited, and related departments.

(l) For QC audits, it is better not to have a standing organization but to form audit teams as required.

(m) Regulations and procedures must not be taken for granted. It is important to examine whether they have been skillfully formulated and whether they are actually being applied systematically. One of the aims of QC audits is to identify the status quo.

Revised 17 June 1980

Item	Checkpoint
6. Standardization.	(1) System of standards. (2) Methods of establishing, revising and discontinuing standards. (3) Track record in establishing, revising and discontinuing standards. (4) Contents of standards. (5) Use of statistical methods. (6) Accumulation of technology. (7) Utilization of standards.
7. Control	(1) Control systems for quality and related costs, quantities, etc. (2) Control points and control items. (3) Use of the statistical approach and statistical tools such as control charts. (4) Contribution of QC circle (small-group) activities. (5) Status of control activities. (6) State of control.
8. Quality assurance	(1) Procedures for new-product development. (2) Quality deployment and analysis, reliability, design reviews. (3) Safety, product liability prevention. (4) Process control and improvement. (5) Process capability. (6) Measurement and inspection. (7) Control of equipment, subcontracting, purchasing and services. (8) Quality assurance system and its auditing. (9) Use of statistical tools. (10) Quality evaluations and audits. (11) Status of quality assurance.
9. Results	(1) Measurement of results. (2) Tangible results: Quality, service, delivery time, cost, profit, safety, environment, etc. (3) Intangible results. (4) Agreement between forecast and actual results.
10. Future plans	(1) Concreteness, and identification of status quo. (2) Strategies for overcoming weaknesses. (3) Future promotion plans. (4) Connection with long-range plans.

(n) One ought to check whether the advice from the previous audit has been implemented.

(o) The advice derived from each audit should be promptly incorporated into the company's quality control plans.

(p) It is necessary to look through any gloss applied for the purposes of the audit and to see how routine work is performed. The best type of investigation starts at one point and bores in deeper and deeper, bringing things to light one after another like pulling up a string of sweet potatoes.

(q) Audits should be carried out with a friendly attitude and without preconceptions.
(r) Advice should be positive and constructive. An audit does not consist of picking away at petty details or pointing out weaknesses and putting the screws on people. Holding an audit should be like a doctor making a diagnosis, curing sickness, and producing a healthy patient.
(s) The department receiving the audit should report on a particular topic, explaining its existing policy, results obtained, problems remaining, future policy, and requests for head office and other departments.
(t) An audit does not consist of simply sitting around a table and talking: it involves finding out what the daily work consists of by actually visiting the workplace, talking to supervisors, foremen, and workers, watching the work being done, and examining documents and data.
(u) It is best to audit individual functions as well as individual departments.
(v) Since audits are concerned with quality control, the emphasis should naturally be on whether the products being made are of a quality that will satisfy consumers and please them so much that they continue to buy them, and on whether quality assurance is sufficient. However, action should be directed not at the products themselves but at the work that produces them, i.e., the process. In a nutshell, the purpose of an audit is to use quality as an indicator for judging whether or not a company is being well managed.

A quality control audit checklist is given in Table 7.2 for reference.

7.12 Policy Management

Management philosophy and methods were discussed in Section 1.5, and it is really sufficient to proceed as explained there. However, as is usual in business management, various expressions have recently become popular, e.g., "management by objectives," "policy management," "priority management," "routine management," etc., and I would like to give my opinion on these.

Since management is impossible without policy, goals, and objectives, all we actually have to do is simply to manage properly. If we talk of management by objectives, policy management, etc., there is a danger that top executives will simply state objectives and policy and then do no more than exhort people to try harder, falling into the trap of managing by exhortation rather than scientifically. This is why the concept of management by objectives, once fashionable in the

United States, has now been discredited. One reason why I used the cause-and-effect diagram and explained the philosophy of management in six steps in Section 1.5.2 was that I wanted to emphasize the need to think of the process (Step 2). However, I believe that management begins with setting policy, and, since the term "policy management" sounds good, I would like to use it here in the sense of "management beginning with policy." We should start by setting policy and then proceed to rotate the control cycle in the order described in Section 1.5.2.

As I will discuss later, long-term plans and annual plans are decided on the basis of long-term policy and annual policy, and the items included in these can be classified as those to be carried out on a priority basis and those to be carried out on a routine basis. If we do use the terms "policy management," "priority management," and "routine management," I think we should use them as follows:

—*Policy management* is management that starts with policy.
—*Priority management* addresses management items that should be implemented on a priority basis by individual departments or by the company as a whole.
—*Routine management* is management that is not a priority but should be carried out naturally as a matter of routine (in connection with QCDS, etc.) by individual departments or the company as a whole.

For methods of deciding on policy, setting plans, and setting objectives and targets, see below and subsections (1) and (2) of Section 1.5.2.

Policies, plans, and objectives are generally decided in the following order: (1) company policy; (2) long-term policy; (3) annual policy; (4) policy for accounting period; (5) monthly policy. Then, (1) long-term plans; (2) annual plans; (3) plans for accounting period; (4) plans for movable blocks of a certain number of months; (5) monthly plans.

A company's annual policy is its policy for the first year of its long-term policy, and its annual plan is its plan for the first year of its long-term plan. Long-term policies and plans should be revised yearly. A long-term plan should usually cover the next five years, but if necessary, it may cover the next ten or fifteen years while a medium-term plan covers the next three to five years.

Some of the benefits of establishing long-term policies and plans are:

(i) The process of creating these policies and plans is inherently valuable.
(ii) It gives management a long-term outlook and makes employees look forward to the future.
(iii) It simplifies the formulation of short-term plans.
(iv) It becomes possible to formulate action plans for each year of the long-term plan.

(v) It creates a framework and vision for the operation of the company.
(vi) It gives pattern to the life of the organization and leads to new-product development.

Below is a list of some of the priorities to be considered when drawing up long-term policies and plans:

- Management decisions should all relate to the future.
- Long-term policies and plans should be expressed in terms of objectives (quality, profit, quantity, capital, manpower). One should only start to consider methods (technology and equipment) after the objectives have been set.
- We make long-term policies and plans in order to execute work smoothly over long spans of time, not so that we can work slowly.
- Policies and plans (both long-term and annual) should generally be expressed as a combination of words (concept) and numerical targets. Either words or numbers alone are not enough.
- We should formulate specific procedures for establishing policies and plans (standards for revising long-term plans, policy management regulations, etc.).
- Sufficient information must be provided for establishing policies and plans, their rationale must be clear, and adequate analysis must be carried out.
- The control cycle must be checked to see that it is being skillfully rotated, taking account of previous policies and plans and the results of implementing these. Problems left over from a previous period should be included in the following period's policies and plans.
- Information and forecasts for setting policies and plans are never perfect. We can regard our management as scientific if it is based on data that are about 70%–95% complete; with less than this, it is unscientific. Managers and others in positions of authority must make bold guesses to cover the missing 5%–30%.
- Policies and plans must be specific and concrete and people must have a means of evaluating their progress. Are these sufficiently closely related to the control items? Have the really important points been decided (priority management)?
- Are the QC policies and plans fully integrated with the management policies and plans? Has QC policy been firmly hammered out?
- Have the policies been deployed through every level of the organization? Are the methods of policy deployment and communication satis-

factory? Does policy become more specific the further down the organization it proceeds? Will top-level policies and plans come to fruition if the policies and plans at the lowest level are put into effect? Are the policies of superiors and their subordinates sufficiently closely related? Is policy consistent throughout the organization?

- Are the policies well formulated and consistent with authority and responsibility?
- Tunnel-type (authoritative, top-down) policy setting is useless. Policy at each level should respect that level's autonomy and include its own ideas.
- Is the scope of authority of each level fully considered and is it given sufficient independence? Have superiors approved the policies and plans of their subordinates?
- Is policy firm, or does it change from day to day? Has a system of policy management been established, and is policy management constantly carried out?
- Policy deployment will not succeed the first year it is started; it will come right gradually as various weaknesses appear, various failures occur, and action is taken to prevent these from recurring.

7.13 CONCLUSION

In conclusion, I would like to re-emphasize the following points:

(1) Quality control should be promoted through the cooperation of all the employees of a company working together to do what should be done. It is not just the job of the TQC promotion office, the quality control department, the quality control staff, or other people or organizations with the words "quality control" in their titles.

(2) The enthusiasm, drive, and leadership of top management, especially the company president, are absolutely indispensable for achieving this.

(3) Quality control should not be practiced simply because it is fashionable. Its purpose is to rationalize industry, establish technology, and enable companies to develop the ability to secure good profits and beat international competition through quality rather than through unfair trading practices such as dumping. QC must be continued throughout the life of a company.

(4) If quality control is practiced in earnest, the money spent on it can be recovered in a matter of months or days. However, if mistakes such as

those described in Section 1.1.3 are made, only a half-hearted attempt is made, or top management does not take the lead, it will be difficult to recover the costs and QC will be a flash in the pan.

Real TQC is a complete revolution in the old-style approach to management. It means that top and middle management, engineers, administrators, and all other employees, as well as affiliated companies, must work together as a team to understand the philosophy of statistical quality control, acquire a feeling for TQC, put it into practice, and build up an effective management organization. This will rationalize every workplace and all types of work, as well as individual companies and industry as a whole. Exports of products and technology will increase, raising the standard of living globally.

At present, some Japanese products are too successful, and this has led to international trade friction. Together with the rise in the yen and the hot pursuit of Japan mounted by the newly industrialized countries, this means that Japanese companies are facing a difficult period in which they must again change their spots. By practicing TQC, CWQC, and GWQC, they must reform their structures and organizations, develop new products, and ride out the storm. Meanwhile, the Japanese are helping to revitalize the advanced nations of the West by disseminating Japanese-style quality control philosophy and methods. We are also helping the developing countries become even stronger. I believe that if every country promotes international specialization and joins the international competition for quality through practicing QC, the result will be world peace. I am promoting QC and TQC in hopes for the happiness of all the people of the world.

Index

H

J

K

P

Q

R

S

T